NEUROMETHODS

Series Editor
Wolfgang Walz
University of Saskatchewan
Saskatoon, SK, Canada

For other titles published in this series, go to
www.springer.com/series/7657

Drug Delivery to the Central Nervous System

Edited by

Kewal K. Jain

Jain PharmaBiotech, Basel, Switzerland

Humana Press

Editor
Kewal K. Jain
Jain PharmaBiotech
Blaesiring 7
4057 Basel
Switzerland
jain@pharmabiotech.ch

ISSN 0893-2336 e-ISSN 1940-6045
ISBN 978-1-4939-6159-7 ISBN 978-1-60761-529-3 (eBook)
DOI 10.1007/978-1-60761-529-3

Printed on acid-free paper

springer.com

Series Preface

Under the guidance of its founders Alan Boulton and Glen Baker, the Neuromethods series by Humana Press has been very successful since the first volume appeared in 1985. In about 17 years, 37 volumes have been published. In 2006, Springer Science+Business Media made a renewed commitment to this series. The new program will focus on methods that are either unique to the nervous system and excitable cells or which need special consideration to be applied to the neurosciences. The program will strike a balance between recent and exciting developments like those concerning new animal models of disease, imaging, in vivo methods, and more established techniques. These include immunocytochemistry and electrophysiological technologies. New trainees in neurosciences still need a sound footing in these older methods in order to apply a critical approach to their results. The careful application of methods is probably the most important step in the process of scientific inquiry. In the past, new methodologies led the way in developing new disciplines in the biological and medical sciences. For example, physiology emerged out of anatomy in the nineteenth century by harnessing new methods based on the newly discovered phenomenon of electricity. Nowadays, the relationships between disciplines and methods are more complex. Methods are now widely shared between disciplines and research areas. New developments in electronic publishing also make it possible for scientists to download chapters or protocols selectively within a very short time of encountering them. This new approach has been taken into account in the design of individual volumes and chapters in this series.

Wolfgang Walz

Preface

The delivery of drugs to central nervous system (CNS) plays an important role in the treatment of neurological disorders. Drugs may be administered directly into the CNS or administered systematically for targeted action in the CNS. The major challenge to CNS drug delivery is the blood–brain barrier (BBB), which limits the access of drugs to the brain substance. Various strategies are used to deliver therapeutics across the BBB, both in experimental animals and humans.

It would be impossible to thoroughly cover this topic in one handy volume. **Chapter 1** is an introduction to the topic with tabulation of various methods and their applications. Important methods form the subject of other chapters for detailed description. A few cover overviews of an innovative method and its applications.

Kewal K. Jain

Contents

Contributors

FARS K. ALANAZI • *Kayyali Chair for Pharmaceutical Industry, Department of Pharmaceutics, College of Pharmacy, King Saud University, Riyadh, Kingdom of Saudi Arabia*

IBRAHIM A. ALSARRA • *Department of Pharmaceutics, College of Pharmacy, King Saud University, Riyadh, Kingdom of Saudi Arabia*

SHAWN BELVERUD • *Department of Neurosurgery, North Shore LIJ Health System, Manhasset, NY*

MARTIN L. BRADY • *Therataxis, LLC, Baltimore, MD*

THOMAS P. DAVIS • *Department of Medical Pharmacology, College of Medicine, The University of Arizona, Tucson, AZ*

REMIGIJUS DIDZIAPETRIS • *Pharma Algorithms Inc., Vilnius, Lithuania*

DALE DING • *School of Medicine, Duke University, Durham, NC*

GAMAL M. EL MAGHRABY • *Department of Pharmaceutics, College of Pharmacy, King Saud University, Riyadh, Kingdom of Saudi Arabia*

JI-MING FENG • *Department of Comparative Biomedical Sciences, School of Veterinary Medicine, Louisiana State University, Baton Rouge, LA*

ITZHAK FISCHER • *Department of Neurobiology & Anatomy, Drexel University College of Medicine, Philadelphia, PA*

REINHARD GABATHULER • *Angiochem Inc., Montreal, QC, Canada*

AMEL Y. HAMED • *Department of Pharmaceutics, College of Pharmacy, King Saud University, Riyadh, Kingdom of Saudi Arabia*

KEWAL K. JAIN • *Jain PharmaBiotech, Basel, Switzerland*

PRANAS JAPERTAS • *Pharma Algorithms Inc., Vilnius, Lithuania*

SIN-SOO JEUN • *Department of Neurosurgery, Seoul St. Mary's Hospital, College of Medicine, The Catholic University of Korea, Seoul, Korea*

SHAILENDRA JOSHI • *Department of Anesthesiology, College of Physicians and Surgeons of Columbia University, New York, NY*

CHARLES W. KANALY • *Division of Neurosurgery, Department of Surgery, Duke University Medical Center, Durham, NC*

SEONG MUK KIM • *Department of Biomedical Science, College of Medicine, The Catholic University of Korea, Seoul, Korea*

KIRIL LANEVSKIJ • *Pharma Algorithms Inc., Vilnius, Lithuania*

JIAN YI LI • *Division of Neuropathology, Department of Pathology and Lab Medicine, Harvey Cushing Brain Tumor Institute, Feinstein Institute for Medical Research, North Shore-Long Island Jewish Health System, Manhasset, NY*

JUNG YEON LIM • *Department of Biomedical Science, College of Medicine, The Catholic University of Korea, Seoul, Korea*

PHILIP C.W. LUI • *Department of Anatomic and Cellular Pathology, Prince of Wales Hospital, The Chinese University of Hong Kong, Shatin, Hong Kong, China*
YAEL MARDOR • *The Advanced Technology Center, Sheba Medical Center, Tel-Hashomer, Israel; Medical School, Tel-Aviv University, Tel-Aviv, Israel*
STEPHAN MITTERMEYER • *BrainLAB AG, Feldkirchen, Germany*
ALON MOGILNER • *Department of Neurosurgery, North Shore LIJ Health System, Manhasset, NY*
BIRGIT NEUHUBER • *Department of Neurobiology & Anatomy, Drexel University College of Medicine, Philadelphia, PA*
ALANAS PETRAUSKAS • *Pharma Algorithms Inc., Vilnius, Lithuania*
RAGHU RAGHAVAN • *Therataxis, LLC, Baltimore, MD*
ZVI RAM • *Medical School, Tel-Aviv University, Tel-Aviv, Israel; Neurosurgery Department, Tel-Aviv Medical Center, Tel-Aviv, Israel*
PATRICK T. RONALDSON • *Department of Medical Pharmacology, College of Medicine, The University of Arizona, Tucson, AZ*
CHUNG HEON RYU • *Department of Biomedical Science, College of Medicine, The Catholic University of Korea, Seoul, Korea*
JOHN H. SAMPSON • *Division of Neurosurgery, Department of Surgery, The Preston Robert Tisch Brain Tumor Center at Duke, Duke University Medical Center, Durham, NC*
KARIN E. SANDOVAL • *Department of Pharmaceutical Sciences, School of Pharmacy, Southern Illinois University Edwardsville, Edwardsville, IL*
MICHAEL SCHULDER • *Department of Neurosurgery, North Shore LIJ Health System, Manhasset, NY*
NATALIA VYKHODTSEVA • *Department of Radiology, Focused Ultrasound Laboratory, Brigham and Women's Hospital, Harvard Medical School, Boston, MA*
YONGTING WANG • *Neuroscience and Neuroengineering Center, Med-X Research Institute, Shanghai Jiao Tong University, Shanghai, China*
KEN A. WITT • *Department of Pharmaceutical Sciences, School of Pharmacy, Southern Illinois University Edwardsville, Edwardsville, IL*
GUO-YUAN YANG • *Neuroscience and Neuroengineering Center, Med-X Research Institute, Shanghai Jiao Tong University, Shanghai, China*
FALEI YUAN • *Neuroscience and Neuroengineering Center, Med-X Research Institute, Shanghai Jiao Tong University, Shanghai, China*

Chapter 1

An Overview of Drug Delivery to the CNS

Kewal K. Jain

Abstract

Limitation of drug delivery to the central nervous system (CNS) is a major problem in development of successful treatment of CNS disorders. Concepts of blood–brain barrier (BBB), its role in transport of various substances from the blood to the brain, as well as strategies to deliver drugs across the BBB have evolved over the past century. This chapter is an overview of the challenges and various approaches to drug delivery in CNS disorders as an introduction to other chapters, which deal with laboratory and clinical methods of CNS drug delivery. A classification of current strategies for drug delivery across the BBB is presented. These include novel formulations of drugs such as nanoparticles and strategies to cross the BBB. Drugs can be introduced directly into the CNS and various devices are used for this purpose. Drugs can be administered systemically by various routes for targeted delivery to the site of action. Various methods of cell and gene therapies are used for drug delivery to the CNS. Finally, methods of delivery are classified according to various neurological disorders.

Key words: Drug delivery, central nervous system, blood–brain barrier, cell therapy, gene therapy.

1. Introduction

The delivery of drugs to the brain is a challenge in the treatment of central nervous system (CNS) disorders. Neurological disorders are a substantial and important part of currently known human diseases. Since the 1950s the number of drugs used for neurological disorders has steadily increased. Currently, about 110 drugs are in use for neurological disorders and over 500 more agents are in development by the pharmaceutical industry. In the past, treatment of CNS disease was mostly by systemically administered drugs and this trend continues. Most of the pharmaceutical research in CNS disorders is directed toward the discovery of drugs and formulations for controlled release; little attention has

K.K. Jain (ed.), *Drug Delivery to the Central Nervous System*, Neuromethods 45,
DOI 10.1007/978-1-60761-529-3_1, © Humana Press, a part of Springer Science+Business Media, LLC 2010

been paid to the method of delivery of these drugs to the brain. Many disorders, such as Alzheimer's disease (AD), have no satisfactory treatment at present, and the evolving treatments require innovative methods of delivery. A detailed description of drug delivery systems is presented elsewhere (1).

Drugs may be administered directly into the CNS or administered systematically (e.g., by intravenous injection) for targeted action in the CNS. The major obstruction to CNS drug delivery is the blood–brain barrier (BBB), which limits the access of drugs to the brain substance. The Chapters 2–4 of this book deal with BBB and strategies for delivery of drugs through this barrier (2–4). This chapter is an overview of drug delivery, whereas other chapters deal with detailed methods of delivery of therapeutics to the CNS.

2. Historical Evolution of Drug Delivery for CNS Disorders

Landmarks in the historical evolution of drug delivery technology to the brain are shown in **Table 1.1**. They are related to the history of the blood–brain barrier (BBB).

Although the cerebral ventricles were tapped for hydrocephalus in ancient times, the first perforation of subarachnoid space by lumbar puncture was made in 1885 to administer cocaine for anesthesia (5). Paul Ehrlich described the concept of the blood–brain barrier in the same year when he observed that dyes injected into the vascular system were rapidly taken up by all the organs except the brain (6). Later, research showed that dyes injected into the cerebrospinal fluid have free access to the neural tissues but do not enter the blood supply of the brain. Coining of the term "blood–brain barrier" to describe this phenomenon is attributed to Lewandowski in 1900 (7). Despite the BBB, intracerebral distribution of various substances was observed. "Barrière hémato-encéphalique" was defined as a cerebral blood vessel compartment in which choroid plexus was semipermeable, facilitating the flow of substances from the blood into the CSF (11). BBB permeability to hexoses, amino acids, amines, and neurotransmitters was demonstrated 50 years later by radiolabeled substances (12).

Broman first observed the transient opening or disruption of the BBB after intracarotid arterial administration of hypertonic solutions in 1941 (13). The first injections into the cerebral circulation were of contrast materials for cerebral angiography (14). The injection of a therapeutic substance (diazepam) into the carotid arteries was not reported until almost a half-century later (15).

Table 1.1
Landmarks in the development of drug delivery to the CNS

Year	Observation/concept/comment
1885	First lumbar puncture to administer cocaine for anesthesia (5).
1885	Concept of BBB indicated by the observation that dyes injected into the vascular system were rapidly taken up by all the organs except the brain (6).
1900	Coining of the term "blood–brain barrier" to describe the phenomenon (7).
1913	BBB observed to be decreased in the choroid plexus (8).
1920	Intracerebral distribution of various substances administered systematically was observed.
1927	First injections into the cerebral circulation: contrast materials for cerebral angiography.
1940	Description of vertebral venous plexus and its connection to blood vessels of the brain laid the anatomical basis for use of epidural venous injection for drug delivery to the CNS (9).
1940s	Tor Broman of Goteborg, Sweden showed that the anatomical substrate of the BBB was the brain capillary wall. This was confirmed by electron microscopic studies a quarter of a century later.
1941	Opening of the BBB by pharmacological means.
1947	Stereotactic equipment for guided placement of instruments at selected targets in the depth of the brain for the treatment of movement disorders.
1950s	Electron microscopy used to demonstrate lack of extracellular fluid compartment between glia and neurons in the brain and this was given as an explanation of failure of substances to enter the brain.
1954	Injection of a mixture of procaine and alcohol into the globus pallidus of the brain for treatment of movement disorders.
1967	Electron microscopy confirmed brain capillary wall to be the BBB (10).
1973	First injection of a therapeutic substance (diazepam) into the carotid arteries.
1978	First implantable pump for intrathecal and intraventricular injection of morphine.
1980s	Studies in molecular biology of the BBB. Cloning and sequencing of glucose transporter gene.
1990s	Further development of direct injections of therapeutic substances, including biologicals (e.g., gene therapy), into the brain or intracranial lesions, and development of strategies to overcome the BBB.
1995	Use of nanoparticles for drug delivery across BBB.

The advent of stereotactic surgery more than 50 years ago opened the way for placing instruments at selected targets in the depths of the brain for the treatment of movement disorders (16). This approach was used some years later to perform chemopallidectomy by injection of a mixture of procaine and alcohol into the globus pallidus (17). The techniques of creating lesions in basal ganglia have been refined, but these principles of localization and injection continue to be used for the introduction of novel therapeutic agents into the brain for the treatment of movement disorders. The first implantable pump for intrathecal and intraventricular injection of morphine for the treatment of cancer pain was described in 1978 (18). During the past 25 years, further progress has taken place with the development of intra-arterial chemotherapy, direct injections of therapeutic substances into intracranial lesions, and strategies to overcome the BBB. Further advances have taken place with the development of cell and gene therapies as well as nanobiotechnology.

3. Passage of Substances Across the Blood–Brain Barrier

Several carrier or transport systems, enzymes, and receptors that control the penetration of molecules have been identified in the BBB endothelium on the basis of physiological and biochemical studies. Passage of substances across the blood–brain barrier is shown in **Fig. 1.1**.

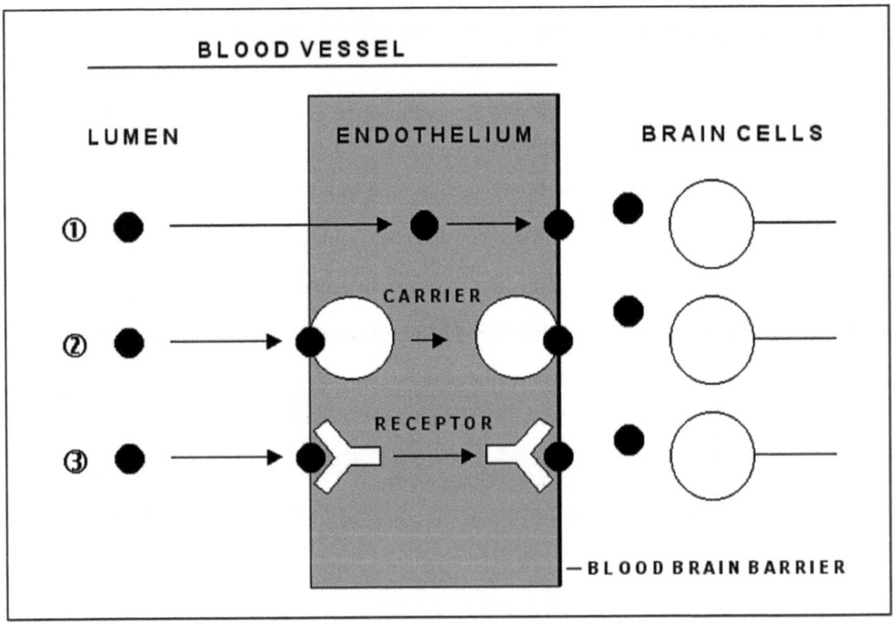

Fig. 1.1. Various forms of passage of substances across the blood–brain barrier. 1. Passive diffusion. Fat-soluble substances dissolve in the cell membrane and cross the barrier, e.g., alcohol, nicotine, and caffeine. Water-soluble substances like penicillin have difficulty in getting through. 2. Active transport. Substances that the brain needs such as glucose and amino acids are carried across by special transport proteins. 3. Receptor-mediated transport. Molecules link up to receptors on the surface of the brain and are escorted through, e.g., insulin.

Various transporters localized in the BBB that control penetration of molecules across the BBB include the following:

- Energy transporters such as glucose transporter
- Amino acid transporters
- Neurotransmitter transporters
- Organic anions and cation transporters
- ABC transporters
- Efflux systems such as P-glycoprotein efflux system
- Miscellaneous transporters such as some hormones and vitamins.

Among these, the action of efflux transporters at the BBB may have clinical significance by reducing the effectiveness of drugs targeted at CNS disorders. Therefore, modulation of these efflux transporters by design of inhibitors and/or design of compounds that have minimal affinity for these transporters may well enhance the treatment of intractable CNS disorders.

Various enzymes control the penetration of molecules across the BBB. An example is monoamineoxidase, which provides an enzymatic barrier and hinders the influx of monoamine precursors into the brain. After their entry into the endothelial cells, monoamines are decarboxylated by cytoplasmic monoamineoxidase, thus, effectively preventing a flood of peripheral monoaminergic neurotransmitters in the neuronal environments.

3.1. Receptor-Mediated Transcytosis

Transcytosis, the transport of peptides and proteins across cellular barriers, has been documented in a number of systems. Examples include the transport of immunoglobulin across the intestinal epithelium and human placenta, the transport of insulin and insulin-like growth factors across the aortic endothelium, and the transport of epidermal growth factor across the kidney epithelium. It is not surprising that transcytosis occurs across the BBB. Factors that participate in receptor-mediated transcytosis across BBB include the following:

- Angiotensin II
- Basic albumin
- Mannose 6-phosphate receptor
- Insulin and insulin-like growth factors-I and II
- Transferrin

3.2. Current Concepts of the Permeability of the BBB

A recent evaluation challenges common opinions about BBB permeation and brain uptake capacity (19). According to this review, BBB does not appear to be as impermeable as often perceived, and is able to absorb compounds of considerably greater size and lower lipophilicity than previously suggested. The absorptive capacity of BBB is similar to that of the highly permeable hepatocytes. The suggestion that only a few drugs enter the brain appears to have been based on studies including a large portion of CNS-inactive drugs with good BBB permeability. The proposed requirement to design CNS compounds for active BBB uptake appears exaggerated.

Data show that compounds with sufficiently good intestinal uptake and no, or minor, BBB efflux and metabolism should have no problems to readily enter the brain. The intestinal uptake could be rate limiting for the absorption rate into the CNS. Intestinal uptake data could, therefore, potentially replace BBB data for compounds with no or negligible active transport.

Unbound compounds are not completely absorbed because of a short transit time through the brain. For high permeability compounds, the uptake of unbound fraction during each transit may reach approximately 50%. This makes brain uptake comparably (e.g., versus intestinal and hepatic uptake) sensitive to involvement and changes of active transport for compounds with high passive permeability. This has been demonstrated in animals and humans in vivo, and in simulations with a human brain uptake PB-IVIV (physiologically based in vitro to in vivo) prediction method.

3.3. Strategies to Cross the BBB

Various strategies that have been used for manipulating the blood–brain barrier for drug delivery to the brain include osmotic and chemical opening of the BBB as well as the use of transport/carrier systems. Other strategies for drug delivery to the brain involve bypassing the BBB. The drawbacks of strategies to open the BBB are damage to the barrier as well as uncontrolled passage of drugs into the brain. The ideal method for transporting drugs across the BBB should be controlled to prevent damage to the barrier.

Potential therapeutic applications of manipulation of the BBB are mostly in use to facilitate drug delivery to brain tumors in clinical trials. Methods of focal delivery of therapeutic and diagnostic substances to the brain across the BBB are in experimental stages for infectious, genetic, and neurodegenerative disorders. A particular advantage will be the delivery of the genes to the required site as opposed to the exposure of the whole brain to the therapeutic agent. Genetic and other defects leading to brain changes in Down's syndrome, Alzheimer's disease, amyotrophic lateral sclerosis, Huntington's disease, Gaucher's disease, hypertension, and other disorders are rapidly being identified. Several effective therapeutic agents are available but their use is limited pending improvement of drug delivery across the BBB.

4. Methods of Drug Delivery to the CNS

A classification of various methods of drug delivery to the CNS is shown in **Table 1.2**.

Table 1.2
Classification of various methods of drug delivery to the CNS

Systemic administration of therapeutics substances for targeted CNS action
Intramuscular
Intravenous injection
Transmucosal: buccal and nasal
Transdermal
Direct administration of therapeutic substances to the CNS
Introduction into CSF pathways: intraventricular, lumbar subarachnoid
Introduction into the cerebral arterial circulation
Introduction into the brain substance
Targeting of CNS by retrograde axonal transport
Intranasal: introduction into the brain along the olfactory tract
Inner ear: intra-tympanic injection into the cochlea
Direct positive pressure infusion
Drug delivery by manipulation of the blood–brain barrier
Drug delivery using novel formulations
Conjugates
Gels
Liposomes
Microspheres
Nanoparticles
Chemical delivery systems
Drug delivery devices
Pumps
Catheters
Implants releasing drugs
Use of microorganisms for drug delivery to the brain
Bacteriophages for brain penetration
Bacterial vectors
Cell/gene therapy

4.1. Delivery of Cell Therapy to the CNS

Various methods for delivery of cell therapies for CNS disorders are shown in **Table 1.3**.

4.2. Gene Delivery for CNS Disorders

Various methods of gene transfer as applied to the nervous system are shown in **Table 1.4**.

Table 1.3
Methods for delivering cell therapies in CNS disorders

CNS implants of live cells secreting therapeutic substances
Introduction into the brain substance at specific locations using stereotactic methods
CNS implants of genetically engineered cells producing therapeutic substances
Introduction into cerebrospinal fluid pathways: intrathecal or intraventricular
Implantation into the brain substance
Genetically engineered stem cells
CNS neo-tissue implant: fetal cells, synthetic matrix elements, and controlled-release protein
Cells for facilitating crossing of the blood–brain barrier
Activated genetically engineered T lymphocytes
Drug delivery devices
Image-guided catheters for delivery of cells into various parts of the CNS
Intrathecal delivery of cells
Subarachnoid delivery of stem cells
Intrathecal implants: encapsulated genetically engineered cells secreting therapeutics
Cells for facilitating crossing of the BBB
Systemic intravenous delivery targeted to the brain

Table 1.4
Methods of gene transfer as applied to neurologic disorders

Indirect gene transfer
Transplantation of genetically engineered cells: neuronal and non-neuronal
Neuronal cell targeting

(continued)

Table 1.4 (continued)

Direct gene transfer	
	Viral vectors: neurotropic and non-neurotropic viruses
	Non-viral vectors
Methods of administration	
	Direct injection into the CNS: specific locations and into lesions
	Systemic delivery, e.g., intravenous injection
	Injection into muscles
	Intranasal instillation for introduction into the brain along the olfactory tract
	Transfer into cerebral circulation
Targeting of CNS by retrograde axonal transport	

5. Drug Delivery in the Treatment of CNS Disorders

This section contains examples of application of the drug delivery methods in the treatment of neurological disorders. Some common and well-known disorders are included such as Parkinson's disease (PD), Alzheimer's disease (AD), and pain.

5.1. Drug Delivery in the Management of PD

Various strategies for drug delivery in the treatment of PD are listed in **Table 1.5**. Some are approved whereas others have been investigated in clinical trials.

Table 1.5
Strategies for the drug delivery in PD

Drug delivery system	Therapeutic agent	Remarks
Special oral formulation	Duodopa® (micronized levodopa-carbidopa)	Duodenal infusion reduces motor fluctuations of PD
Controlled release	Geometrix/Madopar (carbidopa + levodopa)	An oral time-release capsule which expands in the stomach and gradually releases the drug over time to avoid fluctuations of drug serum levels
Sublingual	Apomorphine	For relieving the off periods, particularly if the patients are unable to anticipate such periods

(continued)

Table 1.5 (continued)

Drug delivery system	Therapeutic agent	Remarks
Transdermal	Rotigotine: a D2 agonist	To "smooth out" the levels of the drug in the brain
Intracerebro-ventricular catheter	Glial cell line-derived neurotrophic factor (GDNF)	Discontinued development for lack of efficacy after phase II clinical trials
Direct infusion using a catheter implanted into the putamen	Recombinant human GDNF	In phase II clinical trials to test direct effect of GDNF on dopamine function
Direct infusion into a target identified by CT scan and MRI	Gene therapy: delivery of glutamic acid decarboxylase gene to the brain with AAV vector	Has undergone a pilot clinical trial
Cell therapy: stereotactic image-guided implantation of cells in specific areas of the brain	Spheramine: human retinal pigmented epithelial cells in gel microspheres	Safety and efficacy shown in phase I/II clinical trials
	Encapsulated dopamine-secreting cells	Clinical trials in progress in Japan
	Embryonic nigral cell implants	A double-blind, placebo-controlled trial showed efficacy and safety
	Engineered stem cells to produce and deliver GDNF	Increase the survival and functioning of dopamine neurons shown in clinical trials
Gene therapy: CERE-120	An adeno-associated viral vector encoding neurturin – a neurotrophic factor	Preliminary clinical trials show safety and efficacy
RNAi: administration of siRNAs that are specific to the α-synuclein gene	Silencing of α-synuclein gene, which is associated with PD	Preclinical studies

5.2. Alzheimer's Disease

Some novel drug delivery methods for Alzheimer's disease (AD) therapies are listed in **Table 1.6**. Several of these have been discontinued and therapeutic delivery in AD remains a challenge.

5.3. Drug Delivery for Pain

Pain is a neurological disorder and relief of pain involves delivery of drugs to the nervous system peripheral or central. Methods of delivery of drugs for pain are shown in **Table 1.7**.

Table 1.6
Novel drug delivery methods for AD therapeutics

Drug/delivery methods	Rationale/Comments	Status
Acetylcholinesterase inhibitors: tolserine and thiatolserine	Accentuate the duration of action and can be given by a transdermal patch	Phase I
Acetylcholinesterase (AChE) inhibitor: Debio-9902 SR slow release implant	Dual action: AChE inhibitor and NMDA receptor antagonist	Phase IIa
Arecoline/transdermal	A muscarinic receptor agonist	Phase I completed but no further development
Drugs binding Aβ/nanoparticles	Drug-binding nanoparticles can trace and clear Aβ in the brain	Experimental
Estrogen E_2/chemical delivery system/buccal formulation	Brain targeted delivery of estrogen is safer and more effective in prevention of AD than systemic delivery	Phase II in 2002 but no further development since then
Intracerebroventricular administration of GM1 ganglioside	Neurotrophic/procedure was safe and well tolerated with some beneficial effects	Pilot study but no further development
Intracerebroventricular bethanecol chloride	The procedure was well tolerated but degree of improvement was not sufficient to justify further trials	Multicenter double-blind trial in 1989 but no development since then
Intracerebroventricular NGF	Beneficial effect in a few patients but complication of severe peripheral pain	Open study. No further development
Rivastigmine/transdermal preparation	A cholinesterase inhibitor, approved for AD as an oral preparation.	Approved by the FDA in 2007
Stereotactic intracerebral implantation of engineered cells secreting nerve growth factor (NGF)	NGF can prevent the death of nerve cells affected by AD and enhance the function of remaining brain cells	Phase II
Intranasal administration of NGF	Non-invasive alternative to intra-cerebroventricular injection of NGF	Preclinical
Thyrotropin-releasing hormone (TRH) analogs/molecular packaging	TRH can increase extracellular acetylcholine levels	Preclinical

Table 1.7
A classification of drug delivery methods used in management of pain

Injections
Subcutaneous injection
Local injection in a joint or near a peripheral nerve
Intramuscular injection
Intravenous injection
Administration via gastrointestinal system
Oral: tablets, solutions, and capsules
Rectal suppositories
Topical application with absorption
Ointments and gels
Lotions
Transdermal drug delivery
Patch
Iontophoresis
Electroporation
Ultrasound-mediated
Buccal transmucosal
Intranasal
Pulmonary drug delivery: inhalation
Spinal drug delivery
Intrathecal
Extradural
Intracerebroventricular drug delivery
Pumps
Controlled release technologies
Particulate drug delivery systems
Microspheres
Nanoparticles
Liposomes
Cell and gene therapy

6. Concluding Remarks

This brief overview outlines the scope of drug delivery for neurological disorders and the variety of methods available. Treatment of neurological disorders remains a challenge, not only because of lack of specific treatments for many disorders but also difficulties in delivering it to the CNS. Refinement of drug-delivery systems using new technologies such as nanobiotechnology are contributing to the improvements in clinical applications of some of the new methods of treatment.

References

1. Jain KK. Drug Delivery in Central Nervous System Disorders: Technologies, Companies & Markets. Jain PharmaBiotech Publications, Basel, 2010:1–325.
2. Lanevskij K. Prediction of BBB penetration by drugs. In: Jain KK (ed). Drug Delivery to the Central Nervous System. Springer, New York, 2010.
3. Li JY. Receptor-mediated transport of drugs across the BBB. In: Jain KK (ed). Drug Delivery to the Central Nervous System. Springer, New York, 2010.
4. Vykhodtseva N. Disruption of BBB by ultrasound for drug delivery to the brain. In: Jain KK (ed). Drug Delivery to the Central Nervous System. Springer, New York, 2010.
5. Corning JL. Spinal anesthesia and local medication of the cord. NY Med J 1885;42:483–485.
6. Ehrlich P. 'Das sauerstoffbedürfnis des organismus,' eine farbanalytische studie. Hirschwald, Berlin, 1885.
7. Lewandowski M. Zur Lehre der Cerebrospinalflüssigkeit. Klin Med 1900;40: 480–494.
8. Goldmann E. Vitalfarbungen am Zentralnervensystem. Beitrag zur Physio-Pathologie des Plexus Choroideus und der Hirnhaute (Intravital labeling of the central nervous system. A study on the pathophysiology of the choroid plexus and the meninges). Abhandlungen der konigliche preussischen Akademie der Wissenshaften. Physikalisch-Mathematische Klasse 1913;1:1–64.
9. Batson OV. The function of the vertebral veins and their role in the spread of metastases. Ann Surg 1940;112:138–149.
10. Reese TS and Karnovsky MJ. Fine structural localization of a blood-brain barrier to exogenous peroxidase. J Cell Biol 1967;34: 207–217.
11. Stern L and Gautier R. Rapports entre le liquide céphalo-rachidien et la circulation sanguine. Arch Int Physiol 1921;17: 138–192.
12. Oldendorf WH. Brain uptake of radiolabelled amino acids, amines and hexoses after arterial injection. Am J Physiol 1971;221: 1629–1639.
13. Broman T. The possibilities of the passage of substances from the blood to the central nervous system. Acta Psychol Neurol 1941;16:1–25.
14. Moniz E. L'encéphalographie artérielle, son importance dans la localisation des tumors cérébrales. Rev Neurol 1927;2:72–90.
15. Doppman JL. Intra-arterial valium – its safety and effectiveness. Radiology 1973;106:335–338.
16. Spiegel EA, Wycis HAT, Marks M, and Lee AJ. Stereotactic apparatus for operations on the human brain. Science 1947;106:349–350.
17. Cooper IS. Intracerebral injection of procaine into the globus pallidus in hyperkinetic disorders. Science 1954;119:417–418.
18. Lazorthes Y, Sallerin-Caute B, Verdie JC, and Bastido R. Advances in drug delivery systems and applications in neurosurgery. Adv Tech Stand Neurosurg 1991;18:143–192.
19. Fagerholm U. The highly permeable blood-brain barrier: an evaluation of current opinions about brain uptake capacity. Drug Discov Today 2007;12:1076–1082.

Chapter 2

Receptor-Mediated Transport of Drugs Across the BBB

Ji-Ming Feng, Philip C.W. Lui, and Jian Yi Li

Abstract

Blood–brain barrier (BBB) limits drug delivery to the brain parenchyma. The ultimate goal of brain drug targeting technology is to deliver therapeutic agents across BBB. Insulin or transferrin as well as other endogenous peptides undergo receptor-mediated transcytosis or transport (RMT) across the BBB in vivo. Certain peptidomimetic monoclonal antibodies (mAb) for insulin receptor or transferrin receptor can also cross the BBB via RMT on the endogenous receptors. These mAb can act as molecular Trojan horses to shuttle into the brain a wide range of therapeutics including recombinant proteins, antibodies, RNA interference drugs, or non-viral gene products. During the last two decades, RMT-based brain drug transport techniques have been developed. This chapter will focus on introducing three major technologies in the arena of RMT brain drug delivery: (1) Avidin–biotin technology; (2) Fusion protein technology; and (3) Trojan horse liposomes (THLs).

Key words: Blood–brain barrier, receptor-mediated transcytosis or transport (RMT), molecular Trojan horse, brain drug delivery, avidin–biotin technology, fusion protein technology, Trojan horse liposomes (THLs).

1. Introduction

The bottleneck in the development of new drugs for the brain is the permeability of blood–brain barrier (BBB). There is a widely held misconception that small molecules readily cross the BBB. Unfortunately, only small molecules which are both lipid soluble and have a molecular weight (MW) <400 Da can cross the BBB in pharmacologically significant amounts (1). In the comprehensive medicinal chemistry (CMC) database, <5% of all drugs out of more than 7000 molecules are effective in the brain, which are primarily used to treat four conditions: affective disorders, chronic pain, insomnia, and epilepsy (2). Another study in which affective disorders are excluded demonstrates <1% of all drugs are pharmacologically active in the

K.K. Jain (ed.), *Drug Delivery to the Central Nervous System*, Neuromethods 45,
DOI 10.1007/978-1-60761-529-3_2, © Humana Press, a part of Springer Science+Business Media, LLC 2010

CNS (3). The situation for the larger molecules is even worse including all of the products of biotechnology: recombinant growth factors and enzymes, monoclonal antibodies (mAb), antisense RNA drugs, short interfering RNA (siRNA), or gene therapy. Almost all such large molecule drugs cannot get access to the brain parenchyma due to the presence of BBB. Traditionally, either medicinal chemistry or neurosurgical-based invasive techniques has been applied to achieve the delivery of drugs to the brain with either, which is not as effective as expected.

Table 2.1
Species-specific peptidomimetic monoclonal antibodies for receptor-mediated transport of drugs across the BBB

Species to be targeted	Peptidomimetic monoclonal antibodies
Mouse	8D3 rat mAb to mouse transferrin receptor (TfR)
Rat	OX26 murine mAb to rat TfR
Rhesus monkey	83-14 murine mAb to human insulin receptor
Human	Genetically engineered chimeric mAb to human insulin receptor (HIR)

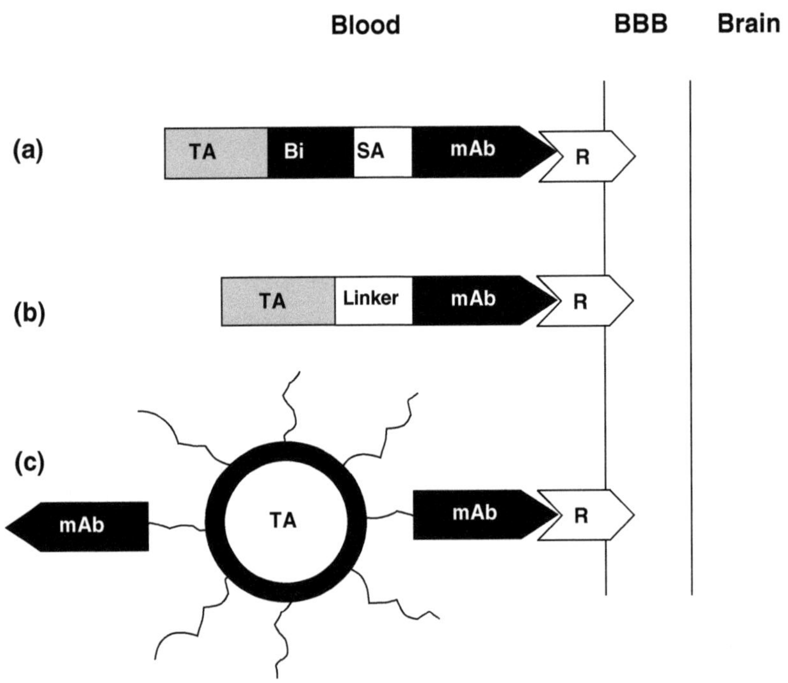

Fig. 2.1. Three technologies for delivery of therapeutics to the brain through RMT. (**a**) Avidin (streptavidin)–biotin technology. (**b**) Fusion protein technology. (**c**) Trojan horse liposomes (THLs). The mAb (monoclonal antibodies) binds the BBB receptor to trigger influx across the BBB. TA (therapeutic agent) follows the transcytosis of mAb and crosses the BBB. SA, streptavidin; Bi, biotin; R, receptor.

Peptidomimetic monoclonal antibodies (mAb) for insulin receptor or transferrin receptor can mimic the way used by insulin or transferrin to bind to their corresponding receptors and induce RMT crossing the BBB. Based on this property, these mAb can be potentially used as the molecular Trojan horses to deliver the macromolecular drugs into the brain parenchyma (**Table 2.1**). The idea has been formulated into one of three technology platforms which can be used to ferry virtually any large molecule pharmaceutical across the BBB in vivo (**Fig. 2.1**). First of all, oligopeptides, antisense RNA, or siRNA can be delivered with avidin (streptavidin)–biotin technology (**Fig. 2.1**). Second, recombinant growth factors and monoclonal antibodies can be delivered with fusion protein technology (**Fig. 2.1**). Finally, non-viral plasmid DNAs or genes encoding shRNAs may be shuttled to the brain with Trojan horse liposomes (THLs) (1, 4) (**Fig. 2.1**).

2. Materials

2.1. Avidin (Streptavidin)–Biotin Technology

1. OX26 hybridoma cell line is originally produced in laboratory of Dr. Alan F. William (University of Oxford) and is obtained from Dr. Arthur Like (University of Massachusetts Medical School).
2. Female Balb/c mice (4–6 weeks) are purchased from Charles River (Wilmington, MA).
3. Cyanogen bromide-activated Sepharose 4B, ABTS, Superose 6HR 10/30 FPLC columns, and 2.6 $\times$ 96 cm column of Sephacryl S300HR are obtained from Pharmacia Biotech., Inc. (Piscataway, NJ).
4. Tris–HCl ready gels (12%), ultra-pure SDS, Coomassie brilliant blue-R250, prestained SDS-PAGE low range and biotinylated SDS-PAGE molecular weight standards, mini-protein electrophoresis apparatus, and mini-transblot apparatus are obtained from Bio-Rad Labs (Hercules, CA).
5. BCA protein assay kit is purchased from Pierce Chemical Co. (Rockford, IL). PVDF Immobilon-P transfer membrane (0.45 μ) is purchased from Millipore Corp. (Burlington, MA).
6. Biotinylated goat anti-rabbit IgG (H+L), biotinylated horse anti-mouse IgG, peroxidase-labeled anti-mouse IgG (H+L), and Vectastain ABC peroxidase kit were obtained from Vector Labs, Inc. (Burlingame, CA).
7. Human recombinant BDNF from *Escherichia coli* is obtained from Amgen (Thousand Oaks, CA).

8. [^{3}H]-biotin (45 Ci/mmol), [^{14}C] sucrose (600 mCi/mmol), and [^{125}I]iodine (17 Ci/mg) are obtained from DuPont-NEN (Boston, MA).
9. Methoxy-PEG-hydrazide (Hz) of molecular weight 2000 Da (designated PEG 2000-Hz) and biotin-PEG 2000-Hz are synthesized by Shearwater Polymers (Huntsville, AL).
10. High-trap copper affinity columns and Superose12HR (10/30) columns are purchased from Pharmaica Biotech., Inc. (Piscataway, NJ).
11. A 7.8 × 300-mm TSK-Gel G2000SW$_{XL}$ HPLC column is obtained from Toso Hass (Montgomeryville, PA).
12. 2-Iminothiolane (Traut's reagent), the bicinchoninic acid assay (BAC) kit, and sulfosuccinimidy 2-(biotinamido) ethyl-1,3′-dithiopropionate (NHS-SS-biotin) are purchased from Pierce Chemical Corporation (Rockford, IL).
13. The G25 Quick Spin column is obtained from Roche Molecular Biochemicals (Indianapolis, IN).
14. Centriprep-10 microconcentrator is obtained from Amicon Corp. (Beverly, MA).
15. Ultrafree MC ultrafiltration units are purchased from Millipore Corporation (Bedford, MA).
16. Male Sprague-Dawley rats (weighing 260–290 g) are obtained from Harlan Sprague-Dawley Inc. (Indianapolis, IN).
17. Lactoperoxidase is obtained from Worthington Biochemical Corporation (Lakewood, NJ).
18. Avidin, rat serum albumin, *N*-methyl-*N*′-3(dimethylaminopropyl)carbodiimide hydrochloride, recombinant streptavidin, and all other chemicals are purchased from Sigma Chemical Co. (St. Louis, MO).

2.2. Fusion Protein Technology

1. 8- to 10-year-old female Rhesus monkeys are purchased from Covance (Alice, TX).
2. Serum-free medium, ultra-pure urea, DMEM, RPMI-1640, Iscove's modified Dulbecco's medium, and fetal calf serum (FCS) are obtained from Invitrogen-Gibco (Frederick, MD).
3. Tris–HCl ready gels (12%), ultra-pure SDS, Coomassie brilliant blue-R250, pre-stained sodium dodecyl sulfate polyacrylamide gel electrophoresis (SDS-PAGE) low range and biotinylated SDS-PAGE molecular weight standards, silver stain plus, mini-protein lectrophoresis apparatus, mini-trans-blot apparatus, and 0.4-cm gap electroporation cuvette are obtained from Bio-Rad Labs (Hercules, CA).
4. BCA Protein assay kit was purchased from Pierce Chemical Co. (Rockford, IL).

5. PVDF Immobilon-P transfer membrane (0.45 mm) is purchased from Millipore Corp. (Burlington, MA).
6. Biotinylated horse anti-mouse IgG, peroxidase labeled anti-mouse IgG (HþL), and Vectastain ABC peroxidase kit are obtained from Vector Labs, Inc. (Burlingame, CA).
7. Oligodeoxythymidine (dT) cellulose, corning spin X filter unit, bovine serum albumin, polyoxyethylenesorbitan monolaurate (Tween-20), 3,3-diamiobenzidine (DAB), L-arginine, imidazole, isopropyl-b-D-thiogalactopyranoside (IPTG), Biomax X-ray film, and all other molecular biology grade reagents are purchased from Sigma Chemical Co. (St. Louis, MO).
8. Chelating Sepharose fast flow, oligo d(T)18, and 100 mM dNTPs is obtained from Pharmacia Biotech., Inc. (Piscataway, NJ).
9. Guanidine hydrochloride is purchased from Fisher Scientific (Tustin, CA).
10. QIAquick Gel Extraction Kit, QIAGEN Plasmid Midi Kit, and QIAquick PCR Purification Kit are obtained from Qiagen Inc. (Valencia, CA).
11. COS cells are purchased from ATCC (Manassas, VA).
12. Pefabloc SC is purchased from Roche Molecular Biochemicals (Indianapolis, IN).
13. Centriprep-10 microconcentrators and Centriprep-30 concentrators are obtained from Amicon Corp. (Beverly, MA).
14. XL1-Blue supercompetent cells, Taq extender, and proteinase K are obtained from Stratagene (La Jolla, CA).
15. All restriction endonucleases (RE) and modifying enzymes are obtained from Promega (Madison, WI).
16. AmpliTaq DNA polymerase is purchased from PE Applied Biosystems (Foster City, CA).
17. Custom oligodeoxynucleotides (ODN) are obtained from Biosource International (Camarillo, CA).

2.3. Trojan Horse Liposomes

1. OX26 hybridoma cell line is originally produced in laboratory of Dr. Alan F. William (University of Oxford) and is obtained from Dr. Arthur Like (University of Massachusetts Medical School).
2. Mouse myeloma ascites IgG2a (k) is from MP Biomedicals (Solon, OH).
3. Protein G Sepharose is obtained from Invitrogen (Carlsbad, CA).
4. [^{125}I]iodine are purchased from Dupont/NEN (Boston, MA).
5. Chloramine T and cholesterol are from Sigma Chemical Co. (St. Louis, MO).

6. 2-Iminothiolane (Traut's reagent) and bicinchoninic acid protein assay kit are obtained from Pierce Chemical Co. (Rockford, IL).
7. Sephadex G-25 gel filtration chromatography, Na-borate buffer, 0.1 mM EDTA, Na-phosphate, Ellmann's reagent Centriprep-30 (molecular weight cut-off: 30,000) concentrators are obtained from Amicon Corp. (Beverly, MA).
8. Distearoylphosphatidylcholine, POPC (1-palmitoyl-2-oleoyl-*sn*-glycerol-3-phosphocholine) and DDAB (didodecyldimethylammonium bromide), and distearoylphosphatidylethanolamine (DSPE) are from Avanti Polar Lipids (Alabaster, AL).
9. 100-nm pore size polycarbonate membrane employing an extruder is purchased from Avestin (Ottawa, Canada).
10. 50-nm polycarbonate membrane and Microtrac Ultrafine Particle Analyzer are obtained from Leeds–Northrup (St. Petersburg, FL).
11. PEG2000 and DSPE-PEG are purchased from Shearwater Polymers (Huntsville, AL).
12. DSPE-PEG-maleimide is custom-synthesized by Shearwater Polymers (Huntsville, AL).
13. The 5.8-kb plasmid, pGL2, encoding for the firefly *Photinus pyralis* luciferase gene, or the 6.8-kb pSV-b-galactosidase plasmid, both under the influence of the SV40 promoter, NickTranslation System, and exonuclease III are obtained from Promega (Madison, WI).
14. QIAfilter Plasmid Maxi kit is obtained from Qiagen (Chatsworth, CA).
15. [a-^{32}P]dCTP (800 Ciymmol) is obtained from Dupont/NEN (Boston, MA).
16. DNA polymerase I, Trichloroacetic acid (TCA), Pancreatic DNase I (from bovine pancreases) with a specific activity of 2000 Kunitz units/mg are purchased from Sigma Chemical Co. (St. Louis, MO).
17. G25 Sephadex column is from Roche Molecular Biochemicals (Indianapolis, IN).
18. Mouse myeloma ascites IgG2a (k) is from Cappel Laboratories, Inc. (Cochranville, PA).
19. Protein G Sepharose CL-4B is from Amersham Pharmacia (Sunnyvale, CA).
20. *N*-succinimidyl[2,3-^{3}H]propionate (^{3}H-NSP) is from Amersham Life Science (Arlington Heights, IL).
21. Male Sprague-Dawley rats (weighing 260–290 g) are obtained from Harlan–Sprague-Dawley Inc. (Indianapolis, IN).

22. Recombinant streptavidin (SA) and m-maleimidobenzoyl *N*-hydroxysuccinimide ester are obtained from Sigma Chemical Co. (St. Louis, MO).
23. Human recombinant BDNF is produced in *E. coli* and is obtained from Amgen (Thousand Oaks, CA).
24. Methoxy-PEG-hydrazide (Hz) of molecular weight 2000 Da (designated PEG 2000-Hz) and biotin-PEG 2000-Hz are synthesized by Shearwater Polymers (Huntsville, AL).
25. EDTA and *N*-methyl-*N*9-3(dimethylaminopropyl)carbodiimide hydrochloride are obtained from Sigma Chemical Co. (St. Louis, MO).
26. High-trap copper affinity columns are obtained from Amersham Pharmacia (Sunnyvale, CA).
27. Centriprep-10 microconcentrator is purchased from Amicon Corp. (Beverly, MA).
28. Tissue-tek OCT embedding medium is obtained from VWR (Batavia, IL).
29. X-GAL histochemistry kit is obtained from Invitrogen (Carlsbad, CA).

3. Method

3.1. Avidin (Streptavidin)–Biotin Technology

We will introduce this technology by the example of delivering BDNF into the brain through linking with the TfR mAb (5–7).

3.1.1. Purification of OX26 Antibody

1. OX26 mAb (monoclonal antibodies against rat TfR) are prepared from the hybridoma cell line as an ascites tumor in pristane-primed Balb/c mice (8).
2. Ascitic antibody is affinity purified with a protein G-Sepharose 4 Fast Flow column.
3. After the column (1.0 × 5.0 cm) is equilibrated with 20 mM phosphate buffer (pH 7.0), 4 ml of ascites are added and followed by a wash with 20 ml of 20 mM phosphate buffer at a flow rate of 1.0 ml/min.
4. Antibodies are eluted with 0.1 M glycine (pH 2.5) and are collected in a fraction collector equipped with an ultraviolet monitor (Pharmacia) set at 280 nm.
5. The IgG fraction is immediately neutralized with 1 M Tris base (pH 9.0) and dialysed against water at 4°C overnight and the precipate is removed by centrifugation.

6. The purified IgG and ascites are analyzed by SDS-PAGE.
7. The protein concentrations are determined by the bicinchoninic acid assay (Pierce Chemical Co.).

3.1.2. Preparation of OX26-SA Conjugate

1. OX26 is thiolated with a 10:1 ratio of Traut's reagent. This ratio generated, on average, two thiol groups per OX26 molecule, which can be determined with 5,5'-dithiobis (2-nitrobenzoic acid) (Ellmann's reagent; Pierce) (9).
2. The recombinant SA is activated with a 20-fold molar ratio of m-maleimidobenzoyl *N*-hydroxysuccinimide ester.
3. Following conjugation of the thiolated OX26 and the activated SA, the OX26-SA conjugate is purified by chromatography on a column (1.6 × 94 cm) of Sephacryl S300HR. The number of biotin binding sites per OX26-SA conjugate can be determined by a [^{3}H]biotin binding assay.

3.1.3. BDNF Radioiodination with Lactoperoxidase

1. Add 3.4 μl of BDNF (50 μg, 2.1 nmol of monomer), 5 μl of lactoperoxidase (0.03 units, 0.15 μg), 2 μCi [^{125}I]iodine (1 nmol), and 3 μl of 0.00075% H_2O_2 (0.022 μg, 0.6 nmol) into a 500-μl Eppendorf tube (10, 11).
2. After 1 min of incubation, another 3 μl of H_2O_2 is added. The step repeats two more times.
3. The reaction is stopped by adding 200 μl of stop solution (2 mg/ml tyrosine, 0.5 M NaCl, pH 7.0)
4. The mixture is applied to a 0.7 × 28 cm column of Sephadex-G25 that is pre-equilibrated with 0.5 M NaCl (pH 6.0).
5. The radiolabeled BDNF is eluted from the column with 0.5 M NaCl (pH 6.0) as a single peak with a specific activity of ~11 μ Ci/μg and a trichloroacetic acid (TCA) precipitability of 99%.
6. The [^{125}I]-BDNF elute from a C4 reverse phase HPLC column has a singe peak using the gradient system, and buffer A containing 0.1% Trifluoroacetic acid (TFA) in water and buffer B containing 90% acetonitrile in 0.1% TFA are used to develop gradients.
7. Labeled BDNF elutes at a slightly longer retention time than unlabeled BDNF. [^{125}I]-labeled BDNF is used within 24 h of labeling.

3.1.4. Preparation of BDNF-PEG 2000-Biotin/OX26-SA Complex

1. [^{125}I]-BDNF(30 μCi, 3 μg of protein) and 6 mg of unlabeled BDNF (0.48 μmol BDNF monomer) are mixed in 200 μl of 0.3 M Pyridine-HCl (pH 4.8).
2. Add 40 mg of PEG 2000-Hz and 18 mg of biotin-PEG 2000-Hz.
3. The reaction is initiated by adding 26 mg of *N*-methyl-*N′*-3(dimethylaminopropyl)carbodiimide hydrochloride and allow to proceed in the dark at room temperature overnight.

4. Adding buffer A (0.05 M Na_2HPO_4/0.5 M NaCl, pH 7.0) to bring the total volume to 1 ml.
5. The mixture is applied to a 5-ml Hi-Trap copper affinity column pre-activated with 0.1 M $CuSO_4$, to remove unreacted PEG 2000.
6. After washing with 20 ml of buffer A, the BDNF-PEG 2000-biotin is eluted with 15 ml of buffer A with 50 mM EDTA, and 1 ml fractions are collected.
7. The blue-colored fractions of copper and pegylated BDNF are pooled by adding buffer A to a volume of 20 ml.
8. The blue-colored fractions of copper and pegylated BDNF are removed to a centriprep-10 microconcentrator to reduce the volume to 2 ml.
9. This cycle is repeated until a final reduction to 1.0 ml
10. Measure protein concentration with the bicinchoninic acid (BCA) assay.
11. Determine the [^{125}I] radioactivity with γ-counter.
12. Store at –20°C until use.
13. In this synthesis, the molar ratio of PEG-Hz/EDAC/BDNF carboxy residues is 22/22/1, and 7.0% of the total PEG-Hz contain biotin at the terminus of the PEG strand opposite of the Hz moiety. The 7.0% biotin content is determined with the HABA assay to yield, on average, of 1.0 biotin residues incorporated per BDNF homodimer.
14. Twenty milligrams (100 nmol) of OX26-SA is added to 2.8 mg of the BDNF-PEG 2000-biotin (100 nmol) containing 12 μCi of [^{125}I]-BDNF-PEG 2000-biotin.
15. The mixture is incubated at room temperature for 30 min.
16. The sample is applied to a 2.6 × 96 cm column of Sephacryl S300HR.
17. Elute with 0.01 M Na_2HPO_4/0.5 M NaCl, pH 7.4 with 0.05% Tween-20 at 30 ml/h.
18. 3-ml fractions are collected.
19. The conjugate peak (fractions 65–90) determined by the 280 nm absorption and [^{125}I] counting, is selected, pooled, and concentrated to 2.0 ml with Centriprep-10 microconcentrator.
20. The final yield of BDNF-PEG 2000-biotin conjugated to OX26-SA is ~65%.
21. The molar ratio of BDNFPEG 2000-biotin to OX26-SA is 1:1. The proportion of the BDNF-PEG 2000-biotin/SA-OX26 that is BDNF homodimer is calculated from the molecular weight (25,000 Da) and the specific activity (mCi/mg) of the BDNF, which is computed from the small amount of

[^{125}I]-BDNF added initially to the unlabeled BDNF. The free [^{125}I]-BDNF-PEG 2000 peak (fractions 91–105) also is collected from the Sephacryl S300 column and concentrated with a Centriprep-10 microconcentrator.

3.1.5. Brain Uptake

1. Male Sprague-Dawley rats are anesthetized i.p. with ketamine (100 mg/kg) and xylazine (2 mg/kg).
2. Administer intravenously (i.v.) through tail vein either (a) [^{125}I]-BDNF (5 μCi/rat), (b) [^{I25}I]-BDNF-PEG2000-biotin (5 μCi/rat), or (c) [^{125}I]-BDNF-PEG2000-biotin (5 μCi/rat) conjugated to OX26/SA (10 μg/rat). These doses are equivalent to 0.06 and 0.05 nmol of BDNFPEG2000-biotin and OX26/SA, respectively.
3. At 30, 60, and 120 min after i.v. injection, animals are sacrificed and brains are removed, weighed, and solubilized.
4. Count the radioactivity using γ-counter.
5. Compute the brain volume of distribution (VD), the blood–brain barrier permeability–surface area (*PS*) product and the brain uptake of the [^{125}I]-BDNF, [^{125}I]BDNF-PEG2000-biotin, and [^{125}I]BDNF-PEG2000-biotin conjugated to OX26/SA.
6. The VD values are calculated by the following equation: VD = μL/g brain tissue. The ratio of the concentration of the BDNF in brain (DPM/g) is divided by the concentration of the BDNF in serum (DPM/uL).
7. The blood–brain barrier permeability–surface area (*PS*) product is expressed as μL/min/g brain tissue, the ratio of the concentration of the BDNF in brain (DPM/g) divided by the concentration of the BDNF in serum (DPM/uL) per min.
8. The brain uptake is expressed as the percentage of injected dose (ID) per gram brain (%ID/g).
9. The brain VD, the BBB PS product, and the brain uptake (%ID/g) of the BDNF-PEG2000-biotin will be expected to increase following conjugation to OX26/SA, which is consistent with drug delivery of the neurotrophin through the BBB mediated by the OX26 mAb delivery system.

3.2. Fusion Protein Technology

We will introduce this technology by the example of delivering anti-Aβ into the brain through fusing the anti-Aβ ScFv into the chimeric human insulin receptor (HIR) mAb (12, 13).

3.2.1. Genetic Engineering

3.2.1.1. Purification of Hybridoma-Generated Monoclonal Antibodies

1. The anti-Aβ ScFv (single chain variable fragment) hybridoma is grown to confluency in RPMI-1640 medium with 10% calf serum (14).
2. The cells are propagated in serum-free medium (SFM).
3. Six liters hybridoma-conditioned SFM are concentrated to

200 ml with an Amicon concentrator (membrane S1Y10) followed by purification of 50-ml aliquots with protein G affinity chromatography.

4. Monoclonal antibodies are purified to homogeneity and only heavy chain and light chains are visible on SDS-PAGE following Coomassie blue staining.

3.2.1.2. Isolation of RNA

1. mRNA is isolated from 2.5 × 10^8 anti-Aβ ScFv hybridoma cells using a single-step method previously described (15, 16).
2. Hybridoma cells are lysed with buffer containing 2% SDS and 400 μg/ml proteinase K.
3. Proteins are completely digested for 45 min at 45°C.
4. Then the concentration of the solution is adjusted to 0.5 M NaCl, which represents the optimal ionic concentration for binding of mRNA to oligodeoxythymidine (dT) cellulose.
5. The selection of mRNA is performed in the same tube by addition of oligo dT cellulose.
6. Following washing of mRNA with 0.5 and 0.1 M NaCl buffers, mRNA is obtained with the aid of a Spin X filter unit and is precipitated with 100% ethanol and glycogen for 2 h at –70°C.
7. The final pellet is resuspended in DEPC-water and stored at –70°C. The final concentration can be measured by a spectrometer and should be adjusted to 1 μg/μl. The ratio of A260 to A280 should be 1.9–2.0, which indicates a good RNA purity.

3.2.1.3. Construction of Anti-Aβ ScFv Expression Vector

1. Reverse transcription of mRNA is performed with oligo $d(T)_{18}$ and AMV-RT (17).
2. PCR amplification of light and heavy chain variable regions (V_L and V_H, respectively) is carried out with 0.5 μg RNA-derived cDNA in a total volume of 100 μl with 2 μM V_L and V_H specific primers (**Table 2.2**), in the presence of 2.5 mM $MgCl_2$, dNTPs and *Taq* polymerase (17, 18).
3. Amplification comprises 35 cycles of denaturation for 1 min at 94°C, followed by annealing for 2 min at 52°C, and extension for 2 min at 72°C.
4. PCR products are analyzed by gel electrophoresis on 1% agarose and identified by UV light after ethidium bromide staining.
5. The expected major cDNA bands corresponding to the PCR amplified ~0.4 kilobase (kb) V_L cDNA and ~0.4 kb V_H cDNA are obtained.
6. The V_H and V_L bands are isolated from the agarose gels with Qiagen QIAquick Gel Extraction Kit and subcloned into a ScFv vector to link the V_H and V_L genes aside a gene encoding

Table 2.2
Light chain and heavy chain cloning and sequencing primers. K, G, or T; M, A, or C; S, G, or C; W, A, or T. Primers Bi4, Bi8b, Bi5c, and Bi3f are from Breitling and Dubel (29)

Cloning primers	
Bi8b (V_L forward)	5′-AATTTTCAGAAGCACGCGTAGATATCKTGMTSAC-CCAAWCTCCA-3′
Bi5c (V_L reverse)	5′-GAAGATGGATCCAGCGGCCGCAGCATCAGC-3′
Bi3f (V_H forward)	5′-CAGCCGGCCATGGCGCAGGTSCAGCTGCAGSAG-3′
Bi4 (V_H reverse)	5′-CCAGGGGCCAGTGGATAGACAAGCTTGGGTGTC-GTTTT-3′
Sequencing primers	
V_H forward	5′-AGCGGATAACAATTTCAC-3′
V_H reverse	5′-GATATCTACGCGTGCTTCT-3′
V_L forward	5′-GACACCCAAGCTTGAAGA-3′
V_L reverse	5′-TCGTTTTATTGATGCCTCA-3′
SA forward	5′-GTGGAGGCACCAAGCTGGAA-3′
SA reverse	5′-TGTACCAGGTGCCGGTGATA-3′

a 17 amino acid linker; the V_L gene is followed by a sequence that encoded the 10 amino acid c-myc epitope (EQKLISEEDL) of the 9E10 mAb, as described previously (14).

7. The anti-Aβ ScFv cDNA is subcloned into the pcDNA3.1 vector to produce the pCD–ScFv plasmid.
8. Anti-Aβ V_L and V_H PCR generated fragments are subcloned into the cohesive ends of the pOPE51 expression vector to produce Anti-Aβ scFv (19).
9. PCR products (i.e., ~0.4 kb V_L and ~0.4 kb V_H) and 1 μg pOPE51-phox are double digested with *Nco*I and *Hin*dIII.
10. Anti-Aβ – V_H and pOPE51 – V_L phox (without V_H phox) are purified by agarose gel electrophoresis and QIAquick Gel Extraction Kit.
11. V_H is ligated into pOPE51-*Nco*I/*Hin*dIII with T4 ligase.
12. Following transformation of *E. coli* JM109 competent cells, clones containing Anti-Aβ–V_H in the open reading frame (ORF) with the c-*myc* tag at the C-terminus are identified by plate-colony screening with the 9E10 mAb following induction with IPTG.
13. Positive clones are further confirmed by *Bgl*II restriction endonuclease mapping.

14. The anti-Aβ scFv cDNA is released from pOPE51 vector with *Nco*I and *Not*I double digestion and ligated into the pcDNA 3.1 expression vector at the same cohesive ends, to form pCD−ScFv plasmid (20).
15. The V_H and V_L genes are aside a gene encoding a 17 amino acid linker; the V_L gene is followed by a sequence that encoded the 10 amino acid c-myc epitope (EQKLISEEDL) of the 9E10 mAb.

3.2.1.4. Construct an Expression Vector for Fusion Protein

1. The pCD−UHC expresses the cDNA encoding for the intronless heavy chain (HC) of the chimeric human insulin receptor (HIR) mAb (13) and has a single *Hpa*I site at the end of the HIRMAb HC open reading frame (ORF) for insertion of genes of interest and expression of HIRMAb HC fusion proteins.
2. For the PCR cloning of the mature anti-Aβ ScFv, the mature ScFv PCR forward primer is designed, so as to delete the human IgG peptide leader sequence from the anti-Aβ ScFv cDNA while maintaining the ORF at the CH_3 region of the HIRMAb heavy chain of the pCD−UHC.
3. This results in the insertion of a Ser−Ser linker between the end of the CH_3 region of the HIRMAb and the amino-terminal part of the anti-Aβ ScFv.
4. The anti-Aβ ScFv reverse PCR primer is designed to delete the c-myc tag and to introduce a stop codon, TAA.
5. Both PCR primers are 5′-phosphorylated for direct ligation into the pCD−UHC at the *Hpa*I site.
6. The PCR cloning of the mature anti-Aβ ScFv cDNA is done using the pCD−ScFv DNA as template.
7. Agarose gel electrophoresis of the PCR products shows the expected single band of ~0.8 kb corresponding to the mature anti-Aβ ScFv cDNA.
8. The engineered ScFv is ligated at the *Hpa*I site in pCD−UHC to form the pCD−HC−ScFv expression vector.
9. The pCD−HC−ScFv clone is validated by DNA sequencing.
10. The deduced amino acid sequence shows that the HC of the fusion antibody is composed of 708 amino acids and included a 18 amino acid signal peptide, a 113 amino acid HIRMAb V_H, a 330 amino acid human IgG1 constant region, a 2 amino acid linker, a 114 amino acid anti-Aβ ScFv V_H, a 17 amino acid linker, and a 113 amino acid anti-Aβ ScFv V_L.
11. The predicted molecular weight of the fusion antibody heavy chain, minus the signal peptide, is 75,594 Da, and the predicted isoelectric point (pI) is 8.81.

3.2.2. Expression of Intact Fusion Antibody in COS Cells

1. COS cells are dual-transfected with pCD-HC-mAβ ScFv and pCD–LC, which is the chimeric HIRMAb LC expression plasmid as follows.
2. Ten million COS cells are centrifuged for 5 min at 4°C and 200 × *g*.
3. The cells are suspended in 900 ml sterile PBS, and transferred to an ice-cold 0.4 cm gap electroporation cuvette (Bio-Rad).
4. Cells are mixed with 5 mg linearized humanized V_L expression plasmid and incubated 10 min on ice.
5. Cells are electroporated at 200 V and 950 mF using a Bio-Rad electroporator.
6. Following incubation on ice for 10 min, cells are diluted with 25 ml IMDM (Iscove's modified Dulbecco's medium)/10% calf serum, penicillin and streptomycin, 100 mg/mL gentamicin SO_4, and plated in 96-well plates.
7. One volume (125 μl) selection medium containing a final concentration of 2 μg/ml mycophenolic acid is added 2 days later.
8. Four days following electroporation, one half of the incubation medium is discarded and 125 μl fresh selecting medium is added (Selecting medium contains: 2 μg/ml mycophenolic acid, 5.25 μg/ml hypoxanthine, and 87.5 mg/ml xanthine).
9. Human kappa LC ELISA is performed using 80 μl conditioned medium and several high producer clones are identified.
10. The myeloma line expressing the humanized light chain is then electroporated with 10 μg linearized humanized HC vector.
11. Clones are selected with 10 mM histidinol and positive transfectants are identified with a human IgG ELISA.
12. After culture of positive transfectants for 3–6 days, conditioned medium is collected.
13. The fusion antibody is purified by protein A affinity chromatography.
14. The purity is assessed with both 12% SDS-PAGE with Coomassie blue staining and Western blotting using an anti-human IgG primary antibody.

3.2.3. Blood to Brain Transport in the Primate In Vivo

1. The fusion antibody is iodinated with [^{125}I]-iodine and chloramine T.
2. At the same time the murine anti-Aβ mAb is tritiated with [^{3}H]-*N*-succinimidyl proprionate.
3. A single intravenous injection a dose of 777 μCi of [^{125}I]-fusion antibody and 888 μCi of [^{3}H]-murine anti-Aβ mAb is administered into an old female rhesus monkey.

4. Collect serum at multiple time points over a 180 min period and count the total radioactivity of serum, and expressed as a percent of injected dose (I.D.) per milliliter of serum.
5. The ^{125}I radioactivity is counted in a γ counter, and the ^{3}H radioactivity is counted in a liquid scintillation counter.
6. The ^{125}I isotope emits radioactivity in the ^{3}H window, and standard curves are prepared to account for ^{125}I emission in the ^{3}H channel.
7. The [^{125}I]-fusion antibody in serum is fit to a two-compartment pharmacokinetic model to yield the pharmacokinetic parameters (21).
8. At 180 min after drug injection, the animal is euthanized. Brain homogenate is prepared and separated into a capillary pellet and a post-vascular supernatant (22).
9. The volume of distribution (VD) of the antibody in the post-vascular supernatant is high, that indicates the antibody has crossed the BBB and entered into the brain interstitial and intracellular spaces.
10. The VD = μL/g brain and is the ratio of the concentration of the antibody in brain (DPM/g) divided by the concentration of the antibody in serum (DPM/uL) at the 180 min terminal time point.
11. The primate brain is divided into frontal, midsection, and caudal parts, and coronal slabs are snap-frozen in dry ice or liquid nitrogen.
12. 20 μm frozen sections are cut with a cryostat.
13. The sections are exposed to Biomax X-ray film for 7–10 days followed by fixation and development.
14. The film is scanned, and the grayscale image is converted into color imagine with *NIH Image* software.

3.3. Trojan Horse Liposomes

In this technique, macromolecules can be packaged into liposomes (as the Trojan horse). With the help of peptidomimetic mAbs, the liposomes can cross the BBB and deliver the encapsulated macromolecules into brain parenchyma. We will introduce this method here by the examples of delivering plasmids encoding either luciferase or beta-galactosidase.

3.3.1. Thiolation of Antibodies

1. [^{125}I]OX26 peptidomimetic mAb to rat TfR or IgG2a isotype control antibody for OX26 are thiolated using 2-iminothiolane (Traut's reagent). mAb is dissolved in 0.15 M Na-borate buffer/0.1 mM EDTA, pH 8.0 (23–28).
2. Addition of Traut's reagent and incubation for 60 min at room temperature using a mAb/Traut's ratio of 1:40 (mol/mol), an average of one primary amine per mAb is thiolated.

3. mAb solutions are concentrated and the buffer exchanged with 0.1 M sodium phosphate (pH8.0) using a Centriprep-30 concentrator.
4. Thiolated mAb is immediately used for conjugation with liposomes (*see* below).
5. Ellmann's reagent is used to determine the number of sulfhydryl groups added by thiolation to mAb.

3.3.2. Plasmid Production

1. The plasmids are amplified in the JM109 strain of *E. coli* (24).
2. DNA is extracted by using alkaline lysis method and purified by precipitation with isopropanol, using QIAfilter Plasmid Maxi kit.
3. DNA is measured by UV absorption at 260 nm and dissolved in TE buffer [10 mM Tris/1 mM EDTA (pH 8.0)].
4. Linearized DNA is obtained by digestion with *Bam*HI.
5. DNA size is confirmed by 0.8% agarose gel electrophoresis and ethidium bromide staining.

3.3.3. DNA Radiolabeling

1. Supercoiled DNA is labeled with [α-^{32}P]dCTP by Nick Translation System using DNA polymerase I and DNase I (24).
2. Unincorporated nucleotide is removed by a G25 Sephadex column.
3. The linearized plasmid is ^{32}P-radiolabeled at both 5′ and 3′ ends with T4 polymerase and purified with a Sephadex G-25 spin column to a trichloroacetic acid (TCA) precipitability of 98%.
4. This material is analyzed with 0.8% agarose gel electrophoresis and film autoradiography and show a single band of 5.8 kb.
5. The ^{32}P-linearized plasmid is used in the pharmacokinetic experiments.
6. The ^{32}P-supercoil plasmid is used as a tracer to measure the incorporation of the unlabeled supercoil plasmid in the liposomes.

3.3.4. Pegylated Liposome Synthesis and Plasmid DNA Encapsulation

1. POPC (19.2 μmol), DDAB (0.2 μmol), DSPE-PEG 2000 (0.6 μmol), and DSPE-PEG 2000-maleimide (30 nmol) are dissolved in chloroformymethanol (2:1, vol:vol) followed by evaporation (24).
2. The lipids are dispersed in 1 ml 0.05 M Tris–HCl buffer (pH 8.0) and sonicate for 10 min.
3. Non-labelled supercoiled plasmid DNA (100 μg) and 1 μCi ^{32}P labelled plasmid DNA are added to the lipids.
4. The liposome/DNA dispersion is evaporated to a final concentration of 200 mM at a volume of 100 μl.

5. The dispersion is frozen in ethanol/dry ice for 4–5 min and thawed at 40°C for 1–2 min, and this freeze–thaw cycle is repeated ten times.
6. The liposome dispersion is diluted 0.05 M Tris–HCl buffer to a lipid concentration of 40 mM.
7. The dispersion is extrude ten times each through two stacks each of 400-, 200-, 100-, and 50-nm pore size polycarbonate membranes, by using a hand-held extruder.
8. The mean vesicle diameters are determined by quasielastic light scattering by using a Microtrac Ultrafine Particle Analyzer.
9. For digestion of the unencapsulated DNA, 5 units of pancreatic endonuclease I and 5 units of exonuclease III are added in 5 mM $MgCl_2$ and 0.1 mM DTT to the liposome/DNA mixture after extrusion.
10. After incubation at 37°C for 1 h, the reaction is stopped by adding 7 mM EDTA (pH 8.0).
11. Aliquots of product taken before and after nuclease treatment are visualized by agarose gel electrophoresis and ethidium bromide staining, to determine the extent to which the nuclease digestion removed the exteriorized plasmid DNA.

3.3.5. Conjugation of OX26 mAb or Mouse IgG2a to the Pegylated Liposome/DNA

1. For conjugation, thiolated antibody is incubated with liposomes containing DSPE-PEG-maleimide prepared as above overnight at room temperature (23, 24).
2. Buffer is exchanged by applying the reaction mix to a 1.6 × 18 cm Sephadex CL-4B column and eluting with 0.001 M PBS.
3. Aliquots of column elute are analyzed by scintillation counting.
4. Efficiency of OX26 coupling is determined by analysis of the column elution profiles. In the case where no ^{125}I-labeled antibody is available the amount of liposome-conjugated protein is determined by a protein assay.
5. The final percentage entrapment of the 100 μg of plasmid DNA in the liposome preparation is computed from the ^{32}P radioactivity and is typically 30% or 30 μg of plasmid DNA.
6. This preparation is then administered to three rats at a dose of 10 μg of plasmid DNA per rat for luciferase gene expression measurements.

3.3.6. Luciferase Gene Expression In Vivo

1. The pegylated liposome/luciferase DNA that is conjugated with either OX26 mAb or mouse IgG2a is injected i.v. into anesthetized rats at a dose of 10 μg of pGL2 DNA per rat (8).
2. Rats are killed at 24, 48, or 72 h after injection.

3. The brain, heart, kidney, spleen, liver, and lung tissues are homogenized in 4 vol of lysis buffer containing 0.1 M potassium phosphate buffer (pH 7.8), 1% Triton X-100, 1 mM DTT, and 2 mM EDTA by using a Polytron (Kinematica Lucerne, Switzerland) homogenizer.
4. The homogenate is centrifuged at 14,000 rpm for 10 min at 4°C.
5. The supernatant is used for measurement of tissue luciferase activity with a Luminometer (Biolumat LB 9507, Berthold, Nashua, NH).
6. 100 μl of reconstituted luciferase substrate is added to 20 μl of tissue extract. Peak light emission is measured for 10 s at 20°C and recorded as relative light units.
7. The background level is determined by measuring a sample containing only lysis buffer.
8. The protein concentration in the tissue extract is determined with the bicinchoninic acid (BCA) protein assay reagent.

3.3.7. β-Galactosidase Histochemistry

1. Expression plasmids encoding for *E. coli* β-galactosidase is an alternative reporter gene other than luciferase. β-galactosidase histochemistry as a reporter gene can be used to identify the cellular location of gene expression.
2. Rat organs are removed and embedded in Tissue-tek OCT in powdered dry ice a various time points after a single intravenous injection PIL with a β-galactosidase gene.
3. 18-μm frozen sections are made with cryostat and stored at –70°C.
4. The sections are fixed at room temperature in 0.5% glutaraldehyde in 0.1 M phosphate buffer (pH 7.0) for 5 min.
5. The β-galactosidase histochemistry is performed with 5-bromo-4-chloro-3-indoyl-β-galactoside (X-gal) with the X-GAL histochemistry kit.
6. The slides are incubated with the X-gal chromagen overnight at 37°C in a humidified chamber.
7. Air dry slides and scan them with a 1200 dpi UMAX flatbed scanner with transluminator and cropped in Adobe Photoshop 5.5 with a G4 Power Macintosh before coverslipping.
8. The slides are coverslipped for light microscopic examination.

Acknowledgments

Authors appreciate many valuable discussions with Dr. William M. Pardridge at UCLA and Dr. Eric V. Shusta at University of Wisconsin-Madison, Madison.

References

1. Pardridge, W.M. (2007) Drug targeting to the brain. *Pharm. Res.*, 24, 1733–1744.
2. Ghose, A.K., Viswanadhan, V.N., and Wendoloski, J.J. (1999) A knowledge-based approach in designing combinatorial or medicinal chemistry libraries for drug discovery. 1. A qualitative and quantitative characterization of known drug databases. *J. Comb. Chem.*, 1, 55–68.
3. Lipinski, C.A. (2000) Drug-like properties and the causes of poor solubility and poor permeability. *J. Pharmacol. Toxicol. Methods*, 44, 235–249.
4. Jones, A.R., and Shusta, E.V. (2007) Blood-brain barrier transport of therapeutics via receptor-mediation. *Pharm. Res.*, 24, 1759–1771.
5. Wu, D., and Pardridge, W.M. (1999) Neuroprotection with noninvasive neurotrophin delivery to the brain. *Proc. Natl. Acad. Sci. U.S.A.*, 96, 254–259.
6. Pardridge, W.M., Wu, D., and Sakane, T. (1998) Combined use of carboxyl-directed protein pegylation and vector-mediated blood-brain barrier drug delivery system optimizes brain uptake of brain-derived neurotrophic factor following intravenous administration. *Pharm. Res.*, 15, 576–582.
7. Pardridge, W.M., Kang, Y.S., and Buciak, J.L. (1994) Transport of human recombinant brain-derived neurotrophic factor (BDNF) through the rat blood-brain barrier in vivo using vector-mediated peptide drug delivery. *Pharm. Res.*, 11, 738–746.
8. Yoshikawa, T., and Pardridge, W.M. (1992) Biotin delivery to brain with a covalent conjugate of avidin and a monoclonal antibody to the transferrin receptor. *J. Pharmacol. Exp. Ther.*, 263, 897–903.
9. Pardridge, W.M., Boado, R.J., and Kang, Y.S. (1995) Vector-mediated delivery of a polyamide ("peptide") nucleic acid analogue through the blood-brain barrier in vivo. *Proc. Natl. Acad. Sci. U.S.A.*, 92, 5592–5596.
10. Sakane, T., and Pardridge, W.M. (1997) Carboxyl-directed pegylation of brain-derived neurotrophic factor markedly reduces systemic clearance with minimal loss of biologic activity. *Pharm. Res.*, 14, 1085–1091.
11. Rosenfeld, R., Philo, J.S., Haniu, M., Stoney, K., Rohde, M.F., Wu, G.M., Narhi, L.O., Wong, C., Boone, T., Hawkins, N.N. et al. (1993) Sites of iodination in recombinant human brain-derived neurotrophic factor and its effect on neurotrophic activity. *Protein Sci.*, 2, 1664–1674.
12. Boado, R.J., Zhang, Y., Zhang, Y., Xia, C.F., and Pardridge, W.M. (2007) Fusion antibody for Alzheimer's disease with bidirectional transport across the blood-brain barrier and abeta fibril disaggregation. *Bioconjug. Chem.*, 18, 447–455.
13. Boado, R.J., Zhang, Y., Zhang, Y., and Pardridge, W.M. (2007) Humanization of anti-human insulin receptor antibody for drug targeting across the human blood-brain barrier. *Biotechnol. Bioeng.*, 96, 381–391.
14. Li, J.Y., Sugimura, K., Boado, R.J., Lee, H.J., Zhang, C., Duebel, S., and Pardridge, W.M. (1999) Genetically engineered brain drug delivery vectors: cloning, expression and in vivo application of an anti-transferrin receptor single chain antibody-streptavidin fusion gene and protein. *Protein Eng.*, 12, 787–796.
15. Li, J.Y., Boado, R.J., and Pardridge, W.M. (2001) Blood-brain barrier genomics. *J. Cereb. Blood Flow Metab.*, 21, 61–68.
16. Boado, R.J., and Pardridge, M.M. (1991) A one-step procedure for isolation of poly(A)+ mRNA from isolated brain capillaries and endothelial cells in culture. *J. Neurochem.*, 57, 2136–2139.
17. Boado, R.J., and Pardridge, W.M. (1994) Measurement of blood-brain barrier GLUT1 glucose transporter and actin mRNA by a quantitative polymerase chain reaction assay. *J. Neurochem.*, 62, 2085–2090.
18. Boado, R.J., Golden, P.L., Levin, N., and Pardridge, W.M. (1998) Up-regulation of blood-brain barrier short-form leptin receptor gene products in rats fed a high fat diet. *J. Neurochem.*, 71, 1761–1764.
19. Dubel, S., Breitling, F., Fuchs, P., Braunagel, M., Klewinghaus, I., and Little, M. (1993) A family of vectors for surface display and production of antibodies. *Gene*, 128, 97–101.
20. Dubel, S., Breitling, F., Kontermann, R., Schmidt, T., Skerra, A., and Little, M. (1995) Bifunctional and multimeric complexes of streptavidin fused to single chain antibodies (scFv). *J. Immunol. Methods*, 178, 201–209.
21. Pardridge, W.M., Kang, Y.S., Buciak, J.L., and Yang, J. (1995) Human insulin receptor monoclonal antibody undergoes high

affinity binding to human brain capillaries in vitro and rapid transcytosis through the blood-brain barrier in vivo in the primate. *Pharm. Res.*, 12, 807–816.

22. Triguero, D., Buciak, J., and Pardridge, W.M. (1990) Capillary depletion method for quantification of blood-brain barrier transport of circulating peptides and plasma proteins. *J. Neurochem.*, 54, 1882–1888.
23. Huwyler, J., Wu, D., and Pardridge, W.M. (1996) Brain drug delivery of small molecules using immunoliposomes. *Proc. Natl. Acad. Sci. U.S.A.*, 93, 14164–14169.
24. Shi, N., and Pardridge, W.M. (2000) Noninvasive gene targeting to the brain. *Proc. Natl. Acad. Sci. U.S.A.*, 97, 7567–7572.
25. Shi, N., Zhang, Y., Zhu, C., Boado, R.J., and Pardridge, W.M. (2001) Brain-specific expression of an exogenous gene after i.v. administration. *Proc. Natl. Acad. Sci. U.S.A.*, 98, 12754–12759.
26. Zhang, Y., Jeong, L.H., Boado, R.J., and Pardridge, W.M. (2002) Receptor-mediated delivery of an antisense gene to human brain cancer cells. *J. Gene Med.*, 4, 183–194.
27. Zhang, Y., Schlachetzki, F., and Pardridge, W.M. (2003) Global non-viral gene transfer to the primate brain following intravenous administration. *Mol. Ther.*, 7, 11–18.
28. Zhang, Y., Calon, F., Zhu, C., Boado, R.J., and Pardridge, W.M. (2003) Intravenous nonviral gene therapy causes normalization of striatal tyrosine hydroxylase and reversal of motor impairment in experimental parkinsonism. *Hum. Gene Ther.*, 14, 1–12.
29. Breitling, F. and Dubel, S. (1997) In Reischl, U. (ed.), *Molecular Diagnosis of Infectious Diseases* in the series *Methods in Molecular Medicine*, Vol. 13. Humana Press Inc., Totowa, NJ, pp. 581–591.

Chapter 3

Disruption of Blood–Brain Barrier by Focused Ultrasound for Targeted Drug Delivery to the Brain

Natalia Vykhodtseva

Abstract

A major challenge to drug delivery in the brain is caused by the blood–brain barrier (BBB), a composite of endothelial structures that exclude over 98% of small-molecule drugs and almost 100% of large-molecule neurotherapeutics from being transmitted to the brain. Current strategies for overcoming the BBB are either invasive, non-targeted, or demonstrating limited carrying capacity. Alternatively, non-invasive, transient, and local, image-guided blood–brain barrier disruption (BBBD) can be accomplished using focused ultrasound (FUS) exposure with intravascular injection of pre-formed microbubbles. Low-intensity FUS administered with microbubble-based ultrasound (US) contrast agents has been shown to transiently disrupt the BBB, allowing agents into the brain over several hours. Evaluation in vivo using MRI-guided FUS showed minimal adverse effects, as compared to invasive interventions. Electron microscopy indicated FUS activated trans-cellular transport, and reversibly affected tight junctions. Transient disruption of the BBB allowed delivery of both small-molecule drugs (Doxorubicin) and large-molecule antibody-based chemotherapeutic (Herceptin). The FUS-induced BBB disruption also allowed delivery of both molecular imaging and therapeutic agents directly to amyloid plaques in Alzheimer's disease mouse models. A detailed description of the method for MRI-guided focal BBBD in animals will be described in this chapter.

Key words: Blood–brain barrier (BBB), ultrasound, BBB disruption, drug delivery to CNS.

1. Introduction

Conventional approaches for the drug delivery to the brain include direct injection, or infusion into the brain substance and use of catheters and implantable devices. The blood–brain barrier (BBB) can be also reversibly disrupted by intra-carotid arterial infusion of hyper-osmolar solutions and vasoactive agents. However, most drugs do not cross the BBB and thus lack efficacy for the treatment of disorders of the central nervous system. Moreover, transcranial

K.K. Jain (ed.), *Drug Delivery to the Central Nervous System*, Neuromethods 45,
DOI 10.1007/978-1-60761-529-3_3, © Humana Press, a part of Springer Science+Business Media, LLC 2010

drug delivery is an invasive procedure that increases the risk of infection and is not very efficient due to limited diffusion of the drugs into brain. Administration via transient osmotic disruption of the BBB is non-targeted and has toxic side effects, including possible neuronal damage. New approaches for drug delivery include design of new drugs (or carriers) such as liposomes and nanoparticles, or conjugation of the drugs with proteins, peptide vectors, or antibodies and use of endogenous BBB carrier systems to ferry drug molecules through BBB. One limitation is that only a finite amount of drug molecules can be delivered due to a limited number of receptors and a discrete quantity of molecules attached to the carriers.

A few years ago, we demonstrated that focused ultrasound (FUS) in conjunction with microbubbles-based ultrasound contrast agents might provide a non-invasive, local, and reversible BBB disruption at the targeted locations in the brain (1). Effectiveness of this method has been verified by further studies (2–10). Since ultrasound, guided by modern diagnostic techniques, can be focused through the skull, this method offers a great potential for investigative and eventually for clinical treatments and diagnosis of brain disorders (11–13).

2. Focused Ultrasound

Focused ultrasound technologies enable concentration of an acoustic energy in a precise volume in the body with minimal effects to the near-field tissues. It is a method of non-invasively inducing local biological effects deep inside the body without surgical intervention. Ultrasound effects can be produced by two major mechanisms: thermal and non-thermal (such as radiation pressure and cavitation). Relatively low-level temperatures (< 43°C), maintained for 30–60 min, can be used for hyperthermia to sensitize tumors to chemotherapy and radiotherapy (14–17). At higher temperatures (>60°C), focused ultrasound can be used for the thermal ablation of tissue. Currently this technique is used clinically for the non-invasive treatment of uterine fibroids (18–22), adenomyosis (23, 24), and myomas (25). The FUS-thermal ablation technique is in clinical evaluation for breast cancer (26–28), renal cancer (29–31), liver carcinoma (32, 33), prostate cancer (34–37), pancreatic tumors (38), and other indications. The MRI-guided FUS is being tested for pain relief in patients suffering from bone metastases (39, 40). There is great flexibility in designing ultrasonic applicators, spherical radiators, lenses, reflectors, and phased arrays to tailor a focal zone to almost any application.

One major shortcoming of ultrasound is that it is strongly attenuated and distorted by bone, and the skull, which induces strong aberrations both in phase and amplitude, resulting in a degradation of the beam shape. Hence, an efficient and non-invasive brain therapy requires an adaptive focusing, taking into account the effects of the skull. This adaptive focusing can be achieved by using high powered transducer arrays with large surface area, the acoustic properties of the skull deduced from high resolution CT scans, and high-performance computers to calculate the corrections necessary to restore a focus in the brain (2, 12, 13, 41, 42). Theoretical and experimental studies have indicated that optimal transcranial focusing can be achieved at frequencies below 1 MHz (2, 41). By reducing the frequency to approximately 250 kHz the need for the adaptive focusing is eliminated with the cost of a larger focal spot and reduced pressure gain (3, 41). Now non-invasive transcranial focusing for the purposes of thermal ablation in the brain has reached the early stages of clinical testing.

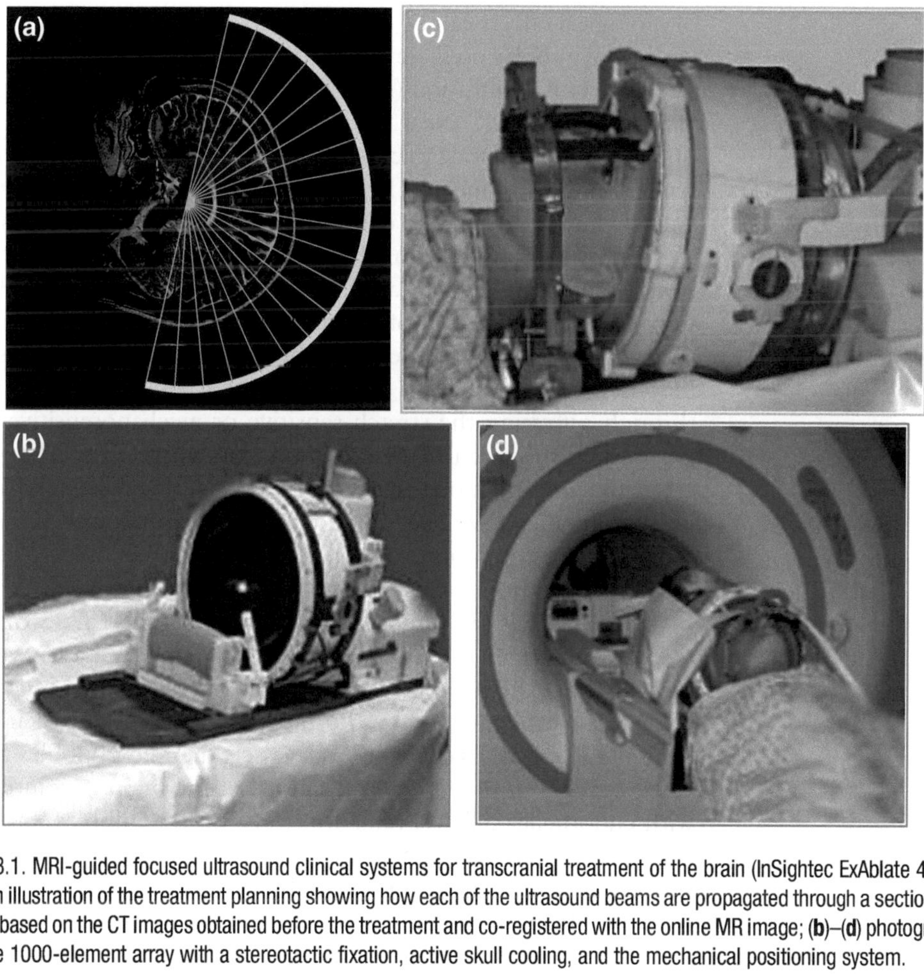

Fig. 3.1. MRI-guided focused ultrasound clinical systems for transcranial treatment of the brain (InSightec ExAblate 4000). (**a**) An illustration of the treatment planning showing how each of the ultrasound beams are propagated through a section of a skull based on the CT images obtained before the treatment and co-registered with the online MR image; (**b**)–(**d**) photographs of the 1000-element array with a stereotactic fixation, active skull cooling, and the mechanical positioning system.

FUS-induced targeted BBB disruption may offer a solution to the problems of the delivery of either small- or large-molecule drugs to the brain, and FUS-based techniques can enter clinical drug development programs for many CNS disorders. **Figure 3.1** shows MRI-guided focused ultrasound clinical systems for transcranial treatment of the brain (InSightec ExAblate 4000). Such a system could be used at low acoustical powers to achieve non-invasive BBB disruption with high spatial resolution for targeted drug delivery.

3. BBB Disruption with FUS and Microbubbles

Transient BBB disruption may be achieved by using low-intensity FUS with intravenously inject ultrasound (US) contrast agents (Optison or Definity), which are encapsulated gas-filled microbubbles less than 5 μm diameter.

When microbubbles interact with an ultrasound beam, a range of biological effects can be induced (43). At low acoustical power levels, the bubbles oscillate within the ultrasound field growing in size via rectified diffusion (44). This type bubble activity is called "stable cavitation." Near an oscillating bubble, the medium undergoes a small-scale micro-streaming that can exert shear stresses on cell membrane and large molecules (43, 45, 46). Vascular endothelium is known to dynamically respond to shear stresses, which play a critical role in BBB homeostasis and vascular pathophysiology (47). Further, acoustic radiation forces cause bubbles to move in the direction of the wave propagation (48) and bring them in contact with vessel endothelial cells. Radiation pressure may be involved in FUS + microbubbles-based BBB opening through activating the stretch-sensitive or mechanosensitive ion channels in the vascular endothelium. The channels can be activated by conformational changes during localized stretching of cell membranes (49, 50). In addition, at high enough acoustic pressures, microbubbles can collapse, a phenomenon known as "inertial cavitation," producing shock waves, high-velocity jets (48), free radicals (51), and high local temperatures (52, 53).

The preformed microbubbles used as US contrast agents presumably can exhibit these effects, either with their shells intact or after being broken apart by the ultrasound beam and their gas contents released. Inside of blood vessels the microbubbles activated by FUS in the targeted locations of the brain may interact with endothelial cells via radiation force on the bubbles, bubble oscillation, micro-streaming of the fluid around the bubbles, and shock waves and jets (if micro-bubbles collapse) (**Fig. 3.2**). Any of these effects may cause BBB disruption by modification of endothelial cell membranes and/or tight junctions; stimulating

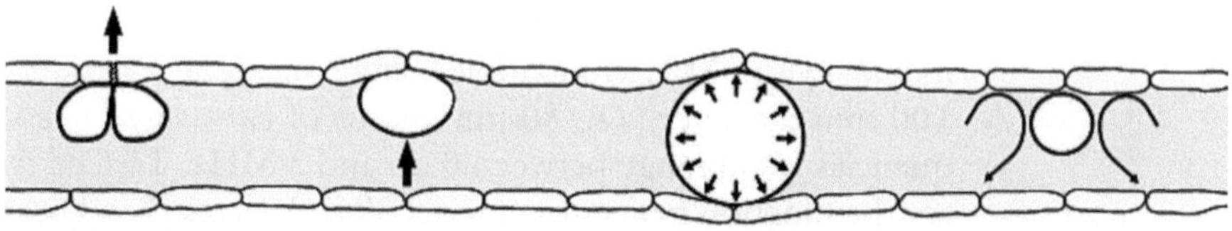

Fig. 3.2. Possible mechanisms for BBB disruption via ultrasound + microbubbles. Assuming that the effect is not due to bubble collapse (inertial cavitation), possible mechanisms include stimulation of the endothelial cells via radiation force on the bubbles, bubble oscillation, and micro-streaming of the fluid around the bubbles.

mechanosensitive channels and triggering a physiological vasoactive response, which can include BBB opening or damaging of blood vessels. Each of these effects could potentially affect the blood vessels and/or blood flow within the microvasculature and be the source of the BBB disruption.

4. Ultrasound Generation

4.1. Basic Quantities Relevant to Therapeutic Ultrasound

In therapy, the most relevant quantities are frequency, pressure, and intensity. Whereas diagnostic ultrasound emits a range of frequencies with each burst, nearly all the therapeutic methods use only a single frequency. The *frequency* (labeled f and reported in MHz) is important for numerous reasons. The energy absorbed by tissue generally increases with frequency, while the cavitation threshold decreases with frequency. Moreover, the focusing ability of a transducer is partially dependent upon frequency, with higher frequencies able to create smaller foci.

Focused ultrasound is a directed pressure wave, and in ultrasound the *pressure* (labeled p and reported in Pa or MPa) is actually the excess pressure present due to the propagating ultrasound wave. Since pressure varies as a function of position, it is common to estimate only the *peak pressure* in the focus of the ultrasound beam.

If pressure is known, the acoustic *intensity* (labeled I and reported in W/cm^2) can be estimated:

$$I = \frac{p^2}{\rho c} \qquad [1]$$

where ρ is the tissue density and c is the sound speed in the tissue. This quantity is important since peak temperature rise is proportional to intensity.

4.2. Ultrasound Transducer

The ultrasound field can be generated by an air-backed spherically curved piezoelectric transducer driven by a function generator. The physical characteristics of the transducer determine the size of the focal spot. For example, we performed successful experiments with MRI-compatible single-element transducers of 100 mm diameter (D), 80 mm radius of curvature (R), and frequencies (f) in range between 0.25 and 2 MHz. To feed transducer the function generator (model 395, Wavetek, San Diego, CA) and an RF amplifier (model 240L, ENI Inc., Rochester, NY) can be used. The electrical impedance of the transducer is matched to the output impedance of the amplifier by an external LC-matching network to allow optimum power transfer. Although we used our own in-house manufactured transducers, custom MRI compatible transducers can be purchased from several sources, for example, from Imasonic, Inc. (Besancon, France).

4.2.1. Ultrasound Transducer Characterization

Whereas approximate tissues properties (sound speed, density, and absorption) may be found in the literature (for example, in (54)), the transducer efficiency and beam-shape are unique to each transducer. It was thus necessary to characterize a transducer so that, given the known quantities of electrical input power, P (W), and driving frequency, f, one can estimate pressure and intensity at the focus.

4.2.2. Ultrasound Transducer Efficiency Measurement

If a certain amount of electrical power is supplied to a transducer, only a percentage of this power is converted to acoustic energy. The fraction of energy converted is the transducer *efficiency*. In fact, most transducers operate at below 50% efficiency. If one knows the transducer's efficiency, the acoustic power can be easily calculated provided that the electrical power measurement is accurate.

So the first step in transducer characterization is a measurement of the ratio of electrical power input to acoustic power output (efficiency). The transducer efficiency can be measured using an acoustic radiation force balance system. The system is based on the principle that an acoustic wave exerts a force F on an ideal absorbing target proportional to the acoustic power P, as described by

$$F = \frac{P \cos \theta}{c} \qquad [2]$$

where c is the sound speed in the medium (usually water) and θ is the angle of incidence between the ultrasound wavefront and the target.

In practice, the transducer is immersed in degassed, de-ionized water (*see* **Note 1**) and driven at a given electrical power, as measured by a power meter. The electrical power is measured in continuous wave (CW) mode with a power meter (model 438A, Hewlett Packard, Palo Alto, CA) and dual directional coupler (model C173, Werlatone, Brewster, NY). The absorbing target is suspended in the water by thin wires attached to an electronic balance, so that its face lay directly opposite the transducer. It is critical that the absorber completely encompasses the acoustic beam emitted by the transducer (*see* **Note 2**). For each voltage, the acoustic power output of the transducer is calculated from the force exerted on the brush, which is measured by the change in weight of the absorbing target (*see* **Note 3**). For increased accuracy output acoustic power is typically averaged over three or more measurements. Efficiency η is simply the mean acoustic power divided by the input electrical power. As an example, the acoustic efficiency of an air-backed single-element transducer for a range of input voltages is presented in **Table 3.1** (55). It is important to note that the electrical power does not directly scale with the input voltage.

Table 3.1
Acoustic efficiency of air-backed single-element transducer; frequency, $f = 1.696$ MHz; radius of curvature, $R = 80$ mm; diameter, $D = 100$ mm; F number, $(R/D) = 0.8$ (55)

Voltage (V)	Electrical power (W)	Acoustical power (W)	Efficiency (%)	Standard error
0.050	1.93	1.04	53.9	6.57×10^{-3}
0.100	7.67	4.17	54.4	4.70×10^{-2}
0.150	17.2	9.30	54.1	3.73×10^{-2}
0.200	30.7	16.6	54.2	7.63×10^{-2}
0.250	48.0	26.4	55.0	1.09×10^{-1}

4.2.3. Ultrasound Pressure Field Measurement

The second step in transducer characterization is the measurement of the ultrasound pressure field (beam-shape). This is generally performed by scanning a needle hydrophone over a plane perpendicular to the ultrasound axis of propagation (*see* **Note 4**).

For this purpose the transducer is placed in a tank filled with degassed water, the walls of the tank are lined with sound absorbing rubber mats. A calibrated membrane hydrophone (spot diameter 0.5 mm, GEC-Marconi Research Center, Chelmsford, UK) is positioned at a focal distance from the transducer. The hydrophone is connected to a standard oscilloscope to measure its voltage. Using pulsed ultrasound and scanning the hydrophone across the focal plane, the amplitudes of the hydrophone signal voltage as a function of the driving radiofrequency (RF) voltage of the frequency generator are recorded. The acoustic pressure amplitude values are calculated using the calibration coefficient of the hydrophone. Example of a spatial distribution of relative

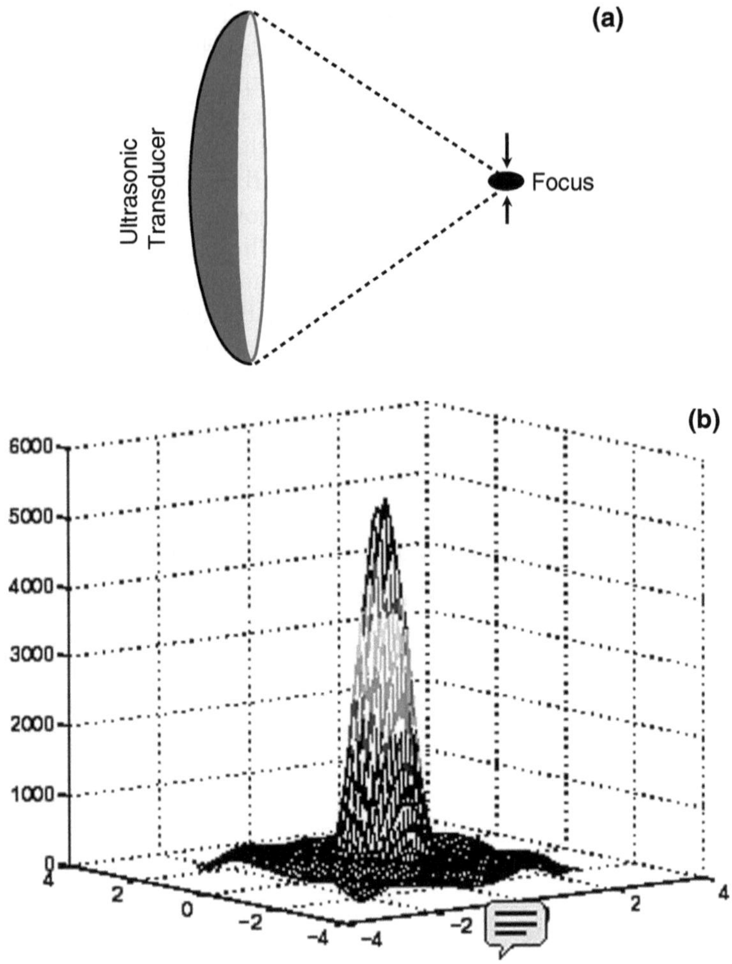

Fig. 3.3. Focused ultrasound transducer. (**a**) Schematic diagram of the focused ultrasound transducer; (**b**) example of a spatial distribution of relative pressure amplitude of a 1.7-MHz transducer at the focal plane parallel to transducer face (measured in water). Reproduced with permission from (55).

pressure amplitude of a 1.7-MHz transducer at the focal plane parallel to transducer face (measured in water) is present in **Fig. 3.3** (55).

To characterize the focal shape of a transducer, it is common to estimate diameter and length of focal spot at half-maximum the pressure amplitude measured with a needle hydrophone (Precision Acoustics, Dorchester, UK) (**Table 3.2**).

4.2.4. Ultrasound Pressure Field: Estimation in Brain

Characterization is performed in water, where absorption of ultrasound is very small and can be neglected over small distances. However, due to the attenuation of ultrasound in the brain tissue and especially in the skull bone, the acoustic energy delivered to the brain in vivo is different from that measured in the water tank.

Table 3.2
Examples of measured focal spot dimensions (50% of the peak pressure amplitude) for single-element transducers with *F* number (*R/D*) = 0.8 (56)

Frequency (MHz)	50% Focus width (mm)	50% Focus length (mm)
0.26	8.0	40.0
0.69	2.3	14.0
1.1	1.7	7.5
1.5	1.0	4.8
3.3	0.6	3.2

To know how a given set of acoustical parameters translates into in vivo effects, the energy losses in the brain tissue and skull bone must be determined.

The pressure amplitude in the brain can be estimated by decreasing the measured water values based on ultrasound attenuation through the brain with an average amplitude attenuation coefficient of 5 Np/m/MHz (57). For example, based on this attenuation coefficient value, the pressure amplitude in the brain at the depth of 10 mm can be obtained by decreasing the measured water values by a factor of 0.92 at 1.63 MHz for the sonication through skulls window.

Energy losses caused by the animal skull, can be determined by performing pressure measurements with the hydrophone in the water tank both with and without an ex vivo animal skull placed in the path of the ultrasound beam. For example, for the rats, the average pressure loss due to the skull over ten individual measurements was found to be 49%, with 51% of the incident energy transmitted through the skull (55). In the mice, minimal attenuation ~18% of the pressure amplitude and minute focus displacement from the sonication through the parietal skull bone has been reported (9). Thus, the pressure amplitude values should be presented "in water" or "in the brain" taking these losses into account.

5. Experimental Set-Up

5.1. Exposure System

Figure 3.4 shows a diagram of the transducer-positioning system.

Ultrasound transducer is submerged in a plastic container filled with degassed water and attached to a XYZ positioning device (*see* **Note 5**). The hole in a plastic platform above the

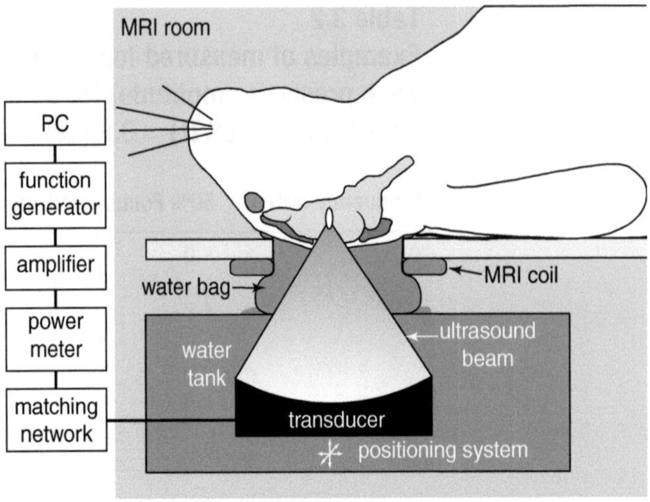

Fig. 3.4. Schematic diagram of the transducer-positioning system. Ultrasound transducer is submerged in a tank filled with degassed water and attached to a XYZ positioning device. The animal is placed in the supine position on the platform above the tank. The hole in the platform allows the ultrasound beam to propagate from the water to the animal brain. A MRI coil is placed around the hole to improve the MRI signal-to-noise ratio. A thin plastic bag filled with degassed water provides acoustic coupling between the transducer and the skin on the head of the animal. The sonication system is placed on the MR table.

container allows the ultrasound beam to propagate from the water to the target placed on the top of the container. A MRI coil is placed around the hole to improve the MRI signal-to-noise ratio. The sonication system is placed on the MR table and moved into a clinical MR scanner so that the transducer focal spot is aimed through a hole to the brain of the animal lying on its back on the platform. The movement of the transducer can be performed with a remote-controlled hydraulic or mechanical computer-controlled system as well as with a manual lead screw-based positioning system.

5.2. Magnetic Resonance Imaging (MRI)

5.2.1. Equipment and Pulse Sequences

Standard MRI scanners that have a space for the sonication system and high enough resolution to image animal brain can be used to target the brain structures and verify the BBB disruption. The standard clinical 1.5 and 3 T magnets (General Electric) with small surface coils that allow good images of the animal brains have been used in our experiments.

Anatomical localization of the brain structures is performed by acquiring three sets of orthogonal T2-weighted fast spin echo (FSE) images (*see* **Table 3.3** for parameters). These images are also used to visualize the craniotomy to ensure that the ultrasound beam is not clipped. The BBB disruption verification and quantification is performed by acquiring a time series of T1-weighted fast spin echo (FSE) images (*see* **Table 3.3** for parameters).

Table 3.3
Parameters used in the MRI imaging

Sequence	TR/TE (ms)	Flip angle (°)	Echo train length	Field of view (cm)	Matrix size	Slice thickness (mm)	Bandwidth
T2-weighted FSE	2000/85	90	8	10	256 × 256	1.5	16
T1-weighted FSE (contrast[a])	500/17	90	4	10	256 × 256	1.5	16

[a]The T1-weighted sequence determines the BBBD after an MRI contrast agent injection.

5.2.2. Localization of Ultrasound Beam Focus with MRI

Before starting the experiments, the focal spot of the ultrasound transducer is localized in the MRI coordinate system. For this purpose the ultrasound beam is aimed on the water surface at a low power level (approximately 1 W), and the focal spot can be visible as a small fountain. A marker is placed on this spot. Using the T2-weigted FSE image of this marker, location of the focal spot is measured in the MRI coordinate system. This location is used as a reference spot when targeting structures in the brain.

6. Animals

Whereas the local BBB disruption in mice and rats is possible using FUS through the intact skull and skin, to achieve the BBB disruption in the deep-situated targeted locations in the rabbit brain, sonications are performed through a cranial window. The craniotomy is performed to simplify the experiments with respect to the ultrasound equipment and to reduce the uncertainty of in situ estimates of acoustic pressure amplitudes.

6.1. Preparation of a Skull Window

The animals are anesthetized with a mixture of 12 mg of sodium xylazine (Xyla-ject; Phoenix Pharmaceuticals, St. Joseph, MO) and 48 mg of ketamine hydrochloride (Abbott Laboratories, North Chicago, IL) per kg of body weight per hour. The hair from the head is removed with clippers and hair removal lotion, and antiseptic lotion is applied to sterilize the skin. The skin is cut over the central line and pushed sideways to expose the skull surface. A piece of skull (up to 18 × 18 mm) is removed. The skin over the skull window is replaced, sutured, and allowed to heal completely at least for 10 days before the experiments.

6.2. Positioning of the Animal for the Experiment

The animals are re-anesthetized and the hair over the top of the head is removed by shaving and by using hair removal lotion. An intravenous catheter is inserted into the ear vein in case of rabbits and into the tail vein (mice and rats). The animal is placed in the supine position on the animal platform with the head fixed in a holder. A receive-only MRI surface coil (7.6 cm diameter, GE Healthcare, Milwaukee, WI) is placed below the head to improve the MRI signal-to-noise ratio. A thin plastic bag filled with degassed water provides acoustic coupling between the transducer and the skin on the head of the animal. Degassed water is injected under the head to improve coupling between the head and the plastic membrane of the sonication system. The body temperature is maintained by a heated pad and monitored with a rectal thermocouple probe. The sonication system is placed on the MR table and moved into a clinical MR scanner for imaging.

6.3. Targeting Locations in the Brain with MRI

Three orthogonal sets of multi-slice T2-weighted FSE images are performed to assure appropriate head position and good acoustic coupling. The target locations are selected from these images. The ultrasound beam focus is aimed to the first targeted location. Before the experiments for the BBB disruption are performed, low-power sonications are delivered to localize the precise position of the focal point in the brain. For example, with the 1.5 MHz transducer used in reference (58) we typically apply a 10-s sonication at an acoustic power of 1.1 W (~90 W/cm^2 peak intensity at the focus). This sonication slightly heats the tissue, and the resulting temperature changes can be visualized using MR thermometry. This MR thermometry uses phase-difference images of a gradient echo sequence to estimate changes in the water proton resonant frequency (59), which changes linearly with temperature (60). In our experience, a focal temperature change on the order of 3°C can be reliably mapped with this imaging; and histological studies have shown that such sonications do not produce damage to the brain (58). The focal coordinate is first found in imaging acquired perpendicular to the direction of the ultrasound beam. Then the procedure is repeated using imaging acquired parallel to the direction of the ultrasound beam. Sometimes a small (about 1 mm) correction is required to make the target location and the focus overlap. After alignment, sonications to disrupt the BBB can be delivered with precise targeting assured (*see* **Note 6**).

7. MRI-Guided BBB Disruption

7.1. Ultrasound Contrast Agents

We used two commercially available ultrasound contrast agents: Optison (Mallinckrodt, St. Louis, MO) containing perfluorocarbon gas-filled albumin shells (mean diameter of 2.0–4.5 μm) and

Definity (Bristol-Myers Squibb, New York, NY) containing perflutren lipid microspheres (mean diameter of 1.1–3.3 μm). Both ultrasound contrast agents are stored at +4°C. Similar results have been achieved with other ultrasound contrast agents, for example with UCA (SonoVue^{TM}, Bracco International, Amsterdam, the Netherlands), which contains phospholipid-coated microbubbles (mean diameter of 2.5 μm, concentration of 1 to 5 × 10^8 bubbles/mL) (10).

7.1.1. Microbubble Injection

Simultaneously with the start of each sonication or 10 s before the ultrasound contrast agent is injected intravenously into the ear vein (rabbit) or tail vein (rat, mouse) at a dosage of 0.05 mL/kg of body weight (Optison) and 0.01 mL/kg (Definity) (**Fig. 3.5**). These dosages are in the ranges recommended for human use by the manufacturer. The injection of the contrast agent is followed by a bolus of a normal saline (approximately 2 mL for rabbit, 0.2 mL for rat, and 0.02 mL for mouse) to flush the agent from an access line.

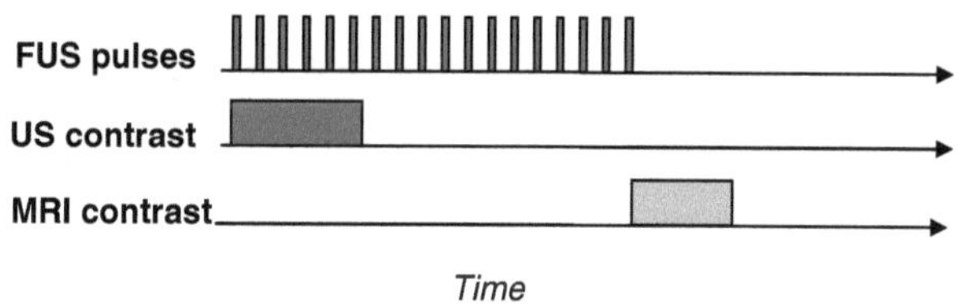

Fig. 3.5. Schematic of protocol used to BBB disruption with FUS and US contrast agent. Simultaneously with the start of pulsed sonication the ultrasound contrast agent is injected intravenously. To confirm the BBB disruption, standard commercial MRI contrast agent is IV administered after sonication.

7.2. Ultrasound Exposure

For the BBB disruption ultrasound energy is delivered in pulsed sonications. For example, at frequencies of 690 kHz suitable for highly localized sonication through human skull (41), successful BBB disruption in the rabbit brain was achieved with burst length of 10 ms and pulse repetition frequency of 1 Hz (duty cycle: 1%); total exposure time of 20 s; peak rarefactional pressure amplitudes less than 1.4 MPa (2).

The sonication can be repeated in another location after the microbubbles are cleared from the circulation. This requires a minimum of 5 min. For next sonication the injection of microbubbles is repeated. For BBB disruptions in a large region, multiple sonications of overlapping locations are spaced also 5–10 min apart and the injections of microbubbles are repeated before each sonication (*see* **Note** 7).

7.3. Detection of the BBB Disruption

The disruption of the BBB is visualized using standard commercial MRI contrast agents, such as gadopentetate dimeglumine (MAGNEVIST, Berlex Laboratories, Inc., Wayne, NJ; molecular weight of 928) that does not penetrate through the BBB. For a visual confirmation of BBB disruption in the harvested brain samples vital dye, trypan blue (ICN Biomedical, Aurora, OH) can be used.

7.3.1. Detection of the BBB Disruption with MRI

To confirm the BBB disruption, contrast-enhanced, T1-weighted FSE images are acquired before and after IV injection of MR contrast agent (Magnevist®, Berlex Laboratories, Inc, Wayne, NJ) at a dose of 0.125 mmol/kg of body weight.

First, the T1-weighted FSE image is acquired in axial planes after the sonication (*see* **Table 3.3** for parameters). This imaging is repeated after injection of the MRI contrast agent. The axial imaging is repeated multiple (7–10) times, and then images in other orientations are acquired.

The average signal intensity within a selected 3 × 3-voxel region of interest (ROI) is normalized to the pre-contrast baseline value of the ROI to calculate the signal enhancement. The percentage of enhancement in a non-sonicated ROI is subtracted from the enhancement at the targeted location to exclude enhancement due to the presence of the contrast agent in the vasculature. The BBB is considered to be opened in the targeted location, if the maximum enhancement of the sonicated ROI, averaged over three consecutive time points, exceeds that of a control ROI outside of the focal target by more than one standard deviation.

Figure 3.6 shows examples of the BBB disruptions in the brains of four rabbits exposed to ultrasound at frequencies of 0.26, 0.69, 1.63, and 2.1 MHz. On the T1-weighted FSE images, focal contrast enhancement detects the areas with disrupted BBB. The size of the BBB-disrupted area depends on US frequency. Lower ultrasound frequency such as 0.69 MHz permits transcranial sonication with a little beam distortion. At 260 kHz, FUS exposure is possible without compensation for any distortion, making a relatively simple ultrasound device possible.

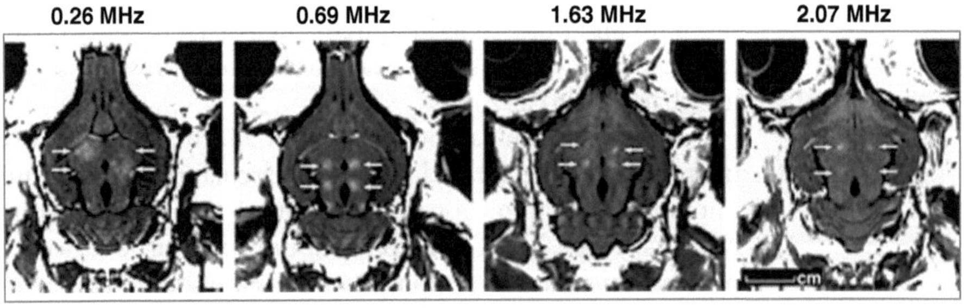

Fig. 3.6. Examples of the BBB disruption in the rabbit brains produced by FUS at four ultrasound frequencies: 2.1, 1.63, 0.69, and 0.26 MHz. The contrast enhanced T1-weighted MR images are acquired perpendicular to the direction of the ultrasound beam. The enhancing spots (*arrows*) detect the areas where the MRI contrast agent (Magnevist, Berlex Laboratories Inc., Wayne, NJ) passed through the FUS-induced BBB disruption. The size of the BBB-disrupted area depends on US frequency, as the frequency decreases, the size of the BBB disrupted area increases. At 260 kHz, exposure will be possible without having to compensate for this distortion, making a relatively simple ultrasound device feasible.

7.3.2. Detection of the BBB Disruption with Trypan Blue

To evaluate the exact location and extent of the area of the BBB disruption a vital dye, trypan blue (ICN Biomedical, Aurora, OH) is administered as an IV bolus injection immediately after sonication. Following Bakay's technique (61) a minimal dose of 0.1 g/kg of body weight in 0.45% NaCl is prepared fresh before injection. Solution is boiled and filtered to avoid formation of microcrystals. After injection the solution is left to circulate for at least 10–15 min before the animal can be sacrificed. Intra-arterial perfusion with isotonic saline followed by 10% formalin is usually used to remove trypan blue from the cerebral blood vessels. The perfusion does not affect the dye deposited in the cerebral tissue. Diffuse light blue staining of the tissue in the sonicated location indicates the penetration of the tissue by trypan blue due to increased permeability of the BBB. Satisfactory concentrations of the dye can be checked on the removed brain: there is distinct blue staining of the dura, hypophysis, choroid plexuses, and area postrema, while the rest brain remains unstained after saline perfusion. Microscopic examination of trypan blue deposition can be carried out on unstained sections of 20–200 μm thickness. On the brain sections, the blue-colored area of the BBBD forms a sharp contrast with the rest of the brain, which shows no trace of coloration after the perfusion of blood vessels (*see* **Note 8**).

8. Delivery of Macromolecular Agent

Molecular imaging and therapeutic agents are IV administered immediately after the contrast-enhanced MR images confirm the BBB destruction in targeted locations. The agents can be also injected simultaneously with the microbubbles during exposure of the targeted volume to the ultrasound or after the sonications. At least for smaller molecules the BBB permeability remains increased up to approximately 6 h after the exposure.

8.1. Examples of Targeted Delivery Across BBB

The high spatial resolution and non-invasive nature of ultrasound-induced BBB disruption make it an advantageous technique for targeted drug delivery to the brain. The feasibility of trans-BBB delivery by FUS has been established for different agents, including chemotherapeutic drugs such as liposomal doxorubicin (DOX) (8, 55, 62, 63) and Herceptin (4) as well as for molecular imaging and immunotherapeutic agents in Alzheimer's disease mouse models (64).

8.1.1. Enhanced Delivery of Chemotherapeutic

In the studies by Treat et al. (8, 55), FUS was applied through the intact skull in normal rat brain and in the brain with implanted 9L gliosarcoma with concurrent IV injections of ultrasound contrast agent and liposomal doxorubicin. Doxorubicin (DOX) is a chemotherapeutic drug with widespread clinical use. DOX showed its potential against glioma (the most invasive form of primary brain tumors) in vitro (65) and in vivo (66). However, it has not been effective in glioma patients. The BBB prevents cytotoxic levels from accumulating in the brain parenchyma. To increase the penetration of DOX into targeted areas of the brain, concurrent FUS treatments and systemic DOX were given with varying doses of ultrasound contrast agent in the normal brain.

The fluorescence image shows focal DOX delivery in the normal rat brain (**Fig. 3.7a**) (8). Concentrations within the human therapeutic range (67) have been achieved without damaging the brain. DOX concentrations in non-treated regions were low, reducing risk of neurotoxic effects (**Fig. 3.7b**). Higher concentrations were possible with some associated tissue damage, which might be acceptable for cancer treatment.

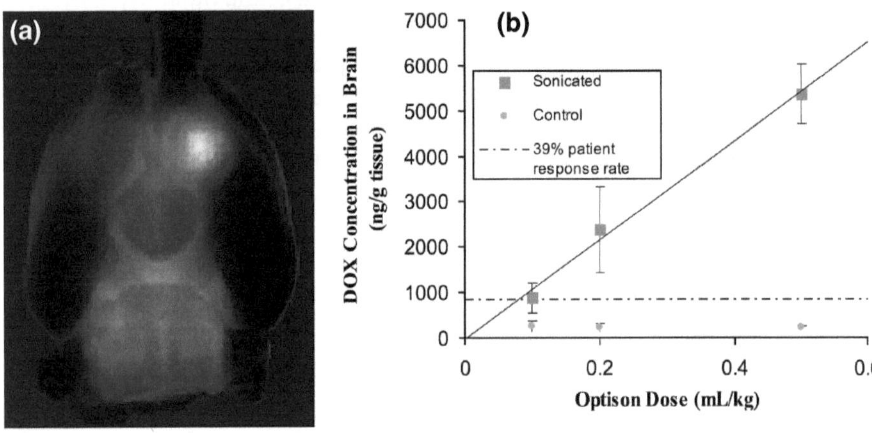

Fig. 3.7. Enhanced delivery of doxorubicin to the normal rat brain by FUS-mediated BBB disruption: (**a**) fluorescence image shows distribution of DOX in the sonicated region after BBB disruption by FUS with Optison and intravenous DOX administration; (**b**) concentration of DOX delivered to the brain as a function of Optison dose. DOX concentration in the sonicated brain tissue met or exceeded the concentration shown to have a clinical response in human (*dotted line*) (67). Reproduced with permission from (8).

The impact of FUS-assisted delivery of DOX to the brain is actively being investigated in glioma models. Preliminary studies in rats with implanted 9L gliosarcoma indicate that concurrent FUS-mediated BBB disruption and IV DOX slowed tumor growth and significantly increased the percentage of long-term (>40 days) survivors (55, 62, 63).

8.1.2. Delivery of Monoclonal Antibody

The delivery of much larger biological molecules, such as antibodies, is also possible. First, experiments by Kinoshita et al. (5) demonstrated that intravenously administered dopamine D4 antibodies crossed the BBB and recognized their antigens when BBB was temporary disrupted with FUS in the presence of Optison. Iimmunohistochemistry confirmed the delivery of dopamine D4 antibodies to the targeted regions of the mouse brain.

Using the same technique Kinoshita et al. (4) delivered antibody-based anticancer agent Herceptin (Trastuzumab, Genentech, San Francisco, CA) through the BBB in the mouse brains. Herceptin is a recombinant humanized IgG monoclonal antibody drug which works on the HER2/neu (erbB2) receptor and is used successfully for treatment of metastatic HER-2-positive breast cancer. This receptor is amplified in approximately 25–30% of breast cancer patients (68). Herceptin targets the tumor cells, which overexpress the HER2/neu (erbB2) receptor. However, in patients with brain metastasis, the BBB limits its use.

Herceptin was successfully delivered into the brains of mice when BBB disruption was performed with the MRI-guided FUS. Herceptin was detected in the sonicated locations, while in regions with an intact barrier its level was below detection limit. Concentrations of Herceptin were found to correlate with the MRI contrast-enhancement, enabling an indirect estimate of the amount of Herceptin delivered.

8.1.3. Targeting Amyloid Plaques in Alzheimer's Disease Mouse Model

Alzheimer's disease (AD) is characterized by the accumulation of small, misfolded proteins, beta-amyloids (Aβ). Therapeutic strategies focus on detecting and reducing Aβ aggregates. One promising strategy for amyloid plaque clearance is immunization against Aβ by administration of anti-β-amyloid antibodies. However, diagnostic and therapeutic agents do not cross the BBB.

Low-intensity FUS with a microbubble contrast agent may be used to transiently disrupt the BBB, allowing non-invasive, localized delivery of imaging fluorophores and immunotherapeutics directly to amyloid plaques (64).

Intravenous administered an amyloid staining red fluorophore (trypan blue) and anti-β-amyloid antibodies were delivered to amyloid plaques when BBB was temporary disrupted with FUS and Optison in the transgenic mouse models of Alzheimer's disease. MRI guidance permitted selective treatment and monitoring of plaque-heavy anatomical regions such as the hippocampus. Treated brain regions exhibited 16.5 ± 5.4-fold increase in trypan blue fluorescence and 2.7 ± 1.2-fold increase in anti-amyloid antibodies that localized to amyloid plaques. However, it is unclear from these experiments if a single FUS-enhanced anti-β-amyloid antibody dose is adequate for amyloid clearance. Future work will quantify antibody dose and amyloid clearance in the treated animals.

9. MRI Free-Guidance Techniques for BBB Disruption

In the animal experiments the BBB disruption in the local regions of the brain can be achieved without MRI guidance.

9.1. Benchtop BBB Disruptions

A simple MRI-free, benchtop sonication system has been developed to allow high-throughput FUS-induced BBB disruption for drug studies (64). Similar to the **Section 5.1**, the transducer is mounted on an XYZ positioning system and submerged in a tank of degassed water. Ultrasound energy is coupled to the animal above via a Mylar window. Before experiments, the transducer is focused onto cross-hairs drawn onto the Mylar window, providing a reference for animal alignment. Mice are placed supine with the shaved head directly contacting a Mylar window and positioned such that the transducer focus is 4 mm caudal to the eyes and centrally aligned. The transducer is then moved to place its focus squarely in the targeted position (in this case, 2–3 mm to the right, in the right hemisphere). Then the mice are treated as described above with FUS exposure and concurrent IV administration of microbubbles. The entire procedure takes approximately 5 min per animal, compared to 30–45 min with the MR-guided procedure. Because benchtop FUS-produced BBB disruption can be performed with a simple setup in a short procedure (5 min), it may aid imaging probe development and pre-clinical drug screening such as an anti-Aβ antibody screening.

However, targeting with the benchtop system is less accurate; it has been found to be within 2.2 mm of the intended location and varied with the system operator. Variability is likely due largely to animal motion and inaccurate animal positioning.

9.2. BBB Disruption: Stereotaxic Approach

A better accuracy can be achieved using a stereotaxic approach with the animal head fixed in ear-bars and transducer attached to a XYZ positioning of a stereotaxic apparatus (Stoelting, Wood Dale, IL) permitting sub-millimeter placement of the ultrasound focus into the target location in the brain (10). A removable cone with a tip capped by a polyurethane membrane was attached to the transducer and filled with deionized and degassed water. The rat with the head fixed in the stereotaxic apparatus was placed prone beneath the cone, and ultrasound gel (Pharmaceutical Innovations, Newark, NJ) was used to maximize the transmission of ultrasound between the transducer and the brain. The sonication was delivered through the cranial window (approximately 10 × 10 mm) to one location in each hemisphere (3.5 mm posterior and 2.5 mm lateral to the bregma and 5.3 mm below the brain surface).

9.3. BBB Disruption: Grid-Positioning Method

The BBB disruption in mice was accomplished by FUS exposure through the intact skull and skin using the diagnostic ultrasound guidance (9). A single-element circular-aperture FUS transducer ($f = 1.525$ MHz; $R = 90$ mm; outer radius: 30 mm; inner radius: 11.2 mm) with a hole in its center for an imaging transducer has been used. **Figure 3.8a** shows experimental set-up. A single-element diagnostic transducer was positioned through the hole so that the foci of the two transducers were aligned. A cone with a tip capped with a polyurethane membrane (Trojan; Church & Dwight

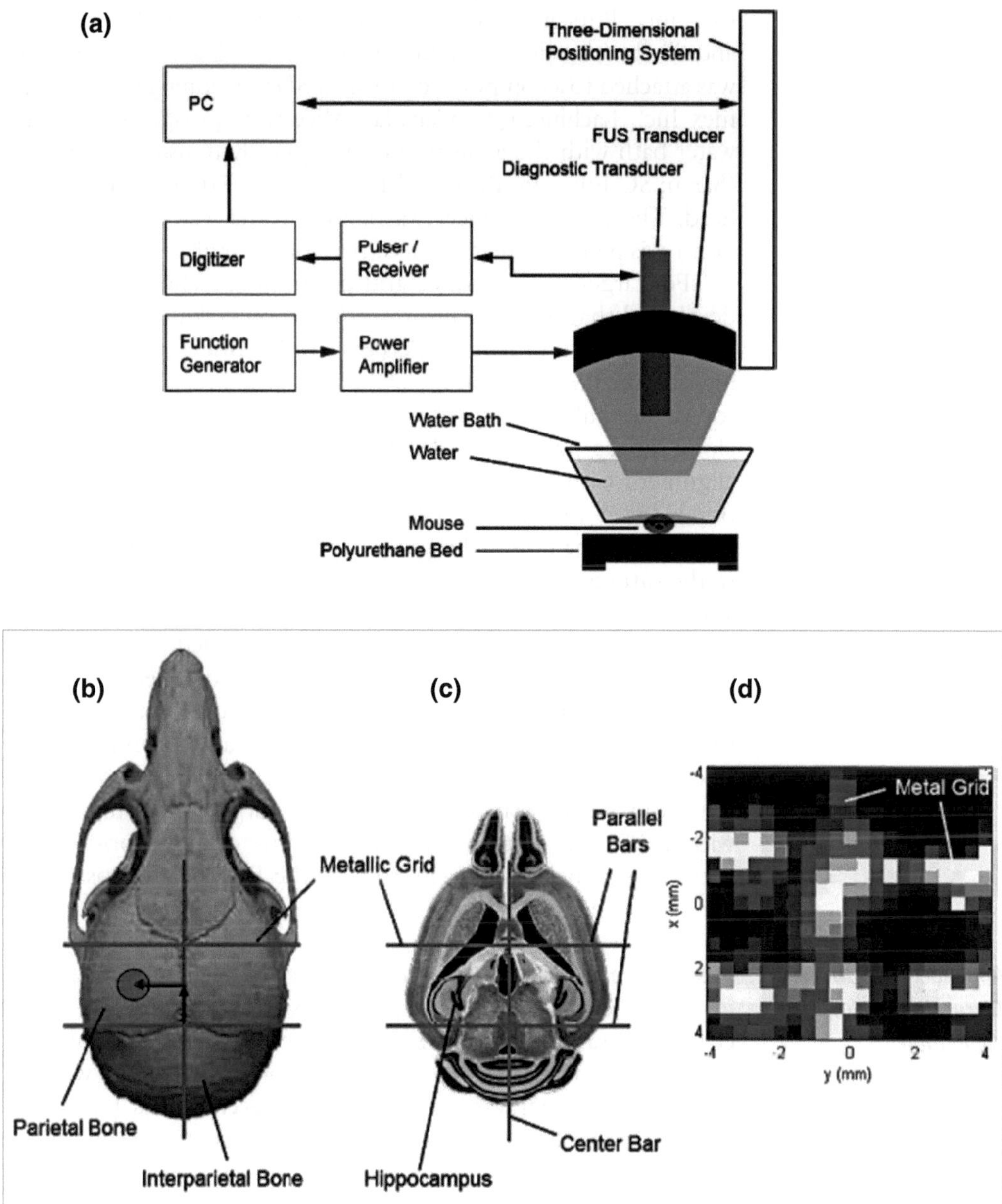

Fig. 3.8. BBB Disruption with a grid-positioning method. (**a**) Experimental set-up: The mouse is laid prone and a water bath with acoustically and optically transparent bottom is suspended over the mouse's head. The ultrasound system, consisting of a FUS transducer and a diagnostic transducer, is inserted in the water bath, and the focus placed in the brain through the intact skin and skull; (**b**) a metal grid is placed in the water bath in alignment with the sutures of the skull seen through the skin of the mouse; (**c**) the location of the intended brain structure, as seen in the histology slice (in this case, hippocampus), is found relative to the grid position; (**d**) a C-scan of the grid is performed using a diagnostic transducer and then the FUS transducer is moved to the targeted location. Reproduced with permission from (9).

Co., Inc., Princeton, NJ) was mounted on the transducer system and filled with deionized and degassed water. The FUS transducer was attached to a computer-controlled 3D positioning system (Velmex Inc., Lachine, QC, Canada). Mice were placed prone and a water bath with the acoustically and optically transparent bottom (Saran; SC Johnson, Racine, WI) was suspended over the mouse's head. The FUS transducer was placed in the water bath with its beam axis perpendicular to the surface of the skull.

For targeting purposes a grid-positioning technique has been developed that uses the sutures of the mouse skull seen through the skin as anatomic landmarks (**Fig. 3.8b**). Briefly, a grid consisting of three 0.3-mm thin metal bars is placed in the water bath on top of the skull and aligned with the skull sutures: (1) parallel and along the sagittal suture; (2) perpendicularly to the first bar and in alignment with the suture between the parietal and inter-parietal bone; (3) 4 mm away from and parallel to the first bar. The locations of the targeted regions of the brain are assumed relative to the sutures on the mouse skull and known brain anatomy. A C-scan of the grid is generated using the diagnostic transducer and the location of the targeted structure is identified relative to this grid. The focus of the FUS transducer is placed beneath the skull by measuring the distance with the diagnostic transducer. Using the grid positioning method and depth calculations, precise, accurate, and reproducible targeting of the local regions of the mouse brain can be performed. The intended target was found to be within 0.5 mm of the actual focus. Thus, the grid-positioning method is sufficiently precise to have the FUS beam consistently overlap the targeted structure of the murine brain.

Appendix 1– BBB Disruption: Histological Evaluation

Histology evaluation shows minimal adverse effects, as compared to invasive interventions. Effects are mostly related to tiny red blood cells extravasations indicating a few affected capillaries. At low ultrasound frequency (0.26 MHz), FUS exposure is possible without any blood cell extravasations (3). No delayed effects have been observed either by MRI or histology up to 4 weeks after sonication.

Appendix 2 – BBB Disruption: Electron Microscopy

Electron microscopy studies have indicated the FUS activates trans-cellular transport and reversibly affects tight junctions (7, 69). In the study using horseradish peroxidase (HRP) (40 kDa) as a tracer, numerous HRP-positive vesicles were observed in the endothelial cells and inter-endothelial clefts immediately after FUS-induced BBB disruption in the rabbit brain At 60 min following BBB disruption, HRP-positive vesicles were visible in the basement membrane and in the interstitial spaces of neuropil (**Fig. 3.9**). Immunoelectron microscopy for tight junction proteins have demonstrated reduced the TJ-specific proteins: occludin, claudin-5, ZO-1, ZO-2, and ZO-3 at 1 and 2 h and restored at 6 and 24 h (69).

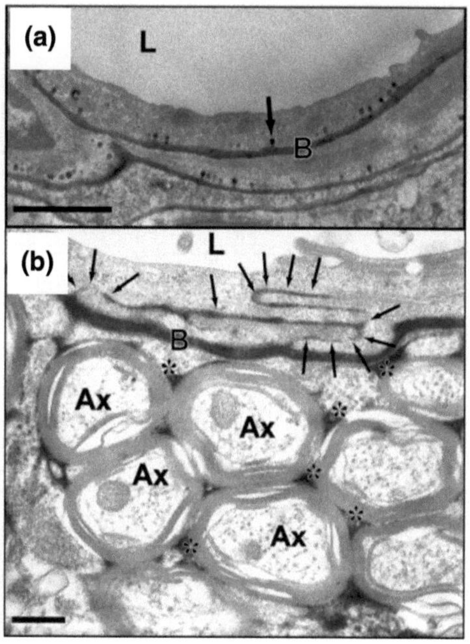

Fig. 3.9. Electron microscopy. BBB permeability for horseradish peroxidase (HRP): (**a**) micrograph showing segment of the arteriole wall 1 h after sonication. An intense vesicular transport is demonstrated by the increased number of HRP-positive caveolae (*arrow*); (**b**) a portion of longitudinally sectioned capillary with adjacent nerve tissue in a sample obtained 2 h after sonication. The tracer, HRP (*black color*) is present in the junctional cleft (*arrows*), the basement membrane and the interstitial spaces between myelinated axons (*asterisks*). (L: lumen; Ax: cross-sectioned axons; B: basement membrane). Scale bars: 200 nm. (**a**) Courtesy of N. Sheikov; (**b**) reproduced with permission from (69).

Appendix 3 – BBB Disruption: In Vivo Multiphoton Imaging

The BBB disruption has been studied in vivo in mice using simultaneous optical imaging and FUS exposure with Optison (70, 71). Exposure to ultrasound was performed through the ventral surface, leaving the dorsal surface of the brain free for imaging through a cranial window (**Fig. 3.10**). For in vivo imaging of blood vessels two fluorescent dyes were used: Alexa fluor (green) and Texas red. Results of observation are present in **Fig. 3.11**. The Alexa fluor IV injected before sonication shows a normal blood flow. The Texas red injected simultaneously with FUS exposure demonstrates an arteriolar vasoconstriction that disrupts blood flow and lasts up to 5 min. The BBB disruption occurs via two distinct processes: (1) perivascular fluorescence gradually increases along the affected vessel or (2) rapidly increases in focal regions. These results have confirmed the electron microscopy studies suggesting activated endothelial transcytosis and affected tight junction. They suggest also that the BBB disruption may be associated with vascular constriction.

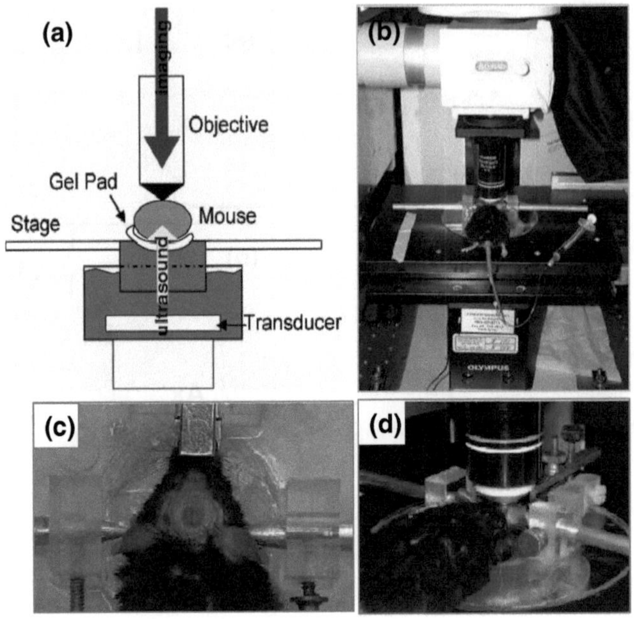

Fig. 3.10. In vivo multiphoton imaging of BBB disruption. (**a**) Schematic of experimental setup: The mouse is positioned to permit optical imaging through a craniotomy from above (via the objective) while sonicating from below. The transducer is held beneath the microscope stage in a cup-like holder which is secured to the microscope's condenser clip, allowing z-translational positioning of the transducer (*arrow*). A modified stereotaxic frame holds the mouse in place for imaging and permits ultrasound transmission to the mouse via a degassed water reservoir; a gel pad is placed between the mouse and the reservoir to ensure efficient coupling; (**b**)–(**d**) photographs of mouse during imaging. Courtesy of S. Raymond.

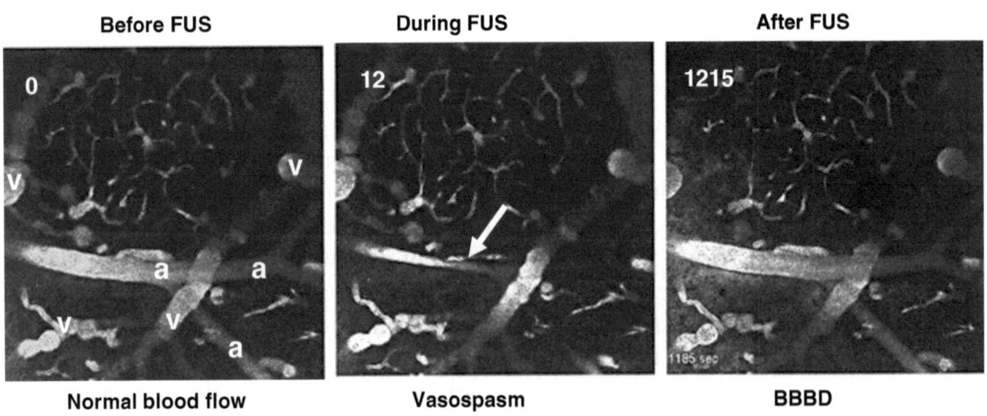

Fig. 3.11 . In vivo multiphoton imaging of the mouse brain vasculature during FUS-induced BBB disruption. The animal received 0.1 mL (2 mg/mL) 10 kDa, dextran-conjugated Alexa Fluor 488 (*green*) intravenously $\sim$ 5 min before imaging. Immediately after the $t = 0$ frame was taken, a 45-s US exposure was initiated and a 0.1 mL bolus (10 mg/mL) of 70 kDa, dextran-conjugated Texas Red was delivered intravenously. Almost total occlusion of the large vessel in the center of the field occurred 12 s after the initiation of US exposure (*arrow*). Beginning at 60 s and by 305 s, leakage in the green channel is apparent in the lower left of the field, and around the central vessel indicating BBB disruption (a = arterioles; v = veins; scale bar is 100 μm). Courtesy of S. Raymond.

10. Concluding Remarks

Low-intensity focused ultrasound administered with a microbubbles-based contrast agent (FUS-MB) may be used to transiently disrupt the BBB, allowing non-invasive, localized delivery of molecular imaging – and therapeutic agents to the brain. At frequencies suitable for non-invasive trans-skull focusing, FUS-MB causes BBB disruption only in the transducer focal volume, which can be selected in real-time by MR imaging for precise anatomical delivery. The BBB disruption lasts a few hours, activates trans-cellular transport, reversibly affects tight junctions, and may be associated with vascular constriction (vasospasm). It is sensitive to US frequency, burst duration, and total exposure duration. Associated effects in the brain are minimal – certainly less than what one would expect from any invasive intervention to deliver drugs. Ultrasound-mediated, transient BBB disruption allows the delivery of chemotherapeutics as well as both molecular imaging and therapeutic agents in Alzheimer's mouse models, which would aid preclinical drug screening and imaging probe development. Furthermore, this technique may be used to deliver a wide variety of small and large molecules to the brain for treatment of brain tumors and for imaging and therapy of neurodegenerative diseases. However, before translation into human use, FUS-microbubbles technique must pass safety and efficacy standards to prove therapeutic benefit and minimize side-effects. Furthermore, the underlying biomechanical mechanisms for FUS-microbubbles therapy must better understand. Ongoing work at our and other laboratories is aimed at understanding the cellular response to FUS-microbubbles and demonstrating the therapeutic effects of FUS-microbubbles-enhanced drug delivery for CNS disorders.

11. Notes

1. The ultrasound beam propagates through a medium from the transducer to the target, and water is usually used for this purpose. Acoustic attenuation in water is small, and the density and sound speed are close to those in soft biological tissues. Since gas bubbles significantly disturb ultrasound propagation, the water is degassed before the use by boiling for 20–30 min or by using degassing systems with a vacuum pump.
2. While absorbing targets are commercially available, often a densely bristled brush is used as an absorbing target.

3. Under conditions of normal incidence in room temperature deionized water the relationship between the mass registered on a scale and power is 68 mg/per 1 W of acoustic power, so that by reading a standard laboratory scale, the acoustic power *in Watts* can be approximated by dividing the mass reading *in milligrams* by 68.
4. Most hydrophones are constructed from a piezoelectric (PVDF) material. The voltage response of this material is linearly proportional to acoustic pressure but rarely absolute, except in the case of high-end calibrated hydrophones. Thus the hydrophone measurement is typically taken as a relative pressure measurement. The scan provides the relative pressure p_{rel} over a plane. If normalized to the peak pressure, $p_{\mathrm{rel}} = p_{\mathrm{scan}}/p_{\mathrm{max}}$, the relative pressure will be related to the true pressure by a scaling factor σ such that

$$p = \sigma p_{\mathrm{rel}}.$$

   If the efficiency of the transducer is known, this scaling factor will be equal to

$$\sigma = \sqrt{\frac{P\eta\rho c}{\sum_{x,y} p_{\mathrm{rel}}^2 \Delta x \Delta y}} \qquad [3]$$

   where P is the electrical power, and Δx and Δy are the scan step size in the x, and y planes, respectively. This scale factor gives the peak pressure as a function of electrical input power. Similarly, the scan may be multiplied this factor to give the pressure anywhere in the field.
5. The transducer is mounted on a mechanical positioning system which has to be MRI compatible; all its parts have to be non-magnetic and made from non-metallic materials such as plastics. Descriptions of positioning systems for experimental (72, 73) and prototype clinical (74) MRI-guided ultrasound systems have been reported.
6. For low-frequency exposures, ultrasound absorption in the brain is very small and it may not be possible to visualize the low-level heating at exposure levels that do not produce cavitation. In addition, in experiments in animals without a craniotomy, it may be difficult to produce heating at the focus without heating the skull bone. In these situations, the focal point is localized in an absorbing phantom before placing the animal on the focused ultrasound device. For trans-cranial application in small animals, even if one can focus through the thin skull bone, the bone may shift the focal point slightly off target. It is important to note that one should not apply

these localization sonications after the contrast agent was delivered, as the exposure levels used can produce tissue damage when microbubbles are present.

7. When fast positioning is possible with a mechanical system or phased array, multiple locations can be sonicated after a single injection of US contrast agent. However, this will require an automated sonication system.
8. Comparison of the tracers P^{32} and trypan blue showed that the area as outlined with P^{32} in radiographs was consistently larger than was the blue-stained area, indicating that the barrier damage when studied with a smaller-particle-sized tracer (P^{32}) is more extensive than it is when outline with a larger-particle-sized trypan blue (61).

Acknowledgments

This work was supported by NIH grants R01EB003268, R33EB000705, U41RR019703.

References

1. Hynynen, K., McDannold, N., Vykhodtseva, N., and Jolesz, F.A., Noninvasive MR imaging-guided focal opening of the blood-brain barrier in rabbits, Radiology, 220 (2001) 640 646.
2. Hynynen, K., McDannold, N., Sheikov, N.A., Jolesz, F.A., and Vykhodtseva, N., Local and reversible blood-brain barrier disruption by noninvasive focused ultrasound at frequencies suitable for trans-skull sonications, NeuroImage, 24 (2005) 12–20.
3. Hynynen, K., McDannold, N., Vykhodtseva, N., Raymond, S., Weissleder, R., Jolesz, F.A., and Sheikov, N., Focal disruption of the blood-brain barrier due to 260-kHz ultrasound bursts: a method for molecular imaging and targeted drug delivery, J. Neurosurg., 105 (2006) 445–454.
4. Kinoshita, M., McDannold, N., Jolesz, F.A., and Hynynen, K., Noninvasive localized delivery of Herceptin to the mouse brain by MRI-guided focused ultrasound-induced blood-brain barrier disruption, Proc. Natl. Acad. Sci. U. S. A., 103 (2006) 11719–11723.
5. Kinoshita, M., McDannold, N., Jolesz, F.A., and Hynynen, K., Targeted delivery of antibodies through the blood-brain barrier by MRI-guided focused ultrasound, Biochem. Biophys. Res. Commun., 340 (2006) 1085–1090.
6. Sheikov, N., McDannold, N., Vykhodtseva, N., Jolesz, F., and Hynynen, K., Cellular mechanisms of the blood-brain barrier opening induced by ultrasound in presence of microbubbles, Ultrasound Med. Biol., 30 (2004) 979–989.
7. Sheikov, N., McDannold, N., Jolesz, F., Zhang, Y.Z., Tam, K., and Hynynen, K., Brain arterioles show more active vesicular transport of blood-borne tracer molecules than capillaries and venules after focused ultrasound-evoked opening of the blood-brain barrier, Ultrasound Med. Biol., 32 (2006) 1399–1409.
8. Treat, L.H., McDannold, N., Vykhodtseva, N., Zhang, Y., Tam, K., and Hynynen, K., Targeted delivery of doxorubicin to the rat brain at therapeutic levels using MRI-guided focused ultrasound, Int. J. Cancer, 121 (2007) 901–907.
9. Choi, J.J., Pernot, M., Small, S.A., and Konofagou, E.E., Noninvasive, transcranial and localized opening of the blood-brain barrier using focused ultrasound in mice, Ultrasound Med. Biol., 33 (2007) 95–104.

10. Yang, F.Y., Fu, W.M., Yang, R.S., Liou, H.C., Kang, K.H., and Lin, W.L., Quantitative evaluation of focused ultrasound with a contrast agent on blood-brain barrier disruption, Ultrasound Med. Biol., 33 (2007) 1421–1427.
11. Hynynen, K., Clement, G.T., McDannold, N., Vykhodtseva, N., King, R., White, P.J., Vitek, S., and Jolesz, F.A., 500-element ultrasound phased array system for noninvasive focal surgery of the brain: a preliminary rabbit study with ex vivo human skulls, Magn. Reson. Med., 52 (2004) 100–107.
12. Aubry, J.F., Tanter, M., Pernot, M., Thomas, J.L., and Fink, M., Experimental demonstration of noninvasive transskull adaptive focusing based on prior computed tomography scans, J. Acoust. Soc. Am., 113 (2003) 84–93.
13. Pernot, M., Aubry, J.F., Tanter, M., Thomas, J.L., and Fink, M., High power transcranial beam steering for ultrasonic brain therapy, Phys. Med. Biol., 48 (2003) 2577–2589.
14. Lele, P.P., Advanced ultrasonic techniques for local tumor hyperthermia, Radiol. Clin. North Am., 27 (1989) 559–575.
15. Hynynen, K. and Lulu, B.A., Hyperthermia in cancer treatment, Invest. Radiol., 25 (1990) 824–834.
16. Diederich, C.J. and Hynynen, K., Ultrasound technology for hyperthermia, Ultrasound Med. Biol., 25 (1999) 871–887.
17. Arthur, R.M., Straube, W.L., Trobaugh, J.W., and Moros, E.G., Non-invasive estimation of hyperthermia temperatures with ultrasound, Int. J. Hyperthermia, 21 (2005) 589–600.
18. Fennessy, F.M. and Tempany, C.M., MRI-guided focused ultrasound surgery of uterine leiomyomas, Acad. Radiol., 12 (2005) 1158–1166.
19. Fennessy, F.M., Tempany, C.M., McDannold, N.J., So, M.J., Hesley, G., Gostout, B., Kim, H.S., Holland, G.A., Sarti, D.A., Hynynen, K., Jolesz, F.A., and Stewart, E.A., Uterine leiomyomas: MR imaging-guided focused ultrasound surgery – results of different treatment protocols, Radiology, 243 (2007) 885–893.
20. Rabinovici, J., Inbar, Y., Revel, A., Zalel, Y., Gomori, J.M., Itzchak, Y., Schiff, E., and Yagel, S., Clinical improvement and shrinkage of uterine fibroids after thermal ablation by magnetic resonance-guided focused ultrasound surgery, Ultrasound Obstet. Gynecol., 30 (2007) 771–777.
21. Zhang, L., Chen, W.Z., Liu, Y.J., Hu, X., Zhou, K., Chen, L., Peng, S., Zhu, H., Zou, H.L., Bai, J., and Wang, Z.B., Feasibility of magnetic resonance imaging-guided high intensity focused ultrasound therapy for ablating uterine fibroids in patients with bowel lies anterior to uterus, Eur. J. Radiol., (2008).
22. Hesley, G.K., Gorny, K.R., Henrichsen, T.L., Woodrum, D.A., and Brown, D.L., A clinical review of focused ultrasound ablation with magnetic resonance guidance: an option for treating uterine fibroids, Ultrasound Q., 24 (2008) 131–139.
23. Fukunishi, H., Funaki, K., Sawada, K., Yamaguchi, K., Maeda, T., and Kaji, Y., Early results of magnetic resonance-guided focused ultrasound surgery of adenomyosis: analysis of 20 cases, J. Minim. Invasive. Gynecol., 15 (2008) 571–579.
24. Yang, Z., Cao, Y.D., Hu, L.N., and Wang, Z.B., Feasibility of laparoscopic high-intensity focused ultrasound treatment for patients with uterine localized adenomyosis, Fertil. Steril. (2008).
25. Funaki, K., Fukunishi, H., Funaki, T., and Kawakami, C., Mid-term outcome of magnetic resonance-guided focused ultrasound surgery for uterine myomas: from six to twelve months after volume reduction, J. Minim. Invasive. Gynecol., 14 (2007) 616–621.
26. Furusawa, H., Namba, K., Thomsen, S., Akiyama, F., Bendet, A., Tanaka, C., Yasuda, Y., and Nakahara, H., Magnetic resonance-guided focused ultrasound surgery of breast cancer: reliability and effectiveness, J. Am. Coll. Surg., 203 (2006) 54–63.
27. Wu, F., Wang, Z.B., Cao, Y.D., Xu, Z.L., Zhou, Q., Zhu, H., and Chen, W.Z., Heat fixation of cancer cells ablated with high-intensity-focused ultrasound in patients with breast cancer, Am. J. Surg., 192 (2006) 179–184.
28. Wu, F., ter Haar, G., and Chen, W.R., High-intensity focused ultrasound ablation of breast cancer, Expert. Rev. Anticancer Ther., 7 (2007) 823–831.
29. Hacker, A., Michel, M.S., Marlinghaus, E., Kohrmann, K.U., and Alken, P., Extracorporeally induced ablation of renal tissue by high-intensity focused ultrasound, BJU Int., 97 (2006) 779–785.
30. Klingler, H.C., Susani, M., Seip, R., Mauermann, J., Sanghvi, N., and Marberger, M.J., A novel approach to energy ablative therapy of small renal tumours: laparoscopic high-intensity focused ultrasound, Eur. Urol., 53 (2008) 810–816.

31. Kohrmann, K.U., Michel, M.S., Gaa, J., Marlinghaus, E., and Alken, P., High intensity focused ultrasound as noninvasive therapy for multilocal renal cell carcinoma: case study and review of the literature, J. Urol., 167 (2002) 2397–2403.
32. Erce, C. and Parks, R.W., Interstitial ablative techniques for hepatic tumours, Br. J. Surg., 90 (2003) 272–289.
33. Okada, A., Murakami, T., Mikami, K., Onishi, H., Tanigawa, N., Marukawa, T., and Nakamura, H., A case of hepatocellular carcinoma treated by MR-guided focused ultrasound ablation with respiratory gating, Magn. Reson. Med. Sci., 5 (2006) 167–171.
34. Barqawi, A.B. and Crawford, E.D., Emerging role of HIFU as a noninvasive ablative method to treat localized prostate cancer, Oncology (Williston. Park), 22 (2008) 123–129.
35. Muto, S., Yoshii, T., Saito, K., Kamiyama, Y., Ide, H., and Horie, S., Focal therapy with high-intensity-focused ultrasound in the treatment of localized prostate cancer, Jpn. J. Clin. Oncol., 38 (2008) 192–199.
36. Koch, M.O., Gardner, T., Cheng, L., Fedewa, R.J., Seip, R., and Sanghvi, N.T., Phase I/II trial of high intensity focused ultrasound for the treatment of previously untreated localized prostate cancer, J. Urol., 178 (2007) 2366–2370.
37. Rouviere, O., Souchon, R., Salomir, R., Gelet, A., Chapelon, J.Y., and Lyonnet, D., Transrectal high-intensity focused ultrasound ablation of prostate cancer: effective treatment requiring accurate imaging, Eur. J. Radiol., 63 (2007) 317–327.
38. Wu, F., Wang, Z.B., Zhu, H., Chen, W.Z., Zou, J.Z., Bai, J., Li, K.Q., Jin, C.B., Xie, F.L., and Su, H.B., Feasibility of US-guided high-intensity focused ultrasound treatment in patients with advanced pancreatic cancer: initial experience, Radiology, 236 (2005) 1034–1040.
39. Catane, R., Beck, A., Inbar, Y., Rabin, T., Shabshin, N., Hengst, S., Pfeffer, R.M., Hanannel, A., Dogadkin, O., Liberman, B., and Kopelman, D., MR-guided focused ultrasound surgery (MRgFUS) for the palliation of pain in patients with bone metastases – preliminary clinical experience, Ann. Oncol., 18 (2007) 163–167.
40. Liberman, B., Gianfelice, D., Inbar, Y., Beck, A., Rabin, T., Shabshin, N., Chander, G., Hengst, S., Pfeffer, R., Chechick, A., Hanannel, A., Dogadkin, O., and Catane, R., Pain palliation in patients with bone metastases using MR-guided focused ultrasound surgery: a multicenter study, Ann. Surg. Oncol., 16 (2009) 140–146.
41. Clement, G.T. and Hynynen, K., A noninvasive method for focusing ultrasound through the human skull, Phys. Med. Biol., 47 (2002) 1219–1236.
42. Pernot, M., Aubry, J.F., Tanter, M., Marquet, F., Montaldo, G., Boch, A.L., Kujas, M., Seilhean, D., and Fink, M., High power phased array prototype for clinical high intensity focused ultrasound: applications to transcostal and transcranial therapy. Conf. Proc. IEEE Eng Med Biol Soc. (2007), 234–237.
43. Nyborg, W.L., Biological effects of ultrasound: development of safety guidelines. Part II: general review, Ultrasound Med. Biol., 27 (2001) 301–333.
44. Hsieh, D.Y. and Plesset, M.S., Theory of rectified diffusion of mass into gas bubbles, J. Acoust. Soc. Am., 33 (1961) 206–211.
45. Nyborg, W.L. and Acoustic, S. In: W.P. Mason (Ed.), Physical Acoustics, Vol. 2B, Academic Press, New York (1965).
46. Miller, D.L., A review of the ultrasonic bioeffects of microsonation, gas-body activation, and related cavitation-like phenomena, Ultrasound Med. Biol., 13 (1987) 443–470.
47. Krizanac-Bengez, L., Mayberg, M.R., and Janigro, D., The cerebral vasculature as a therapeutic target for neurological disorders and the role of shear stress in vascular homeostatis and pathophysiology, Neurol. Res., 26 (2004) 846–853.
48. Leighton, T. The Acoustic Bubble, Academic Press, San Diego, CA (1994).
49. Mihran, R.T., Barnes, F.S., and Wachtel, H., Transient modification of nerve excitability in vitro by single ultrasound pulses, Biomed. Sci. Instrum., 26 (1990) 235–246.
50. Hamill, O.P., Twenty odd years of stretch-sensitive channels, Pflugers Arch., 453 (2006) 333–351.
51. Edmonds, P.D. and Sancier, K.M., Evidence for free radical production by ultrasonic cavitation in biological media, Ultrasound Med. Biol., 9 (1983) 635–639.
52. Apfel, R.E., Acoustic cavitation: a possible consequence of biomedical uses of ultrasound, Br. J. Cancer. Suppl., 45 (1982) 140–146.
53. Flynn, H.G., Generation of transient cavities in liquids by microsecond pulses of ultrasound, J. Acoust. Soc. Am., 72 (1982) 1926–1932.
54. Duck, F.A. Physical Properties of Tissue, Academic Press Inc., New York (1990).

55. Treat, L.H.. Advances in targeted chemotherapy using MRI-guided focused ultrasound to disrupt the blood-brain barrier. PhD Thesis, Massachusetts Institute of Technology, USA. 2009.
56. Hynynen, K., Macromolecular delivery across the blood-brain barrier, Methods Mol. Biol., 480 (2009) 175–185.
57. Goss, S.A., Johnston, R.L., and Dunn, F., Comprehensive compilation of empirical ultrasonic properties of mammalian tissues, J. Acoust. Soc. Am., 64 (1978) 423–457.
58. Hynynen, K., Vykhodtseva, N.I., Chung, A.H., Sorrentino, V., Colucci, V., and Jolesz, F.A., Thermal effects of focused ultrasound on the brain: determination with MR imaging, Radiology, 204 (1997) 247–253.
59. Ishihara, Y., Calderon, A., Watanabe, H., Okamoto, K., Suzuki, Y., and Kuroda, K., A precise and fast temperature mapping using water proton chemical shift, Magn. Reson. Med., 34 (1995) 814–823.
60. Hindman, J.C., Proton resonance shift of water in the gas and liquid states, J. Chem. Phys., 44 (1966) 4582–4592.
61. Bakay, L., Hueter, T.F., Ballantine, H.T. Jr., and Sosa, D., Ultrasonically produced changes in the blood-brain barrier, Arch. Neurol. Psychat., 76 (1956) 457–467.
62. Treat, L.H., Zhang, Y., McDannold, N., and Hynynen, K.. Impact of focused ultrasound-enhanced drug delivery on survival in rats with glioma. Proceedings of the 8th International Symposium on Therapeutic Ultrasound./Minneapolis, Minnesota, USA. Ebbini, E. S. 2008.
63. Treat, L.H., Zhang, Y., McDannold, N., and Hynynen, K., MRI-guided focused ultrasound enhanced chemotherapy of 9L rat gliosarcoma: Survival study. Proceedings of the 16th Scientific Meeting and Exhibition of the International Society for Magnetic Resonance in Medicine. Toronto, Ontario, Canada. 2008:71.
64. Raymond, S.B., Treat, L.H., Dewey, J.D., McDannold, N.J., Hynynen, K., and Bacskai, B.J., Ultrasound enhanced delivery of molecular imaging and therapeutic agents in Alzheimer's disease mouse models, PLoS. ONE., 3 (2008) e2175.
65. Stan, A.C., Casares, S., Radu, D., Walter, G.F., and Brumeanu, T.D., Doxorubicin-induced cell death in highly invasive human gliomas, Anticancer Res., 19 (1999) 941–950.
66. Gupta, B. and Torchilin, V.P., Monoclonal antibody 2C5-modified doxorubicin-loaded liposomes with significantly enhanced therapeutic activity against intracranial human brain U-87 MG tumor xenografts in nude mice, Cancer Immunol. Immunother., 56 (2007) 1215–1223.
67. Cummings, J. and McArdle, C.S., Studies on the in vivo disposition of adriamycin in human tumours which exhibit different responses to the drug, Br. J Cancer, 53 (1986) 835–838.
68. Benz, C.C., O'Hagan, R.C., Richter, B., Scott, G.K., Chang, C.H., Xiong, X., Chew, K., Ljung, B.M., Edgerton, S., Thor, A., and Hassell, J.A., HER2/Neu and the Ets transcription activator PEA3 are coordinately upregulated in human breast cancer, Oncogene, 15 (1997) 1513–1525.
69. Sheikov, N., McDannold, N., Sharma, S., and Hynynen, K., Effect of focused ultrasound applied with an ultrasound contrast agent on the tight junctional integrity of the brain microvascular endothelium, Ultrasound Med. Biol., 34 (2008) 1093–1104.
70. Raymond, S.B., Skoch, J., Bacskai, B.J., and Hynynen, K., Modular design for in vivo optical imaging and ultrasound treatment in the murine brain, IEEE Trans. Ultrason. Ferroelectr. Freq. Control, 54 (2007) 431–434.
71. Raymond, S.B., Skoch, J., Hynynen, K., and Bacskai, B.J., Multiphoton imaging of ultrasound/Optison mediated cerebrovascular effects in vivo, J. Cereb. Blood Flow Metab., 27 (2007) 393–403.
72. Cline, H.E., Schenck, J.F., Watkins, R.D., Hynynen, K., and Jolesz, F.A., Magnetic resonance-guided thermal surgery, Magn. Reson. Med., 30 (1993) 98–106.
73. Hynynen, K., Darkazanli, A., Unger, E., and Schenck, J.F., MRI-guided noninvasive ultrasound surgery, Med. Phys., 20 (1993) 107–115.
74. Cline, H.E., Hynynen, K., Watkins, R.D., Adams, W.J., Schenck, J.F., Ettinger, R.H., Freund, W.R., Vetro, J.P., and Jolesz, F.A., Focused US system for MR imaging-guided tumor ablation, Radiology, 194 (1995) 731–737.

Chapter 4

Prediction of Blood–Brain Barrier Penetration by Drugs

Kiril Lanevskij, Pranas Japertas, Remigijus Didziapetris, and Alanas Petrauskas

Abstract

This chapter presents a mechanistic approach to in silico prediction of blood–brain barrier penetration by drugs and drug-like molecules. Focus is brought to critical analysis of experimental data used for modeling; several factors affecting data quality are discussed. The main experimental techniques used for measuring brain uptake are also briefly reviewed. The chapter provides a theoretical background for obtaining physicochemically reasonable and easily interpretable predictions of passive diffusion across blood–brain barrier, as well as some general advices regarding descriptor selection and model development.

Key words: Blood–brain barrier, drug transport, passive diffusion, permeability, brain uptake index, in situ perfusion, non-linear regression, QSAR.

1. Introduction

The ability to predict blood–brain barrier (BBB) penetration potential of drugs and drug-like molecules has always been of major interest to scientists dealing with CNS drug design. Due to its unique characteristics blood–brain barrier constitutes a major obstacle for delivery of therapeutic compounds to the brain, preventing a wide range of molecules from reaching their target areas in brain. Many drug candidates targeted for CNS activity are rejected because of low brain penetration, despite being sufficiently potent in vitro (1). Nowadays, in silico predictive tools gain increasingly greater roles in discovery decision making (2). A model able to provide a reliable first estimate of brain uptake for novel molecules could be very helpful for guiding drug discovery toward compounds with acceptable permeability profiles. Although first attempts to obtain predictive computational models of brain penetration date back to several decades ago (3, 4),

K.K. Jain (ed.), *Drug Delivery to the Central Nervous System*, Neuromethods 45,
DOI 10.1007/978-1-60761-529-3_4, © Humana Press, a part of Springer Science+Business Media, LLC 2010

advances in this field remain quite modest, primarily due to the lack of high-quality experimental data. Recently, we addressed this issue by performing an exhaustive analysis of original publications on brain uptake of drugs and by compiling a larger data set compared to those used in the previous studies. Based on these data we developed a mechanistic ionization-specific model of blood–brain permeability by passive diffusion with this main objective – to provide a clear theoretical basis behind the predictions (5). In this chapter we present our modeling approach and data verification procedures focusing mainly on practical details and aspects that often receive little attention but are essential for building a robust model.

The first step toward development of a QSAR model is selection of the particular parameter that the model will predict. The ultimate goal of modeling BBB penetration is to predict whether the compound of interest will be active in CNS. However, presence or absence of CNS activity is an overall outcome of a multitude of different processes, which is highly unlikely to be meaningfully described by a single model. In addition to qualitative classification, two quantitative measures characterizing different aspects of brain uptake are usually distinguished – log BB and log *PS* constants.

- log BB or *b*rain to *b*lood distribution ratio is a thermodynamic constant defining the extent of brain penetration (6). It is given by the ratio of drug concentrations in brain tissue and plasma at steady-state conditions (c_{SS}):

$$\log \mathrm{BB} = \log\left(\frac{c_{SS}^{\mathrm{brain}}}{c_{SS}^{\mathrm{plasma}}}\right)$$

- *PS*, defined as membrane *p*ermeability–*s*urface area product, is a kinetic parameter describing the rate of BBB permeation by a compound (7). *PS* relates to other capillary transport parameters according to Kety–Renkin–Crone equation, and by its physical sense corresponds to unidirectional influx rate constant corrected for cerebral blood flow rate (F):

$$K_{\mathrm{in}} = F \cdot \left(1 - e^{-PS/F}\right)$$

The major advantage of log *PS* as a determinant of brain uptake is that it represents an intrinsic characteristic of compounds' diffusion across cerebral endothelium cell membranes. *PS* is only physicochemically meaningful when it is unaffected by any side processes other than passive diffusion. Oppositely, log BB does not provide a measure of blood to brain transport process as such. If only passive transport is involved, brain/plasma partition ratio at equilibrium does not depend on the rate of crossing blood–brain barrier, but rather indicates the relative strengths of drug binding to brain tissue and plasma constituents (6). Despite this fact, very few QSAR studies on BBB permeability (i.e., log *PS*) have been published (8–13). The vast majority of earlier modeling attempts (*see* (14) for a review) including the most recent studies (15–17) deal with either prediction of log BB constants or CNS activity classification. In the latter case a popular approach is to produce empirical models based on complex, hardly interpretable descriptors without any consideration of actual brain penetration mechanisms. On the other hand, studies based on simple physicochemical relationships often result in heavily oversimplified considerations. Failing to understand the nature of the analyzed process together with the lack of publicly available experimental data suitable for modeling constitutes the main obstacles toward development of robust and widely applicable predictive algorithms of brain penetration.

Our preliminary analysis involving blood–brain distribution data in addition to permeation rates showed that if both quantitative characteristics of brain uptake are considered, a simple combination of the respective models allows obtaining a reasonably accurate estimate whether brain uptake of a compound is sufficient for being active in CNS (18). Any single parameter alone does not suffice for correct assignment of activity category. Yet, penetration rate is of primary importance since a compound cannot achieve steady-state conditions if it is not permeable to a considerable extent. On the other hand, there are many drugs that are largely restricted to plasma (and exhibit low log BB) due to high protein binding, but are still active in CNS. Therefore, this chapter will focus on the prediction of blood–brain permeability (log *PS* constants).

2. Experimental Data

A crucial aspect of every QSAR model is the quality of experimental data that it is based on. A computational prediction cannot be better than the underlying data. Critical verification of available experimental values is therefore of extreme importance for the development of accurate and reliable predictive models. In vivo blood–brain permeability of drugs is usually measured using one of the following experimental techniques: initial rate evaluation after intravenous administration, brain uptake index (BUI) studies, and in situ brain perfusion (7). In this section the respective methods will be briefly reviewed, and their advantages and potential pitfalls as well as the impact they pose on data quality will be discussed in detail.

2.1. Methods of Experimental Determination

Intravenous administration technique is one of the simplest approaches utilized for brain uptake studies. The procedure typically involves a single injection of labeled test substance to rats, arterial blood sampling at several time points after the administration, and subsequent decapitation followed by determination of brain tissue concentration at the terminal time point (C_{br}). Influx rate constant is then calculated as $K_{in} = C_{br}/AUC_0{}^T$, where $AUC_0{}^T$ is the area under concentration–time curve obtained from plasma data (7). Despite its technical simplicity, intravenous technique potentially offers the highest sensitivity of measurements since it is a multi-passage method (19). It is thus suitable for analysis of compounds with very low permeability. However, data coming from these experiments should be used with caution due to several apparent shortcomings. Intravenous experiments are conducted under physiological conditions that may result in flawed data for metabolically unstable substances or compounds highly bound to plasma proteins. The equation for calculating K_{in} relies on the assumption of unidirectional uptake, which may not hold throughout the whole experiment as exposure times are relatively high and significant backflux may occur at later time points (7, 19).

Brain uptake index (BUI) technique introduced by Oldendorf (20, 21) is a single-passage method. BUI experiment implicates rapid intracarotid injection to anesthetized rats followed by decapitation within 5–15 s of bolus administration (corresponding to 1–2 s passage time through brain endothelial capillaries (7) while the substance remaining in blood is washed out of brain circulation at 15 s (20)). Brain transport of the compound of interest is then quantified as first-pass extraction (E) and BUI is defined as relative extraction of test compound compared to freely diffusible internal reference (E_{ref}), usually tritiated water or [^{14}C]-labeled lipid-soluble compound such as butanol or antipyrine:

$$\mathrm{BUI\%} = \frac{E}{E_{ref}} \cdot 100\%$$

As in the case of K_{in} values, first-pass extraction can be converted to permeability–surface area product using Kety–Renkin–Crone equation:

$$E = 1 - e^{-\mathrm{PS}/F}$$

The possibility to freely manipulate the composition of the injection fluid (including drug concentration, pH) is often outlined as one of the main advantages of the BUI method (7, 19). The injected solution is not completely free of mixing with endogenous blood stream, which may lead to biased results for highly protein-bound drugs. Another drawback of the method is low sensitivity. Due to the short exposure time absolute extraction values of solutes entering the brain at slower rates cannot be reliably evaluated. Therefore, BUI measurements should only be expected to produce accurate results for moderate- to high-permeability compounds while applicability of the method for hydrophilic compounds is limited.

A more advanced method for measuring blood–brain permeability is in situ brain perfusion, first suggested by Takasato in 1984 (22). In these experiments common carotid artery of anesthetized rats is perfused with oxygenated physiological buffer containing test substance for up to 10 min. Calculation of the unidirectional influx rate constant (K_{in}) in this case is simplified compared to intravenous studies as perfusate concentration is maintained constant during the experiment (7). This method resolves most of the issues associated with other techniques and at the same time gives the researcher full control over the experimental conditions such as buffer composition, flow, or perfusion duration. It is though more difficult to perform due to involvement of a complex experimental setup, as well as accomplishing a surgical procedure (19). Most earlier in situ perfusion studies were conducted with rats, while in 2000 Dagenais and coworkers modified the method for use with mice (23). Since then a significant amount of high-quality mice brain uptake data have become available (24–26).

2.2. Data Analysis

Interspecies variation. Once the complete set of experimental log *PS* values is ready, the first thing to consider is the necessity to bring data determined in different rodent species on a single scale. For this purpose, in our recent study (5) we selected 25 compounds for which log *PS* values determined in both rats and mice were available and evaluated the interspecies correlation as shown in **Fig. 4.1**. The slope in the obtained scatter plot is close to 1,

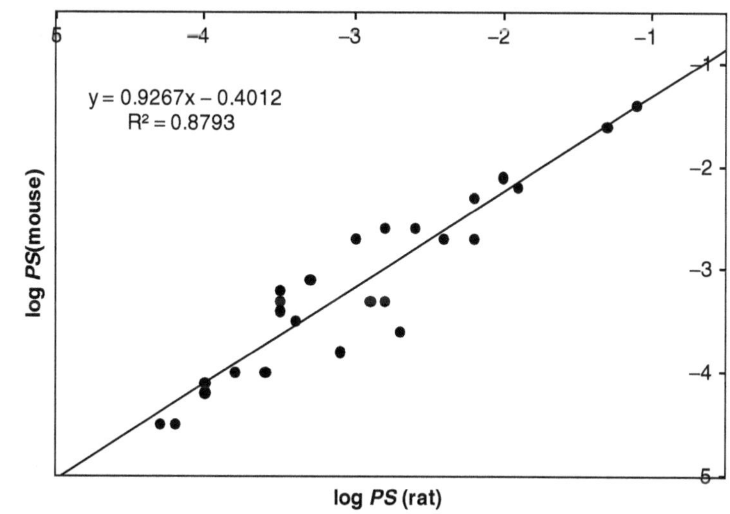

Fig. 4.1. Interspecies correlation of log *PS* constants (rat versus mouse data, $N = 25$).

indicating that blood–brain permeability (*P* term of the *PS* product) is equal in both species, while non-zero intercept may originate from the differences in the surface area of cerebral endothelium per gram of brain tissue (*S* term of the *PS* product). A simple equation ($\log PS_{rat} = \log PS_{mouse} + 0.4$) was used in our study to bring these data on a single scale yielding an overall data set of 280 compounds with quantitative log *PS* data.

Plasma protein binding. As already mentioned in **Section 2.1**, intravenous administration and BUI methods involve mixing of test solution with endogenous plasma. Since only free (i.e., unbound) drug molecules are available for diffusion across blood–brain barrier (27), the effective plasma concentration of drugs in these experiments is lowered by the amount bound to plasma proteins. This may not pose considerable impact on measured brain uptake values for hydrophilic compounds as the majority of them exhibit low affinities to human serum albumin and other carrier proteins (α_1-acid glycoprotein, lipoproteins, etc.) and circulate mostly in the unbound form. However, brain penetration of many lipid-soluble molecules that tightly bind to albumin may be restricted to a significant extent under in vivo conditions. Recently it was shown that blood–brain permeability of a few specific albumin ligands (28) as well as non-steroidal anti-inflammatory drugs (29) is markedly decreased if albumin is present in the perfusion buffer. The unidirectional influx rate constant could then be expressed by modified Kety–Renkin–Crone equation, where intrinsic permeability–surface area product (marked PS_u to emphasize that it represents a "true" value unaffected by protein binding) is corrected for the compound's unbound fraction in plasma (f_u):

$$K_{in} = F \cdot \left(1 - e^{-f_u \cdot PS_u}\right)$$

The results reported in (28) demonstrate that whole blood perfusion or other brain transport experiments involving drugs' interactions with albumin may underestimate intrinsic *PS* values by 1–3 orders of magnitude for highly protein-bound compounds. There is evidence that in some cases apparent exchangeable fraction between plasma and brain is higher than plasma f_u, and the simple equation presented above may not adequately describe the influence of protein binding on BBB permeation rate (30, 31). Despite such uncertainty regarding the extent to which protein binding affects *PS* product, it is obvious that if brain uptake of compounds exhibiting high affinities to plasma proteins is assayed by intravenous administration or BUI techniques, the measurements are likely to result in significantly biased data that should be excluded from modeling.

Facilitated diffusion and active efflux. As mentioned in Introduction, we suggest that blood–brain permeability should be modeled using a theoretically based physicochemical approach. A simple physicochemical model would though only describe BBB permeability driven by passive diffusion and could not account for the contribution of carrier-mediated processes. A variety of xenobiotics as well as substances of endogenous origin are transported by carrier proteins expressed at the blood–brain barrier (32, 33). These include both influx carriers such as transport systems specific for amino acids (System A, L1, n, y^+, x^- (34)), various organic cation and anion transporters (OCTN, OATPs, etc. (32, 35)), and active efflux pumps (P-gp, BCRP, MRPs (36)). Measured permeation rates for transporter substrates usually lack clear physicochemical meaning as they represent a cumulative effect of several different processes. All such values should be removed from the data set intended for modeling purposes as presence of data points apparently affected by enzymatic mechanisms would not give any valuable information, but would rather introduce additional noise and hinder the real dependencies between passive membrane diffusion and physicochemical parameters. Yet, the fact that a compound may undergo enzymatic influx or efflux does not mean that it is not suitable for analysis. Some transporter substrates may be left in modeling if log *PS* values representing passive diffusion are available for these compounds.

When brain uptake of a solute is characterized in concentration-dependent manner, linear dependence of permeation rate on solute concentration in the perfusion buffer is observed in those cases when transport is governed by passive diffusion, whereas involvement of carrier-mediated mechanisms in addition to the purely passive route manifests itself as saturable uptake. In the latter situation there are several possibilities to obtain log *PS* values unaffected by enzymatic processes:

1. If test compound is known to be a substrate of influx transporter expressed at BBB, additional experiments are usually conducted with excess of unlabeled substrate present in the perfusion fluid. Under these conditions transporter eventually becomes saturated, and it can be assumed that measured uptake rate of labeled solute corresponds to passive permeability. Generally speaking, presence of saturable component may be disregarded in those cases when it can be considered insignificant compared to overall transport rate.
2. A more accurate approach is to describe saturable transport in terms of modified Michaelis–Menten kinetic equation:

$$PS = \frac{V_{\max}}{K_{\mathrm{M}} + C_{\mathrm{test}}} + K_{\mathrm{d}}$$

Here K_{M} and $V_{\max}$ are standard Michaelis–Menten parameters, C_{test} is the concentration of test compound in the perfusion buffer, and additional term K_{d} corresponds to non-saturable transport component (7). Diffusion constant K_{d} represents uptake clearance driven solely by passive transport and the value log $PS \approx \log K_d$ should be included in the data set instead of the overall log *PS* constant.

Nevertheless, it should be kept in mind that K_d values for compounds crossing BBB by facilitated diffusion are estimated indirectly, by decomposition of overall transport rate to saturable and non-saturable components. Quality of such data is usually lower than that for passive permeants. For instance, suspicions regarding reliability of reported K_d values were raised for several influx transporter substrates during our QSAR analysis. In particular, K_d constants determined by first method should be used with caution as even at large substrate concentrations carrier-mediated transport component may be relatively significant (*see* **Section 2.3**).

The following considerations apply to the analysis of compounds subject to P-gp or other efflux mechanisms:

1. Log *PS* values supposedly driven by passive diffusion for active efflux substrates can be estimated from in situ perfusion experiments conducted with inclusion of specific transport inhibitors in the perfusion buffer. Verapamil, quinidine, cyclosporine A, as well as specific inhibitor valspodar (PSC833) are frequently used to block P-gp activity, while more recent studies also utilize inhibitors of other efflux carriers expressed at BBB (GF120918 inhibits BCRP (37), probenecid, MK-571, celecoxib – MRPs (38)).
2. In situ brain perfusion of *mdr1a*-knockout mice, first described by Dagenais and coworkers (23), remains the best source of blood–brain permeability data in those cases when disposition of the compounds of interest is restricted by P-gp. In contrast to experiments with addition of P-gp efflux inhibitors, brain uptake clearance measured in mice lacking the corresponding gene may be considered completely free of P-gp-mediated transport effect. It is now recognized that P-gp is not the only protein limiting brain entry of xenobiotics and other lines of transporter-deficient laboratory mice are available (39, 40). A number of studies utilizing *bcrp*(−/−) and *mrp*(−/−) mice to study the influence of the corresponding protein on brain penetration have been recently published, but most of these works focus on the determination of brain/plasma partitioning ratio. High-quality permeability data for BCRP and MRP efflux substrates are still lacking.

Zwitterionic drugs. An additional note should be made regarding zwitterions and other types of polyelectrolytes. Up to date, most available literature data for zwitterions represent amino acids that enter the brain via specific transporters (34), quinolone antibiotics that are substrates for P-gp or other efflux pumps (41), and other compounds susceptible to carrier-mediated transport. Although it can be assumed that some of the 40 zwitterionic compounds present in our log *PS* data set cross blood–brain barrier by passive diffusion (e.g., imipenem, morphine glucuronides), their permeability is so low (log $PS < -4$) that quality of reported values remains questionable, as additional concerns could be raised regarding sensitivity of measurements. Consequently, it was not possible to derive a predictive model for blood–brain permeability of zwitterions from currently available data. More studies on brain uptake of polyelectrolytes are needed with particular focus on mechanistic understanding of the analyzed process before any definitive conclusions could be made for this class of compounds.

2.3. Application Examples

Protein binding. As mentioned above, concerns regarding protein binding could be raised for compounds assayed using intravenous administration or BUI techniques. Several such compounds along with reported BBB permeability data and values predicted by the final model are shown in **Table 4.1**. The first three entries represent mustard derivatives studied by Levin (3). Despite having similar physicochemical characteristics, spiromustine is considerably less permeable than lomustine owing to much lower free fraction in plasma. Although other mustards are also bound to plasma proteins to a significant extent, the respective log *PS* constants conformed to the general trend observed in preliminary modeling; thus, only spiromustine was excluded from the data

Table 4.1
Examples of compounds for which suspicions regarding the influence of plasma protein binding on measured blood–brain permeability data were raised during data verification. Log *PS* values predicted by final model and key physicochemical properties are provided along with experimental values

No.	Name	Log *PS* exp.	Log *PS* pred.	Excluded	Log $P_{o/w}$	$f_{u,\ plasma}$	Acid pK_a	Base pK_a	References
1	Carmustine	−1.7	−1.8	No	1.33	0.23	–	–	(3)
2	Lomustine	−1.9	−1.5	No	2.52	0.50	–	–	(3)
3	Spiromustine	−2.4	−1.5	Yes	2.65	0.07	–	6.6	(3)
4	PNU-87663	−2.2	−1.1	Yes	4.30	0.01[a]	–	7.1	(51)
5	PNU-89843	−1.9	−1.2	Yes	3.18	0.07[a]	–	7.3	(51)
6	S-312	−1.9	−1.1	Yes	4.95	0.02	–	–	(30)
7	S-145	−2.6	−2.1	Yes	4.06	0.05	4.9	–	(30)

[a]Values predicted using ACD/ADME Suite software (Advanced Chemistry Development, Inc. (ACD/Labs), http://www.acdlabs.com/products/admet/adme/)

set. One should also pay attention that in situ perfusion experiments are sometimes conducted with serum albumin added to the perfusion buffer (e.g., entries from (30) in **Table 4.1**). This allows achieving more physiological conditions but yields permeaxbility data not suitable for modeling.

Influx. Decomposition of overall brain uptake clearance to saturable facilitated diffusion component and passive diffusion constant K_d can be illustrated using concentration-dependent BBB permeability data for pentazocine (42). In the original study brain uptake was measured in the presence of excess unlabeled pentazocine (PTZ concentrations ranging from 0.01 to 30 mM). Applying non-linear optimization procedure to *PS* product values obtained at different substrate concentrations based on modified Michaelis–Menten equation (*see* **Section 2.2**) yields an estimate of all three principal parameters: K_M, V_{max}, and K_d. In **Fig. 4.2** blood to brain flux values ($J = PS \cdot C_{PTZ}$) corresponding to overall, saturable, and non-saturable transport rates are plotted against the respective PTZ concentrations. The presented plot also highlights another important aspect: one should be particularly cautious to assume that observed *PS* values are equal to K_d even when excess of unlabeled substrate is present. When transporter becomes fully saturated as is the case in the provided example from (42), facilitated diffusion may still account for a considerable part of overall transport rate, especially when passive permeability is relatively low. Therefore, performing thorough Michaelis–Menten kinetic analysis is always desirable for obtaining reliable K_d estimates if sufficient data are available.

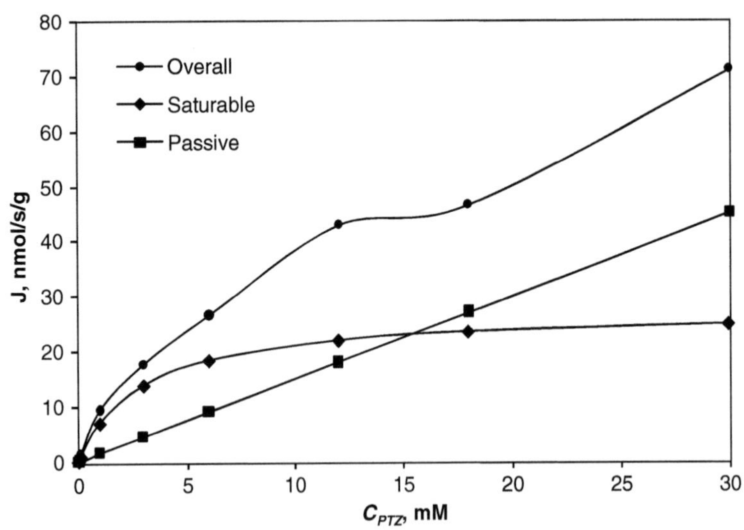

Fig. 4.2. Blood to brain flux of pentazocine and contributions of facilitated and passive transport at different concentrations of unlabeled substrate. Experimental data are taken from (41).

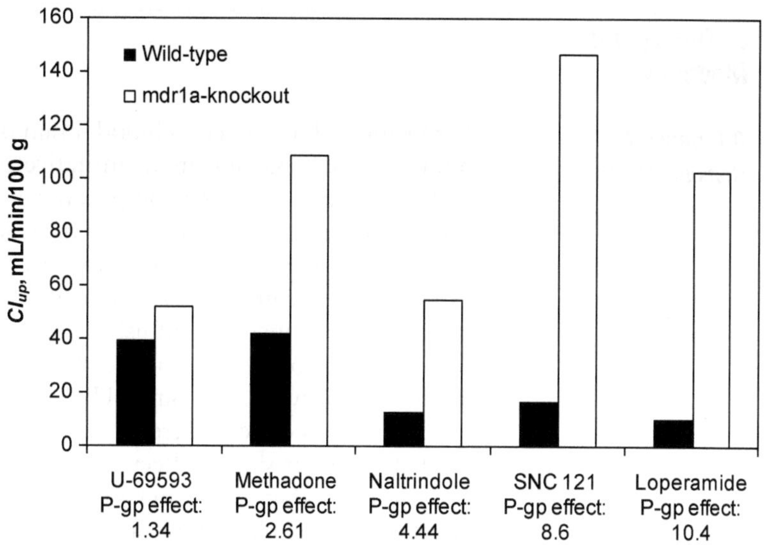

Fig. 4.3. Initial uptake clearances of selected P-gp substrates determined by in situ perfusion in wild-type and *mdr1a*-deficient mice. Experimental values taken from (9).

Efflux. **Figure 4.3** presents experimentally determined values of brain uptake clearance (the same as influx rate constant K_{in}) for selected P-gp efflux substrates extracted from (26). As one can see in the figure, P-gp effect (the ratio between brain uptake rates in P-gp-deficient and wild-type mice) varies quite significantly among studied molecules. For example, U-69593 exhibits very weak efflux transport, with KO/WT permeability ratio only about 1.3, showing that transport of this compound across BBB occurs predominantly by passive diffusion. On the contrary, efflux of naltrindole, SNC 121, and loperamide is much more marked, resulting in permeability ratios up to 10 and higher. Inclusion of the original values for these compounds measured in presence of P-gp in log *PS* data set would introduce discrepancies of up to 1 log unit compared to compounds possessing similar physicochemical properties but not interacting with P-gp.

A good example of involvement of efflux transporters other than P-gp is given in the study of BBB permeability of several flavonoids (37). The reported K_{in} values for quercetin were lower than 0.3 μL s^{-1} g^{-1} in control experiments. Those were not significantly different from the values obtained in presence of P-gp inhibitor valspodar (PSC833), indicating that P-gp does not have a profound influence on brain uptake of quercetin. After pretreatment with BCRP inhibitor GF120918, almost tenfold increase of quercetin uptake rate was observed in all regions of rat brain. Only the latter value of ca. 3 μL s^{-1} g^{-1} therefore truly reflects passive transport and is suitable for modeling.

3. Theory and Modeling

3.1. General Considerations

Permeation of drugs across blood–brain barrier can be viewed as a multi-step process, and its quantitative measure – permeability–surface area product – is a complex parameter that can be decomposed into a few simpler terms. In absorptive systems, overall system resistivity to diffusion (inverse of permeability) can be expressed as a sum of resistances of all particular components, which typically includes diffusion through unstirred water layer (characterized by P_{UWL}) prior to diffusion through phospholipid membrane bilayer (P_{m}). Since UWL thickness in brain capillary endothelium is close to zero (43), the former term can be disregarded. Another similar effect – flow-limited permeability (P_{F}) – occurs at blood–brain barrier (44), and we assume that limiting effect of cerebral blood flow rate on brain uptake may be modeled in a similar manner as P_{UWL}:

$$\frac{1}{P_e} = \frac{1}{P_{\mathrm{UWL}}} + \frac{1}{P_{\mathrm{m}}} \quad \frac{1}{PS} = \frac{1}{P_{\mathrm{F}}} + \frac{1}{P_{\mathrm{m}}S}$$

P_{F} term was introduced to account for the behavior of lipid-soluble drugs such as tricyclic antidepressants or benzodiazepines that readily cross blood–brain barrier and actually diffuse faster than they are delivered to the diffusion place. Inclusion of an additional term in the equation presented above is associated with limited instrumental sensitivity of available techniques rather than with an actual physiological process. Theoretically, PS product expressed from Kety–Renkin–Crone equation represents influx rate constant already corrected for flow (F). In practice, experimentally determined uptake clearance approaches F with increasing permeability, and it becomes virtually indistinguishable from flow ($K_{\mathrm{in}} \approx F$) when $P_{\mathrm{m}} \cdot S$ exceeds K_{in} more than about twice (7). Given a value of F ca. 0.05 mL s^{-1} g^{-1} typical for in situ perfusion experiments (8, 10, 22), the upper limit of reliably measurable PS values does not exceed $P_{\mathrm{F}} \approx 2 \cdot F \approx 0.1$ mL s^{-1} g^{-1}. The proposed assumption becomes necessary to ensure that no meaningless predictions ($\log PS > -1$) are produced by the model.

Membrane permeability ($P_{\mathrm{m}} \cdot S$). From the modified Fick's law and other considerations regarding solute partitioning between phospholipid bilayer and water, transmembrane flux of a drug can be expressed as a product of its membrane permeability P_{m} and concentration gradient ΔC_{m}, where P_{m} depends on kinetic diffusivity D and a thermodynamic diffusion constant K_{d} corresponding to membrane/plasma partitioning ratio.

$$J = P_{\mathrm{m}} \cdot \Delta C_{\mathrm{m}} \quad P_{\mathrm{m}} = C_{\mathrm{mem}} \cdot D \cdot K_{\mathrm{d}}$$

$C_{\mathrm{mem}} = S/h$ is a constant that represents certain properties (surface area and thickness) of the particular membrane under analysis. Kinetic diffusivity can be expressed from the Stokes–Einstein equation as a function of temperature (T), solution viscosity (η), and molecular radius of the solute (r):

$$D = \frac{k_{\mathrm{B}} T}{6\pi\eta r}$$

Temperature and viscosity are system-specific parameters that do not differ among solutes. These may be replaced by a constant (C_{diff}), while the remaining non-constant parameter r can be expressed by the cubic root of McGowan characteristic volume (V_x) yielding $D = C_{\mathrm{diff}} \cdot V_x^{-1/3}$. It is noteworthy that D only slightly varies among molecules of different size, and if K_{d} is assumed constant, increase in MW from 100 to 1000 would only explain about 0.3 log units decrease in permeability. This observation illustrates that major part of log *PS* variations is determined by K_{d}, whereas low permeability of large molecules is not due to weak diffusivity as suggested by earlier models (3), but is rather determined by their hydrophobicity (*see* below) or other properties such as high hydrogen-bonding potential.

3.2. Thermodynamic Diffusion (K_d)

All subsequent considerations will apply to the prediction of diffusion constant K_{d}. According to thermodynamic partitioning theory of electrolytes, K_{d} should depend on a compound's ionization state. If all types of ionic species are assumed to partition independently of each other, the following expression could be obtained:

$$K_{\mathrm{d}} = \sum_{i=0,+,-,\pm} f_i \cdot K_{\mathrm{d}}^i = \sum_{i=0,+,-,\pm} f_i \cdot 10^{\log K_{\mathrm{d}}^i}$$

Here i denotes the particular type of ionic species (neutral, mono-cation, mono-anion, and zwitterions species were considered in our study), f_i the fraction of the ith species at given pH, and $K_{\mathrm{d}}^{\,i}$ the respective characteristic partitioning constant of the ith species. Note that in the right side, $K_{\mathrm{d}}^{\,i}$ constants are converted to logarithmic form to ensure that they uphold LFER (*l*inear *f*ree *e*nergy *r*elationship) principle. The provided equation has the same form as the equation widely used to calculate pH-dependent octanol/water distribution coefficient log $D_{\mathrm{o/w}}$ at given conditions from the corresponding log *P* values of different ionic species:

$$\log D_{\mathrm{o/w}} = \log \sum_i f_i \cdot 10^{\log P^i_{\mathrm{o/w}}}$$

Octanol/water log D is frequently used for modeling such processes as blood–brain permeability (12) or intestinal absorption (45) as a descriptor that combines the effect of lipophilicity and ionization – two main physicochemical determinants of membrane penetration in a single parameter. The major drawback of this approach is that ionization dependence of the analyzed process is implicitly assumed to be the same as that of 1-octanol/water partitioning. This assumption is not permissible since in vivo membrane/plasma partitioning observed at BBB may differ from any known in vitro system. Instead, an independent evaluation of the ionization influence should be performed for the particular process under study. BBB-specific values of electrostatic charge parameters Δ_i representing the differences between permeabilities of neutral and charged species could then be obtained:

$$\log K_{\mathrm{d}}^{i} = \log K_{\mathrm{d}}^{\mathrm{o}} + \Delta_i$$

For a further discussion of ionization effect and comparison of our ionization-specific BBB model versus extrapolation from log $D_{\mathrm{o/w}}$, please refer to **Note 3**.

Characteristic constants log $K_{\mathrm{d}}{}^{i}$ may be described in terms of lipophilicity (given by octanol/water log *P*) and hydrogen-bonding capacity. The following equation proved to yield the best prediction results for non-electrolytes:

$$\log K_{\mathrm{d}}^{\mathrm{o}} = c_{\mathrm{o}} + c_1 \cdot \log P_{\mathrm{o/w}} + c_2 \cdot \left(\mathrm{HD} + \frac{1}{2}\mathrm{HA}\right)$$

Since passive diffusion is a completely non-specific process, only the intercepts c_i ($\Delta_i = c_i - c_{\mathrm{o}}$) should differ in the corresponding equations for electrolytes.

Non-linear log P *dependence*. Further analysis has shown that simple equations presented above are not sufficient for comprehensive description of blood–brain permeability of charged species due to the fact that BBB diffusion rate does not uphold linear relationship with lipophilicity throughout the whole range of log *P* values. **Figure 4.4** illustrates the relationship between experimental log *PS* and octanol/water log *P* for neutral molecules and mono-anions. Permeability of neutral species increases linearly with log $P_{\mathrm{o/w}}$ to a certain point when cerebral blood flow rate rather than diffusion becomes the limiting factor, and log *PS* levels off at maximum measurable value of −1. The corresponding curve for acidic drugs is shifted to the right by the value of Δ_{-1}. As one can see in **Fig. 4.4**, acids never reach the flow-limited plateau of non-electrolytes, but a sharp decrease in permeability is observed when lipophilicity of the analyzed compounds exceeds certain "optimum" log $P_{\mathrm{o/w}}$ range. Data points on the falling part of the curve represent fatty acids, which are typical components of cell membranes. The observed behavior can thus be explained by hydrophobic entrapment in phospholipid bilayer slowing down the passage of highly lipophilic molecules across the membrane. This type of dependence is best described by the bilinear equation of Kubinyi (46, 47).

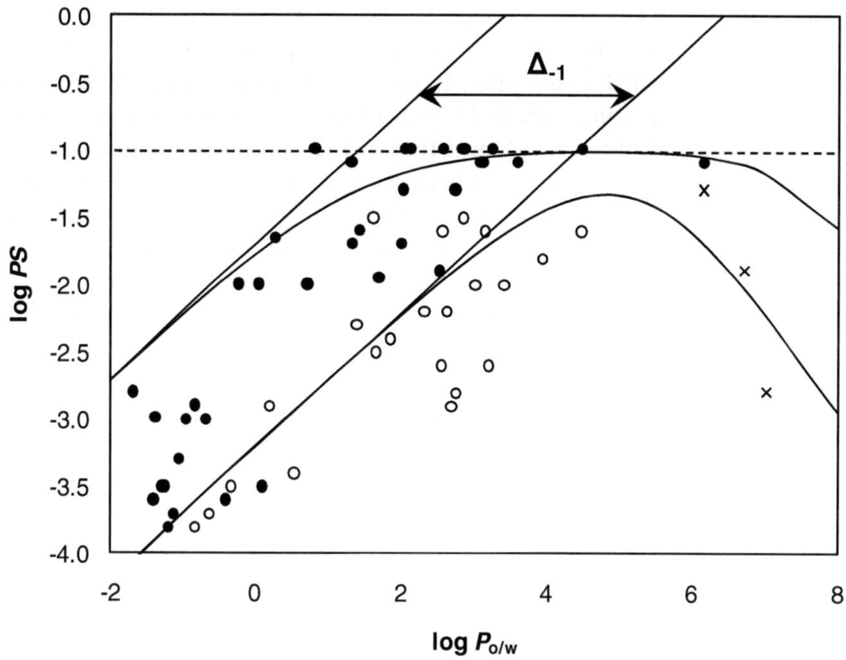

Fig. 4.4. Bilinear relationship between log *PS* and log $P_{o/w}$. Filled data points correspond to acidic drugs, unfilled points represent non-electrolytes, and data points marked by crosses represent fatty acids.

To account for bilinear transformation of log *P* values, an additional member was introduced in the expression for characteristic diffusion constants log $K_d{}^i$:

$$\log K_d^i = c_i + c_1 \cdot \log P_{o/w} + \alpha \cdot \log\left(10^{\log P_{o/w} - \beta} + 1\right) + c_2 \cdot \left(\mathrm{HD} + \frac{1}{2}\mathrm{HA}\right)$$

Here α controls the steepness of the falling part of the permeability – log *P* curve – and β is close to the optimum log $P_{o/w}$ value corresponding to the highest achievable permeability. As mentioned previously, only the intercepts c_i vary among considered ionic species, while slopes c_1, c_2, α, and β should remain the same for all compounds.

3.3. Overall Model

Combining all factors mentioned above into a single expression yields the following overall equation for the prediction of log *PS* constants:

$$\log PS = -\log\left(\frac{1}{P_F} + \frac{C \cdot V_x^{1/3}}{\sum_i f_i \cdot 10^{\log K_d^i}}\right)$$

Here P_F is fixed at 0.1 mL s^{-1} g^{-1} (*see* **Section 3.1**), log $K_d{}^i$ constants are calculated as outlined in **Section 3.2**, and *C* is a product of all constant terms introduced in the equations above

($C = C_{mem} \cdot C_{diff}$). It can be considered a normalizing factor between the rates of cerebral blood flow and diffusion across BBB membranes. In our recent study (5) log *PS* data set cleaned of values affected by side processes was randomly split into training and test sets (consisting of 125 and 78 compounds, respectively). Overall physicochemical equation was fitted to experimental blood–brain permeability data for training set compounds using the non-linear least squares (NLS) optimization procedure in the R program (48). Optimization was performed in two stages: the first stage involving only non-electrolytes, while electrolyte-specific electrostatic parameters were determined in the second stage. The necessity for such multi-step fitting is discussed in **Note 5**. The obtained model showed very good predictivity achieving $R^2 > 0.80$, and RMSE < 0.5 log units on both training and test sets, while consistency of statistical parameters among the sets proves the intrinsic correctness of the model and assures that it is not overfitted. Since the model does not represent a fully empirical correlation between experimental data points and a set of descriptors, but is based on theoretical background, its prediction accuracy should not be very sensitive to the amount of experimental data used for modeling (*see* also **Note 1**). Parameterizing the model using newly available data would only lead to slight changes in parameter values while underlying theoretical considerations would remain the same.

4. Notes

1. *Mechanistic versus empirical approach.* Quite often QSAR analysis of biological properties employs a formal statistical approach. The main objective raised in such studies is to build a model possessing good statistical characteristics without any concern about understanding the factors governing the analyzed process. While such approach is feasible for modeling simple physicochemical properties such as log *P*, usually characterized by huge amounts of experimental data for a wide range of chemically diverse compounds, it is clearly not suitable for the analysis of complex biological systems. As evident in the case of blood–brain transport, quite few experimental values determined in vivo are generally available compared to in vitro systems. Moreover, measured parameters may represent a superposition of several different processes, which is the case for log BB (*see* **Section 1**), making mechanistic analysis essential. Under these circumstances even if conventional statistical methods produce satisfactory results on existing data sets, it is very likely that the resulting models

accommodate to the small regions of chemical space represented by these data. Their performance on external data and extensibility to other classes of drugs and drug-like compounds (especially considering newly synthesized molecules) remains highly questionable.

2. *Linear versus non-linear fitting*. The model described in this chapter may appear quite complex, and it could seem reasonable to describe blood–brain permeability as a simple linear combination of physicochemical descriptors (log $D_{o/w}$, TPSA, MW). However, this approach leads to heavily oversimplified considerations and should be avoided. A particular concern applies to ionization, which is usually either completely neglected in simple linear models (3) or accounted by octanol/water log D (9, 12) which does not correctly reflect the actual ionization–permeability relationship observed in the in vivo BBB system (*see* below and also **Section 3.2**). Furthermore, these models cannot capture non-linear effects specific for blood–brain barrier transport such as flow-limited permeability or log PS decrease due to hydrophobic entrapment (which is then arguably attributed to large MW rather than log P (3, 13)).
3. *Ionization dependence*. Inappropriateness of using log D to model in vivo ionization-partitioning dependence is best illustrated by comparing BBB-specific Δ_i values obtained in our study (5) to literature data for 1-octanol and liposome systems (43) presented in **Table 4.2**. As one can see from the table, the respective values for both biological systems are qualitatively similar and significantly lower than Δ_i in octanol, meaning that biomembranes are much more susceptible to charged species. Extrapolation of octanol data to other

Table 4.2
Optimized electrostatic parameters (Δ_i) for blood–brain barrier permeability characterized by log *PS* constants compared to liposome/water and 1-octanol/water partitioning

System	References	Electrostatic parameters	
		Mono-cation effect $\Delta_+ = (c_+ - c_0)/c_1$[a]	Mono-anion effect $\Delta_- = (c_- - c_0)/c_1$
Blood–brain barrier (BBB)	(5)	−1.8	−3.0
Liposomes (LIP)	(43)	−1.25	−2.5
1-Octanol (OCT)	(43)	−3.1	−4.1

[a]Provided Δ_i values are normalized by the value of lipophilic increase parameter (c_1) for the respective partitioning system in order to bring the data from different systems on a single scale.

systems would result in a model producing significant systematic deviations for ionizable molecules, thus we would strongly recommend using both log P and pK_a as descriptors instead of single log D parameter.

4. *Choice of descriptors* is closely related to the modeling approach. Obviously, mechanistic models should only use principal physicochemical properties to describe the property of interest. Prevalence of scientific background over statistical considerations constitutes the major advantage of mechanistic models compared to formal statistical methods. Theoretical reasoning and interpretability are the main factors to consider when selecting appropriate descriptors for your model. Additional terms should never be included in model equations just for the sake of slight improvement of prediction accuracy. The five descriptors used in our study were log $P_{o/w}$, pK_a, N_{HD}, N_{HA}, and McGowan volume (V_x). These are understandable to a medicinal chemist and may be helpful in guiding drug discovery and design. In principle, log $P_{o/w}$ can be replaced with experimental log P_x^o from some other relevant systems that may prove to be a better approximation of phospholipid bilayer/plasma partitioning than octanol/water system. Molecular weight (MW) can be used instead of V_x as these two descriptors of molecular size are closely related. There is also a variety of parameters suitable for the description of hydrogen-bonding capacity of the molecule, including Abraham solvation parameters (A, B, E, S (8)) and topological polar surface area (TPSA (49)). A simple count of hydrogen bond donors (N_{HD}) and acceptors (N_{HA}) may be preferable for modeling as it is very simple to calculate and gives direct insight on the parts of the molecule responsible for low permeability and how the structure could be changed to achieve desired permeability characteristics.
5. *Multi-step modeling.* Due to the complexity of the analyzed process it is not advisable to determine all parameter values in a single step. Our practice shows that more reliable results are obtained when modeling is performed in a multi-step approach. First, it is desirable to generate a simple model for non-elecrolytes. Isolating non-ionizable molecules from the rest of the data set and describing log K_d^o in terms of log P and hydrogen-bonding potential allow separating the effects of key physicochemical properties on BBB permeability and give the opportunity to estimate the influence of lipophilicity unaffected by ionization. In the second step, the obtained parameter values should be fixed and ionization-specific Δ_i parameters for all considered classes of electrolytes as well as bilinear log P dependence could be modeled.

6. *In combo approach.* Apart from ease of interpretation, use of main physicochemical determinants as descriptors of blood–brain permeability provides another important advantage – the ability to improve prediction quality by using experimentally determined log *P* and pK_a instead of their calculated values. Such integration of in vitro measurements into in silico predictions called in combo approach (50) is particularly useful for assessment of brain uptake of novel compounds. Predictions of log *P* and pK_a for newly synthesized molecules potentially possessing unknown functional groups may introduce significant errors, which can be avoided if experimentally measured values are available.

References

1. Pardridge, W. M. (2005) The blood-brain barrier: bottleneck in brain drug development. *NeuroRx* **2**, 3–14.
2. Goodwin, J. T. and Clark, D. E. (2005) In silico predictions of blood-brain barrier penetration: considerations to "keep in mind." *J Pharmacol Exp Ther* **315**, 477–83.
3. Levin, V. A. (1980) Relationship of octanol/water partition coefficient and molecular weight to rat brain capillary permeability. *J Med Chem* **23**, 682–4.
4. Young, R. C., Mitchell, R. C., Brown, T. H., Ganellin, C. R., Griffiths, R., Jones, M., Rana, K. K., Saunders, D., Smith, I. R., Sore, N. E., and Wilks, T. J. (1988) Development of a new physicochemical model for brain penetration and its application to the design of centrally acting H2 receptor histamine antagonists. *J Med Chem* **31**, 656–71.
5. Lanevskij, K., Japertas, P., Didziapetris, R., and Petrauskas, A. (2009) Ionization-specific prediction of blood-brain permeability. *J Pharm Sci* **98**, 122–34.
6. Hammarlund-Udenaes, M., Fridén, M., Syvänen, S., and Gupta, A. (2008) On the rate and extent of drug delivery to the brain. *Pharm Res* **25**, 1737–50.
7. Bickel, U. (2005) How to measure drug transport across the blood-brain barrier. *NeuroRx* **2**, 15–26.
8. Abraham, M. H. (2004) The factors that influence permeation across the blood-brain barrier. *Eur J Med Chem* **39**, 235–40.
9. Bodor, N. and Buchwald, P. (1999) Recent advances in the brain targeting of neuropharmaceuticals by chemical delivery systems. *Adv Drug Deliv Rev* **36**, 229–54.
10. Gratton, J. A., Abraham, M. H., Bradbury, M. W., and Chadha, H. S. (1997) Molecular factors influencing drug transfer across the blood-brain barrier. *J Pharm Pharmacol* **49**, 1211–16.
11. Lane, M. E., Levis, K. A., and Corrigan, O. I. (2006) Effect of intestinal fluid flux on ibuprofen absorption in the rat intestine. *Int J Pharm* **309**, 60–6.
12. Liu, X., Tu, M., Kelly, R. S., Chen, C., and Smith, B. J. (2004) Development of a computational approach to predict blood-brain barrier permeability. *Drug Metab Dispos* **32**, 132–9.
13. Luco, J. M. and Marchevsky, E. (2006) QSAR studies on blood-brain barrier permeation. *Curr Comp Aided Drug Des* **2**, 31–5.
14. Norinder, U. and Haeberlein, M. (2002) Computational approaches to the prediction of the blood-brain distribution. *Adv Drug Deliv Rev* **54**, 291–313.
15. Fu, X. C., Wang, G. P., Shan, H. L., Liang, W. Q., and Gao, J. Q. (2008) Predicting blood-brain barrier penetration from molecular weight and number of polar atoms. *Eur J Pharm Biopharm* **70**, 462–6.
16. Kortagere, S., Chekmarev, D., Welsh, W. J., and Ekins, S. (2008) New predictive models for blood-brain barrier permeability of drug-like molecules. *Pharm Res* **25**, 1836–45.
17. Zhang, L., Zhu, H., Oprea, T. I., Golbraikh, A., and Tropsha, A. (2008) QSAR modeling of the blood-brain barrier permeability for diverse organic compounds. *Pharm Res* **25**, 1902–14.
18. Lanevskij, K., Japertas, P., Didziapetris, R., and Petrauskas, A. (2009) Ionization-specific QSAR models of blood-brain penetration of drugs. *Chem Biodiv*, In press, doi: 10.1002/cbdv.200900079.

19. van Bree, J. B. M. M., de Boer, A. G., Danhof, M., and Breimer, D. D. (1992) Drug transport across the blood-brain barrier. II. Experimental techniques to study drug transport. *Pharm Weekbl [Sci]* **14**, 338–48.
20. Oldendorf, W. H. (1970) Measurement of brain uptake of radiolabeled substances using a tritiated water internal standard. *Brain Res* **24**, 372–6.
21. Oldendorf, W. H. (1971) Brain uptake of radiolabeled amino acids, amines and hexoses after arterial injection. *Am J Physiol* **221**, 1629–39.
22. Takasato, Y., Rapoport, S. I., and Smith, Q. R. (1984) An in situ brain perfusion technique to study cerebrovascular transport in the rat. *Am J Physiol* **247**, H484–93.
23. Dagenais, C., Rousselle, C., Pollack, G. M., and Scherrmann, J. M. (2000) Development of an in situ mouse brain perfusion model and its application to mdr1a P-glycoprotein-deficient mice. *J Cereb Blood Flow Metab* **20**, 381–6.
24. Cisternino, S., Rousselle, C., Dagenais, C., and Scherrmann, J. M. (2001) Screening of multidrug-resistance sensitive drugs by in situ brain perfusion in P-glycoprotein-deficient mice. *Pharm Res* **18**, 183–90.
25. Cisternino, S., Rousselle, C., Debray, M., and Scherrmann, J. M. (2004) In situ transport of vinblastine and selected P-glycoprotein substrates: implications for drug–drug interactions at the mouse blood-brain barrier. *Pharm Res* **21**, 1382–9.
26. Dagenais, C., Graff, C. L., and Pollack, G. M. (2004) Variable modulation of opioid brain uptake by P-glycoprotein in mice. *Biochem Pharmacol* **67**, 269–76.
27. Smith, D. A., van de Waterbeemd, H., and Walker, D. K. (2001) *Pharmacokinetics and Metabolism in Drug Design*, Wiley-VCH Verlag GmbH, New York.
28. Mandula, H., Parepally, J. M., Feng, R., and Smith, Q. R. (2006) Role of site-specific binding to plasma albumin in drug availability to brain. *J Pharmacol Exp Ther* **317**, 667–75.
29. Parepally, J. M., Mandula, H., and Smith, Q. R. (2006) Brain uptake of nonsteroidal anti-inflammatory drugs: ibuprofen, flurbiprofen, and indomethacin. *Pharm Res* **23**, 873–81.
30. Tanaka, H. and Mizojiri, K. (1999) Drug-protein binding and blood-brain barrier permeability. *J Pharmacol Exp Ther* **288**, 912–8.
31. Videbaek, C., Ott, P., Paulson, O. B., and Knudsen, G. M. (1999) Blood-brain barrier transport and protein binding of flumazenil and iomazenil in the rat: implications for neuroreceptor studies. *J Cereb Blood Flow Metab* **19**, 948–55.
32. Tsuji, A. (2005) Small molecular drug transfer across the blood-brain barrier via carrier-mediated transport systems. *NeuroRx* **2**, 54–62.
33. Tsuji, A. and Tamai, I. (1999) Carrier-mediated or specialized transport of drugs across the blood-brain barrier. *Adv Drug Deliv Rev* **36**, 277–90.
34. Hawkins, R. A., O'Kane, R. L., Simpson, I. A., and Viña, J. R. (2006) Structure of the blood-brain barrier and its role in the transport of amino acids. *J Nutr* **136**, 218S–26S.
35. Ohtsuki, S. and Terasaki, T. (2007) Contribution of carrier-mediated transport systems to the blood-brain barrier as a supporting and protecting interface for the brain; importance for CNS drug discovery and development. *Pharm Res* **24**, 1745–58.
36. Zhou, S. F. (2008) Structure, function and regulation of P-glycoprotein and its clinical relevance in drug disposition. *Xenobiotica* **38**, 802–32.
37. Youdim, K. A., Qaiser, M. Z., Begley, D. J., Rice-Evans, C. A., and Abbott, N. J. (2004) Flavonoid permeability across an in situ model of the blood-brain barrier. *Free Radic Biol Med* **36**, 592–604.
38. Yu, X. Y., Lin, S. G., Zhou, Z. W., Chen, X., Liang, J., Yu, X. Q., Chowbay, B., Wen, J. Y., Duan, W., Chan, E., Li, X. T., Cao, J., Li, C. G., Xue, C. C., and Zhou, S. F. (2007) Role of P-glycoprotein in limiting the brain penetration of glabridin, an active isoflavan from the root of *Glycyrrhiza glabra*. *Pharm Res* **24**, 1668–90.
39. Kruh, G. D., Belinsky, M. G., Gallo, J. M., and Lee, K. (2007) Physiological and pharmacological functions of Mrp2, Mrp3 and Mrp4 as determined from recent studies on gene-disrupted mice. *Cancer Metastasis Rev* **26**, 5–14.
40. Vlaming, M. L., Lagas, J. S., and Schinkel, A. H. (2009) Physiological and pharmacological roles of ABCG2 (BCRP): recent findings in Abcg2 knockout mice. *Adv Drug Deliv Rev* **61**, 14–25.
41. Terasaki, T. and Hosoya, K. (1999) The blood-brain barrier efflux transporters as a detoxifying system for the brain. *Adv Drug Deliv Rev* **36**, 195–209.
42. Suzuki, T., Oshimi, M., Tomono, K., Hanano, M., and Watanabe, J. (2002) Investigation of transport mechanism of

pentazocine across the blood-brain barrier using the in situ rat brain perfusion technique. *J Pharm Sci* **91**, 2346–53.

43. Avdeef, A. (2003) *Absorption and Drug development. Solubility, Permeability and Charge State*, Wiley-Interscience, Hoboken, NJ.
44. Summerfield, S.G., Read, K., Begley, D.J., Obradovic, T., Hidalgo, I.J., Coggon, S., Lewis, A.V., Porter, R.A., Jeffrey, P. (2007) Central nervous system drug disposition: the relationship between in situ brain permeability and brain free fraction. *J Pharmacol Exp Ther.* **322**, 205–13.
45. Hou, T., Wang, J., Zhang, W., and Xu, X. (2007) ADME evaluation in drug discovery. 7. Prediction of oral absorption by correlation and classification. *J Chem Inf Model* **47**, 208–18.
46. Kubinyi, H. (1979) Lipophilicity and biological activity. Drug transport and drug distribution in model systems and in biological systems. *Arzneimittelforschung* **29**, 1067–80.
47. Kubinyi, H. (1977) Quantitative structure-activity relationships. 7. The bilinear model, a new model for nonlinear dependence of biological activity on hydrophobic character. *J Med Chem* **20**, 625–9.
48. R, program for statistical computing, Department of Statistics and Mathematics, WU Wien, Wien.
49. Ertl, P., Rohde, B., and Selzer, P. (2000) Fast calculation of molecular polar surface area as a sum of fragment-based contributions and its application to the prediction of drug transport properties. *J Med Chem* **43**, 3714–17.
50. van de Waterbeemd, H. (2005) Which in vitro screens guide the prediction of oral absorption and volume of distribution? *Basic Clin Pharmacol Toxicol* **96**, 162–6.
51. Sawada, G. A., Williams, L. R., Lutzke, B. S., and Raub, T. J. (1999) Novel, highly lipophilic antioxidants readily diffuse across the blood-brain barrier and access intracellular sites. *J Pharmacol Exp Ther* **288**, 1327–33.

Chapter 5

Intraparenchymal Delivery and Its Discontents

Raghu Raghavan

Abstract

In the treatment of central nervous system diseases, therapeutic particles need to breach the blood–brain barrier (BBB) to reach their intended target, and it has long been known that this is a difficult barrier to breach. The barrier is both passive with tight junctions hindering passage of even rather small molecules, as well as active with transporters that can pump select molecules back into the bloodstream. The diseases include both those where this barrier has been compromised such as brain tumors, hemorrhagic stroke, and neurotrauma; as well as neurodegenerative disorders where the compromise is substantially less and often unnoticeable. In addition to the pharmaceutical solution which involves design of small molecules that may be administered orally or at least systemically, there is also a compelling need at present to design devices for direct delivery of therapeutics into the central nervous system. Such delivery methods are unpopular because of their intrusiveness; however, the lack of success of the systemic route in the treatment of many of the most severe brain diseases has established a need for such devices, and the success and popularity of deep brain stimulation (DBS) has lowered the psychological barrier to these. Many clinical trials are under way which demand direct delivery of therapeutics to the brain. Such therapeutics range from small molecules, large proteins, nanoparticles, and viral carriers for gene therapies, to cells. This chapter is a review of intraparenchymal delivery of such therapies for brain diseases and of the devices used for such purposes.

Key words: Intracerebral drug delivery, convection-enhanced delivery, infusion, brain, intraparenchymal catheters, catheters.

1. Introduction

Direct delivery into brain tissue, as described further below, often involves pumping therapeutic particles in solution, or in suspension, directly into the brain through a catheter or cannula, which is usually the device in question. Alternatively, they may be simply injected into the brain tissue, relying thereafter on simple diffusion, intercellular (axonal) transport, or other assists for

K.K. Jain (ed.), *Drug Delivery to the Central Nervous System*, Neuromethods 45,
DOI 10.1007/978-1-60761-529-3_5, © Humana Press, a part of Springer Science+Business Media, LLC 2010

dissemination of the therapy if needed. The devices themselves are, with rare exceptions duly noted later, simply tubes with variations and hence rather simple, even crude. However, before focusing on these, we need to discuss the difficulties of reproducible intraparenchymal delivery. Many of these difficulties are due to tissue characteristics that strongly affect the flow of infusates. The plan of the chapter therefore is as follows. We continue in this introductory section first to indicate the problems standing in the way of getting drugs into the central nervous system, and to the reasoning in favor of intraparenchymal infusions. This reasoning justifies, if we may say so, the exploration of such a "barbarous" method (1). The following section expands on the tissue characteristics, which make reliable delivery difficult. The third section then takes up the theme proper of this chapter, namely the devices. It will turn out that several rather simple device-related questions have not yet been answered. The reason is not that these are hard problems, but rather that the community has simply not gotten around to solving them. It would be nice if some of us find the time to do so.

Toward the end, we take a brief look at other technologies that are being attempted to avoid the need for direct delivery. In the conclusion, we also refer briefly to "smart" catheters, which combine delivery with the need for monitoring how the delivery is progressing to reach better outcomes.

1.1. Intraparenchymal Delivery: Early Promise and Difficulties

The blood—brain barrier (BBB) (see (2) for a recent comprehensive review) has been a core problem in getting otherwise promising therapies into the central nervous system, even for small molecules (3). Many methods for overcoming this barrier have been proposed and tried — see (3, 4) for somewhat recent reviews; an older review is (5). These include the strong reliance on small molecules by the pharmaceutical industry (despite the dismal record), sonoporation (6–8), the use of active transporter proteins (see the reviews cited), the use of (rabies) viruses for transvascular delivery of small interfering RNA (9), as well as what is the subject of much of this chapter, namely the direct insertion of a catheter into the brain parenchyma followed by pressure-driven infusion of a solution containing the therapy. Reviews of the penetrability of the BBB to molecules include (10, 11), while predictive models of the transport of molecules across the BBB are reviewed in, for example, (12–15). However, these predictive models are largely confined to small and often lipophilic molecules, and additionally to brains where the BBB is not compromised, e.g., by cancer. Psychiatric and pain medications are often usefully of this form, but the serious neurodegenerative diseases and brain cancer are better served by protein-based toxins that are too water soluble and too big to cross the BBB. The use of pressure-driven infusions for proteins of molecular weight around 60 kDa was pioneered by researchers at the NIH (16), and called convection-enhanced delivery (CED) by them. They and others have extended this to a large number of therapeutic particles of various sizes, from those which are quite small and have significant penetration across the BBB to delivery of large viral vectors which produce therapeutic protein products. It has been almost two decades since this method was proposed and explored, and yet it still has not found its way to clinical practice, though there is a very large number of papers that report on its performance and advocate its use. Searching with key words such as "convection enhanced delivery" will yield a large number of references. The NIH group which pioneered the method, has been prolific in this area, a sample of their work may be found in (17–21) (we omit references to considerable work directed at infusions to treat particular diseases for the moment since our focus is on the engineering science of the infusion).

Intraparenchymal delivery suggested itself to its founders at the *NIH* (22), as an inference they made from the even then well-known observation that significant edema — visible in *T2*-weighted *MRI* — accompanies primary brain cancers. In fact, Rakesh Jain and his coworkers had estimated that interstitial fluid is produced by an active tumor of radius 2 cm or so at a rate upto 10 μL/ min, which is quite substantial. The edema is a result of the leakiness of the BBB, and the investigators were aware that such edema spread the serum protein albumin, with a molecular weight of about 66 kDa, to at least the full extent of the edema. In normal brain, this water-soluble protein is found only at very low concentrations within the interstitial spaces of the brain (about 1/200th of its concentration in the bloodstream ((23), Appendix B), but with a compromised BBB, it leaks out of the blood vessels. Reasoning that the albumin was carried in the interstitium by the flow of the fluid, which formed a more efficient mechanism to spread it than diffusion which would be very slow for such large a molecule, they proposed, and then carried out, such a method for distributing macromolecules artificially. It is obvious that if the therapy can be carried by fluid flow under a pressure gradient, it will reach distances somewhat proportional to the time for which it is infused. (This is just the meaning of "convection" as opposed to "diffusion": the latter process tends to reach distances proportional to the *square root* of the time elapsed.) *Furthermore, when there is edema in the extracellular space, a small pressure gradient can drive a large flux.* Technically, the hydraulic conductivity of the extracellular space will have increased as a consequence of the edema. We should point out that the consequences of the leakiness of the BBB and the concomitant edema are still issues at the forefront of clinical researches in drug delivery, cancer dissemination, and other topics that are outside the scope of this review. **Figure 5.1** and its caption illustrates the hope of convection-

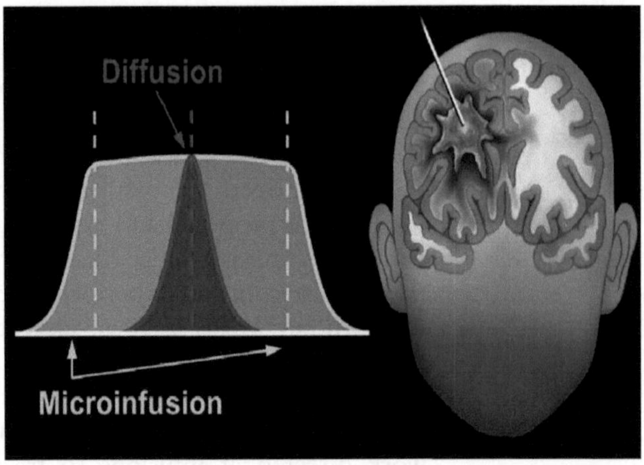

Fig. 5.1. Direct intraparenchymal delivery is illustrated by the figure on the *right*. A catheter is introduced through a small (~3 mm) burr hole made in the skull and the drug particles in solution pumped into tissue. The figure on the *left* illustrates the processes of convection and diffusion. Ideally, the transport by the fluid of the molecule results in a spread at full concentration through the tissue. Diffusion processes, even when the drug concentration is kept fixed at the source, will decay rapidly with distance. This illustration is courtesy of Dr. John Sampson at Duke University.

enhanced delivery. For the moment, we assume that the therapeutic particles will transport easily through the fluid-filled interstitial spaces in the brain. Later, we will examine the size restrictions this assumption entails.

At the time of this writing, it has been over 14 years since this pioneering application, and yet not a single therapy is being delivered this way in clinical practice! We may cite several factors which, acting together perhaps, have so far prevented use of this technique. First, a reason that should emphatically *not* be discounted, is the invasiveness of this method which has to be performed with care in a hospital, so the big pharmaceutical companies have, with reason, shunned it. Obviously, without an adequate supply of candidate drugs or therapies required to be delivered intraparenchymally, the technique will not be used. However, this has not been the most serious drawback, since the smaller biotechnology companies have come up with more than an adequate number of drug candidates in clinical trials to test the concept of intraparenchymal delivery. This leads us to a second reason for the lack of success of the method. There has been even within the community of its advocates, a blitheness about the difficulties of getting reliable infusions in brain tissue. The brain medium is inhomogeneous (its fluid conductivity varies from region to region); anisotropic (a pressure gradient in a given direction may cause the fluid to flow in a different direction); and poro-elastic (which means here roughly that the fluid flow causes the tissue to expand or contract, which can radically alter the fluid conductivity in highly expandable regions such as white matter). In addition, the brain has special channels that favor fluid flow. These are known as the perivascular spaces because they track blood vessels, particularly the significant ones. We shall discuss these issues in more detail below, but the net result of such complexity is that reliable infusions are not to be had by simply sticking a cannula in, and infusing. At the very least, careful planning and device placement are essential. Third, the devices themselves have been sometimes poorly chosen, and even when difficulties have been identified, the suggested corrections have sometimes been made based on unverified folklore. In fact, we hope to convince the reader that several questions, to which the answers undoubtedly depend on simple and quite classical physical principles, still remain to be explored or answered. Finally, it has not helped the cause of intraparenchymal delivery that there have been some spectacular failures of clinical trials involving these, and for which there is some evidence that the delivery itself failed to provide an appropriate dose to reach a therapeutically effective response.

In the next section, we continue to review the second difficulty mentioned, namely the complexity of the tissue. It is not our purpose to be comprehensive about this, since the mandate of the article is on the devices. Rather, the discussion of the tissue environment is to set the stage for a discussion of the device–tissue interaction which can help us decide on better devices, and better protocols to allow for successful intraparenchymal infusions.

2. Difficulties with Delivery: The Tissue Environment

The brain is complex even from a purely "hydraulic" point of view, which makes reliably reproducible delivery difficult. We now take up a more detailed presentation of these complexities. The extracellular, or better, the interstitial, pathways are the conduit for the transport of the infused particles and fluid. (The interstitium comprises the extracellular, extravascular space of the brain.) It is readily appreciated that only the interstitium "counts" for the transport of the active, or free, drug: once it either acts upon a cell or is taken up into the cell, or traverses the BBB back into the vasculature, it is essentially removed from consideration as a free drug. These pathways are inhomogeneous and anisotropic due to both the axonal and white matter pathways, as well as the vasculature. In addition, they may alter their "channel capacities" in response to external influences, such as these fluid infusions, or pathological states in the brain, such as primary tumors, that result in vasogenic edema. Thus the routes taken by the molecules in our application are dominated arguably less by the original pathways, as by the altered,

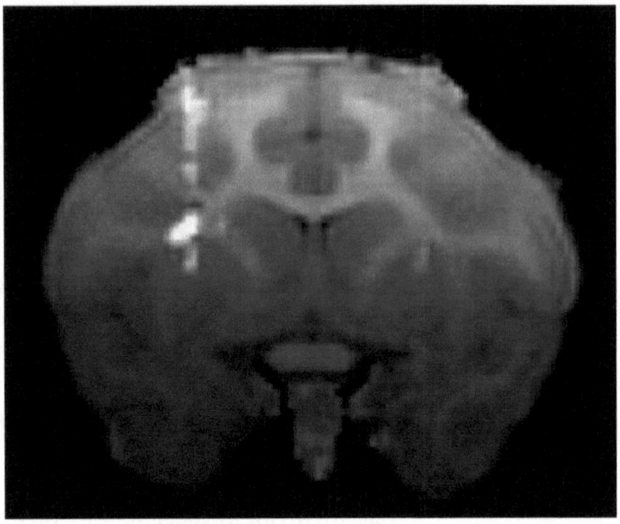

Fig. 5.2. The loss of an infusion. The bright band ends within the center of the putamen (in a monkey). However, the infusion backflows (see later text for explanation) and then spreads along the white matter fibers of the capsule. This preferential spread takes away the infusion, preventing the intended filling of the putamen. The experiment illustrated was conducted in Dr. Krys Bankiewicz' laboratory at the University of San Francisco.

expanded routes primarily in the major white matter tracts. An example of the effect of this, fairly typical is shown in **Fig. 5.2**, which represents an infusion from a single end port catheter into the putamen of a monkey. As the caption explains, the infusion flowed back into white matter tracts, which expanded and provided a low resistance path for the fluid so that the infusion failed to fill the putamen as intended. The perivascular spaces along the major blood vessels are also an effective conduit for flow of infusate. An example showing the possible importance of these spaces, subject to further studies, is shown in **Fig. 5.3** and **5.4** below (see the captions for explanation).

The distribution of sinks for fluid flow such as the ventricles and sulci, and cysts in the case of brains following surgery or disease, are also hazards that can compromise an infusion. In the example shown in **Fig. 5.5**, we have a situation in which both the white matter and a sulcus play their parts in siphoning off the infusate. There is infusion within the putamen around the port of the catheter, but the loss is considerable. Further the distribution or concentration of the molecules in parenchyma is very significantly influenced by the loss mechanisms for the molecule: other than the obvious binding with receptors and metabolization, the distributed sinks in the vasculature for the smaller, and crossing into the subpial layers or into the ventricles for the larger molecules are significant in deciding the dose that reaches a desired target. These sinks also strongly influence the flow of the fluid, and hence

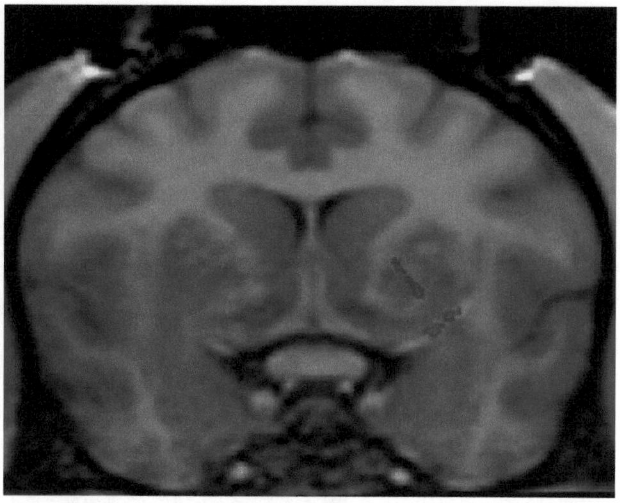

Fig. 5.3. A frontal slice of a cynomolgous monkey brain in which time of flight angiography has been used to delineate two major blood vessels within the putamen, a target for infusion of therapeutics in Parkinson's disease. The experiment was performed at the University of Wisconsin, Madison, at Dr. Marina Emborg's laboratory: the imaging was conducted by Dr. Andrew Alexander.

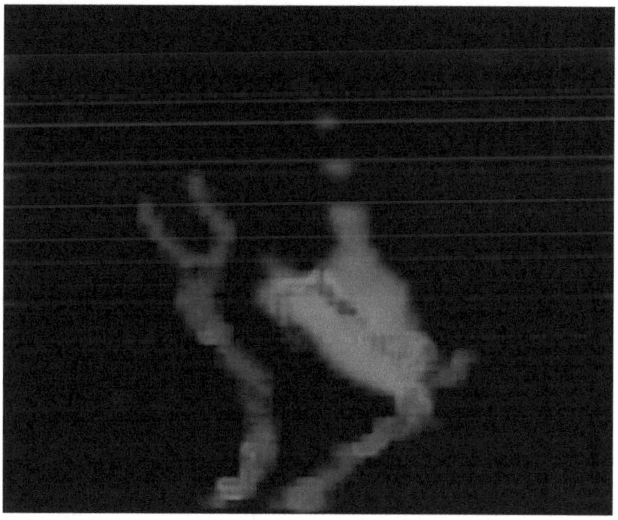

Fig. 5.4. An enlarged volume rendered view of the same blood vessels in the putamen. The volume rendering better displays the 3D spatial positioning of the vessels. The overlaid blob is the actual distribution of a tracer (Gadodiamide) visible in MR, indicating a preference for flowing along the vessel tract. The particular vessel was nearer the catheter port. The simulation and image processing were by Dr. Martin Brady of Therataxis.

the corresponding advective flow for any particle that is carried by the fluid. Thus, while the interstitial spaces are the desired pathways for fluid and therapeutic particle, the other parts of the brain including blood vessels and capillaries, perivascular and Virchow-

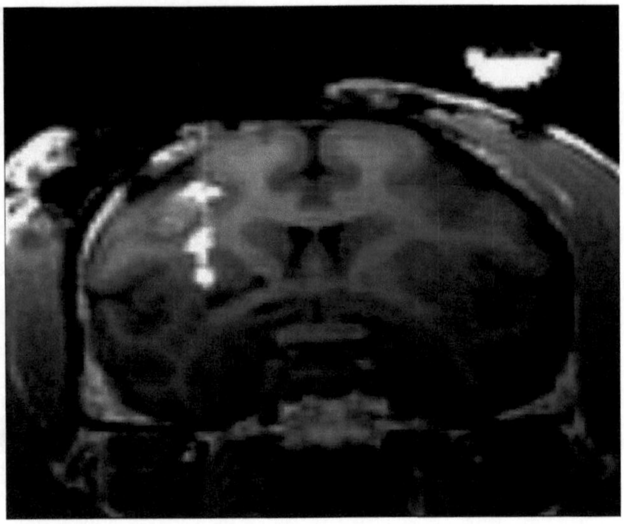

Fig. 5.5. A different infusion from Dr. Bankiewicz' laboratory behaves similar to the one illustrated earlier, but further losses occur at a sulcal boundary to the cerebrospinal fluid.

Robin spaces, and of course ventricles and CSF all act against confining the transport to within the interstitium. Proceeding optimistically, one generally assumes that the interstitial spaces are connected, so that fluid or particles able to traverse in these spaces can move between any two points in the tissue each within the pores. However, how well the free molecule is carried by the fluid does depend on its size, its physical adsorption to the cellular membranes, and also whether they have quite different densities than fluid, i.e., not of "neutral buoyancy" in the fluid. The drug particle interacts with the fluid medium in the interstitium which includes many polymers, extracellular matrix molecules, and the like, and, furthermore, the solute particles can themselves be of sufficient size that the walls (cell membranes) are sufficiently near to provide a drag force on the particles. In other words, what we are faced with is transport of a particle in connected tubular (in a generalized sense) regions of fluid containing suspensions of other large particles, with the solute particle sizes not necessarily being negligible compared with the widths of the regions. By measuring diffusivities, and using the results of microhydrodynamics, the widths of the interstitial spaces have been inferred in particular by (24). They find that the interstitial spaces have a width of about 50 nm. This means that proteins of about 60 kDa can easily fit, and indeed even adeno-associated viruses (AAV) with characteristic sizes of about 24 nm do not encounter an insuperable barrier from the widths alone, while larger viruses, even the adenoviruses with sizes of about 70 nm seem to get stuck and not convect. (AAV does not convect well for other reasons: they bind to sites which are ubiquitous in the extracellular matrix of the brain. We shall not discuss such issues here.)

Thus, in addition to the devices which are the principal subject of this chapter, a host of other criteria must be met during the planning to have a truly successful infusion. Given the complexity of the pathways of the infusions, it is obviously valuable to obtain tracers such as Gadodiamide for magnetic resonance that can be tracked. (This tracer in the region of moderate concentrations, i.e., below a saturation level, shows up as high intensity in a T1-weighted MR image, with intensity increasing within concentration, within a range of concentrations.) This has been, and continues to be, extensively used in animal experiments. However, this has not been used in intraparenchymal infusions in humans except by obtaining special permission from the FDA in select clinical trials (see, for example, (20, 21)), so that obtaining permission for routine use of such surrogate markers carried by fluid flow remains quite uncertain at this time. Further, the use of such tracers so far has been confined to reporting so-called volumes of distribution obtained by simply thresholding an intensity. It is obvious that such approaches have large uncertainties. For example, the doubling of a ball from 200 to 400 μL results in only the slight increase of radius from 3.57 to 4.62 mm. Yet the root mean square distance of diffusion of a molecule the size of the Gadodiamide (~500 Da), is about 5 mm in tissue in 2 h. Thus statements about the volume of distribution supposedly obtained due to convection at a flow rate of say 2 μL/min over such times are quite suspect to say the least. It is important, we feel, to get beyond the "volumes of distribution" often reported, and display the concentration profile so that realistic estimates can be made about the distribution of surrogate tracers. (The issue of using tracers raises the larger issue of *monitoring* infusions as they happen, which, in turn, raises the issue of "smart catheters" which are equipped with sensors for a variety of purposes. We make some brief remarks about these issues in the conclusions.)

We have mentioned that most boundaries of fluid-filled spaces can become sinks and take away from effective distribution of the infusate. These include the CSF boundaries (25–27): the pial-CSF barrier, the ventricular-CSF barrier, and the surfaces of any resection cavities from prior surgeries. If we were to consider other routes of entry of the drug molecule such as from the ventricles or intrathecally, such as mentioned in (28) or proposed in (29), we may need to incorporate fluid flow within the ventricles themselves as dramatically illustrated in (30). However, in intraparenchymal delivery, the ventricles provide only an exit route for both fluid and drug molecules smaller than viral size. In such cases, we do not need to account for advective transport of the agent within the ventricles and out into the sub-arachnoid space due to CSF flow. The pial surfaces do act somewhat as barriers especially to the large molecules we are concerned with. However, the pia is not chemically attached to the cortex, and thus allows a very thin layer of fluid between cortex and pia which may serve as a fast flow path tangent to the cortical surfaces and carry away the molecule to distant regions.

The routes for particle transport have been assumed to be within the interstitial space. Axonal transport, assisted by microtubules, is an important route for some particles (some proteins, viruses) and naturally complicate the delivery of infusate to predictable volumes. (The mathematics of axonal migration is described in (31, 32).) Finally, physical adsorption (33) at the interface between tissue and extracellular space is a *surface* phenomenon distinct from the volumetric losses we have discussed. Thus the transport processes in the brain are indeed complicated!

2.1. Fluid Flux in Porous Media

We now recapitulate and extend the above discussion by providing a somewhat more mathematical description. The description of fluid flow in porous media usually begins with D'Arcy's law, illustrated in a spherical geometry in **Fig. 5.6**, where we consider the tip of the catheter as a small spherical source with radius equal to the radius of the catheter.

D'Arcy's law posits a linear relationship between the flow of fluid across an oriented area element, in the limit as the area becomes vanishingly small, and a pressure gradient which describes a force per unit volume due to hydrostatic pressure:

$$\mathbf{v} = -K\cdot\nabla p \qquad [1]$$

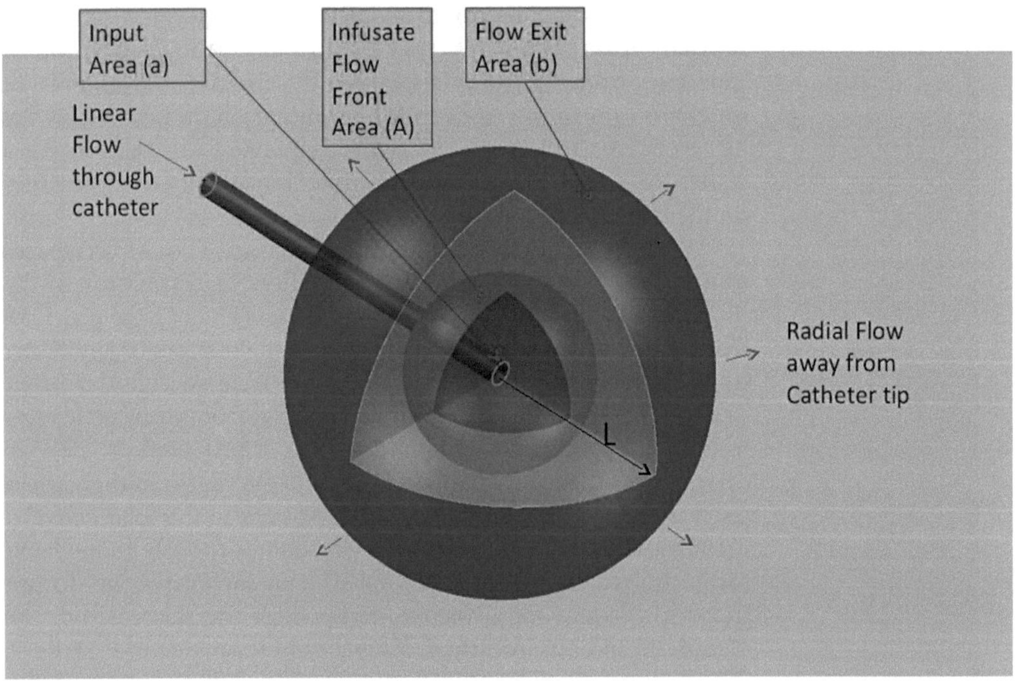

Fig 5.6. An ideal spherically symmetric infusion from the end port of a catheter. Drawing courtesy of Chris Ross of the Engineering Resources Group, Florida.

If **v** is interpreted as a fluid velocity (see below, though) in laboratory-fixed coordinates, D'Arcy's law is a *steady-state homogenized* picture of *rigid porous media* (see (34) where extensive discussion of these concepts are presented). For such a medium, what the above equation says in words is the following. The left hand side is a vector which is a pseudo-velocity vector which is defined to be the volume of fluid that crosses a unit area *fixed in space* and oriented normally to the vector **v**, in one unit of time. The right hand side is the "vector" which is force per unit volume described by the gradient of the hydrostatic pressure and denoted by ∇p, multiplied on the left by the matrix K at each point of a medium. If there are other forces acting on each fluid particle, such as gravity, these should be included additively to the force terms on the right hand side: we are confining ourselves to the simplest situation where this is due to hydrostatic pressure gradients. The matrix K is the expression in a particular coordinate system of a tensor field known as the hydraulic conductivity tensor. The point is that it is that the flux, or velocity, of fluid is proportional to a force, which is characteristic of a friction-dominated situation. (Newton's laws would say that the acceleration is what is proportional to the force: here we assume all accelerations are small and that viscosity dominates over inertia.) In the case illustrated by the figure, with a spherical source and a perfectly homogeneous and isotropic medium, the matrix K has only diagonal entries which are all equal, and so can be treated as a single constant K. In this over-simple case, the pressure at the tip of the catheter will be

$$p_0 = \frac{Q}{4\pi KR} \qquad [2]$$

where R is a radius of the catheter. Several points should be noted about D'Arcy's law as it stands.

- First, if there are solutes in the interstitial space that have strong osmotic effects, the pressure gradients must also take these into account so that the law gets modified (as with accounting for gravity). Such is indeed the case for a brain with a highly compromised blood—brain barrier, in which the serum proteins such as albumin are found in high concentration in the interstitium in contrast to the case for intact brains.
- D'Arcy's law may be combined with allowance for compressibility and shear of the medium: of both the fluid and the solid constituents if need be. Its extension to such *poroelastic* media and time-dependent phenomena is available but will not be discussed here beyond the following brief comments on each. Poroelasticity is the subject of Biot's theories and its generalizations: see, e.g., (35, 36), or (37), for recent textbooks in this field. The extension to dynamic phenomena, within the realm of linear response, is as usual to regard equation [1] as referring to individual Fourier components of the quantities,

in which case the response function, which is the hydraulic conductivity, becomes frequency dependent and complex, with the imaginary part referring to attenuation; see, for example, (38). However, the following caveat should be kept in mind. Biot's theories, as they stand, are incomplete for application to the brain environment. The reason is that while allowing for deformation of tissue, Biot's theories assume the hydraulic conductivity as input. However, as discussed below, the hydraulic conductivity particularly of white matter can change dramatically from its resting value as a result of such deformation. This makes the description of fluid flux in the brain quite nonlinear and predictable delivery more difficult.

- When an infusion is commenced (with the flow rate from the catheter held steady), Biot's theory of poroelasticity correctly predicts that the interstitial pressure does not immediately reach its steady-state value, but does so with some characteristic time. The process (for the flow rates we might be concerned with) is largely diffusive. In other words, the pressure creeps up to its final "infinite" time value. In fact, the diffusivity of the pressure may be estimated to be very roughly the product of the hydraulic conductivity of the tissue and its shear modulus. (The actual formula is more complicated and is available, for example, in (39).) In white matter, where this rough approximation should hold, this works out to be less than $0.1 \text{cm}^2/\text{s}$. (We may, for example, use the values quoted in (40).) In a "nice" infusion such as shown by the behavior of the pressure in **Fig. 5.7**, which was obtained during an experiment in gels, it is seen that after the pump is turned on, the pressure rises over several minutes to its steady value (with noise in the transducer showing fluctuations), and again when the pump is turned off, it decays to its background value over time. What is more frequent, however, is that the pressure needs to rise above the value needed to maintain infusion, and then drop precipitately. This "discontinuity", shown in **Fig. 5.8**, is usually attributed to tissue occluding the catheter, but can also be due to bubbles. We will have more to say about this in the section on the devices themselves.

The characteristic times for the pressure to equilibriate seem to depend surprisingly strongly on seemingly small differences in conditions as **Figs. 5.9** and **5.10**, obtained in early gel experiments, indicate. This phenomenon is neither well studied nor understood if it is real, and needs to be further examined.

When we account for the elasticity and compressibility of the solid or fluid constituents of the brain or other porous medium (or both), then, for situations in which the pressure is slowly varying, D'Arcy's law as described above still holds, with the following (important!) re-interpretation. The area element we used in

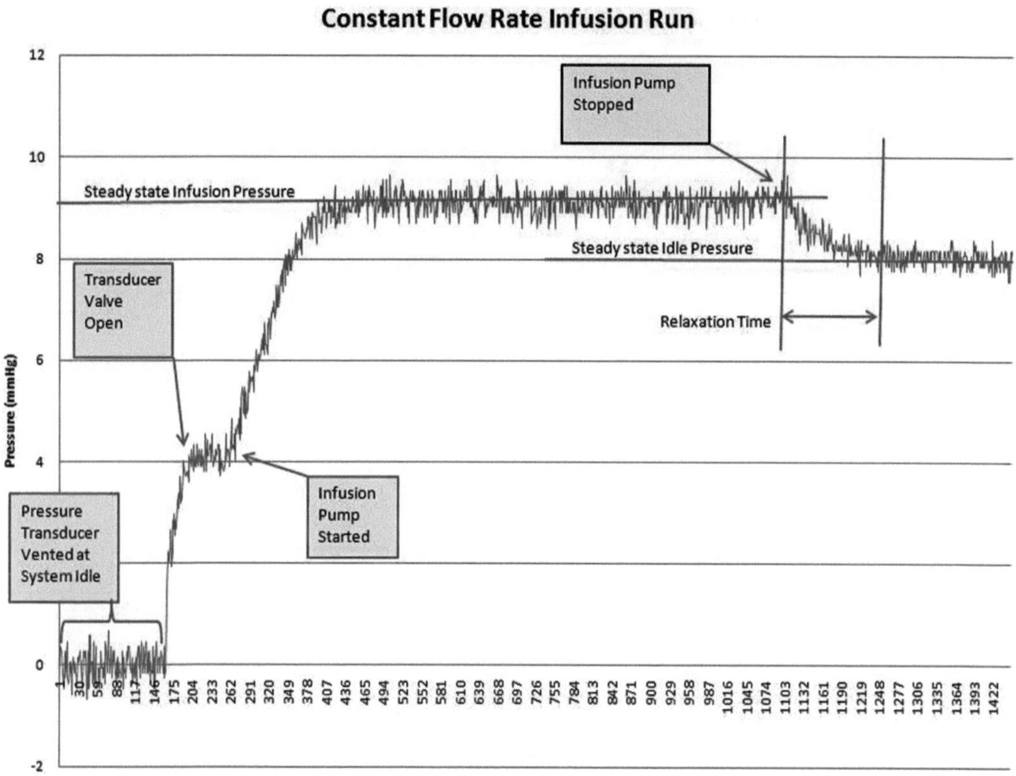

Fig 5.7. The time dependence of the pressure during a well-behaved infusion. When the flow commences, the pressure rises more or less slowly to its steady-state value: when the flow is turned off it similarly decays to a background pressure. The rise and decay times are determined by the various lines as well as the fluid conductivity and the elastic coefficients of the medium, in this case a 0.2% agarose gel. The results reported here and in the next figure were funded at Therataxis by the Kinetics Foundation, and the experiment performed by Chris Ross of the Engineering Resources Group.

defining the velocity is now anchored as to the solid constituent: so the velocity is a *relative* velocity of fluid with respect to the solid. Thus, we can have a situation in which the diffusivity of pressure is very low: indeed for low concentration polyacrilamide gels, the diffusivity has been calculated and measured to be of the order of 10^{-7} cm^2/s (39). If we infuse fluid rapidly into such a medium, the solid framework must move and almost keep pace with the fluid so that there is no contradiction between D'Arcy's law and the fluid infusion rate. The former says that the velocity can be negligible a little distance away since the pressure has not had time to diffuse far enough, while the rapid infusion means the fluid is moving rapidly. Contradiction is avoided by recognizing that the solid must move too, so that the relative velocity is as negligible as D'Arcy's law, using the low pressure diffusivity, calls for. It is clear from this discussion that such movements and expansion of the solid constituent are important in many cases of infusion of fluids into the medium. However, gels are orders of magnitude different in their diffusivity compared even to the most expandable tissue, so again

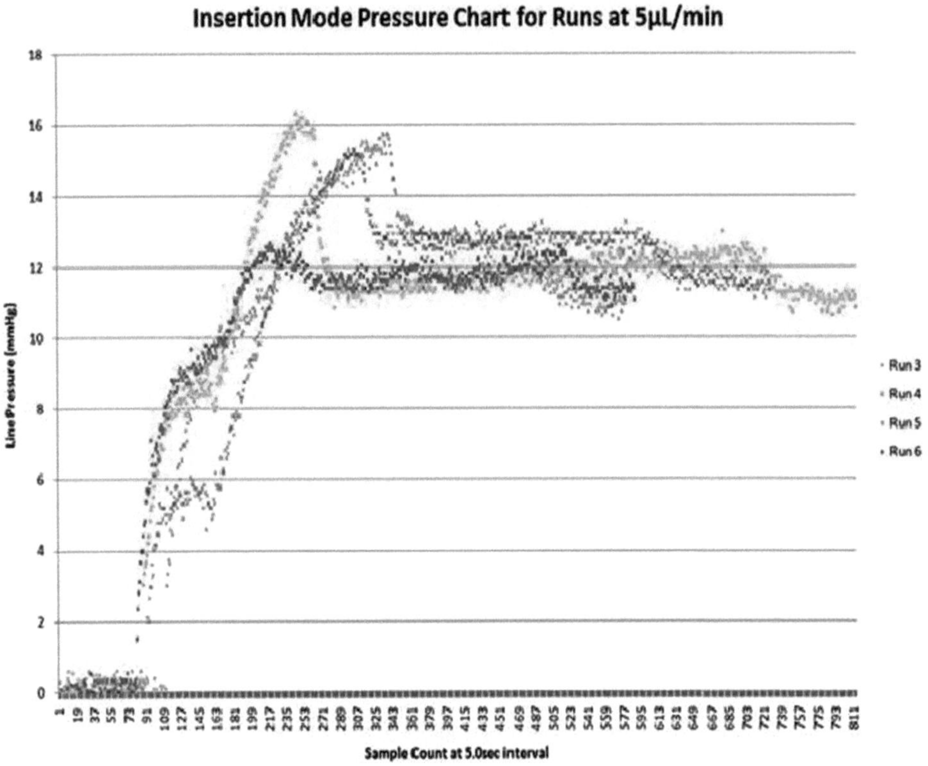

Fig 5.8. Typical behaviors of the pressure during infusions. The pressure almost always rises to a higher value than required to maintain infusion, and then precipitately drops. It is usually conjectured that this is due to an occlusion that is removed. A bubble would have a similar discontinuous effect on the pressure. In the previous figure, care was taken to avoid such occlusions.

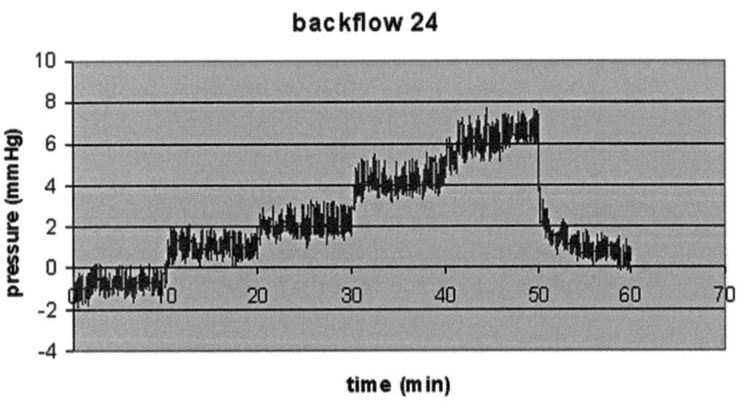

Fig 5.9. The pressure measured at the pump plotted against time. There are five distinct levels of pressure observed which correspond to no flow, and then to flows set to 1,2,5, and 8 μL/ min, respectively. At the highest flow rate, it seems clear that the pressure takes longer to equilibriate. The pressure readings are quite noisy, in part due to the power supply of the transducer used to record pressure. The catheter used had an outer diameter of 0.66 mm, and the medium was gel with 0.6% agar. Experiment conducted by Dr. Zhijian Chen of Virginia Commonwealth University. The research was funded by the NIH, with the author as PI.

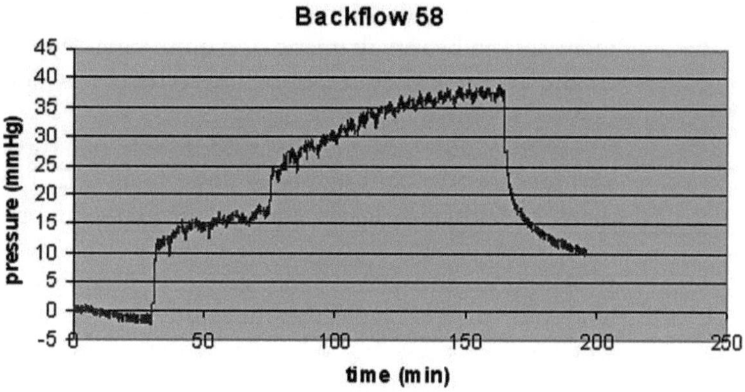

Fig 5.10. In contrast to the previous graph, this used a denser gel (2% agar) and a wider catheter (1 mm diameter). Even at only 2 μL/ min flow rate (the third level) the time to reach a steady pressure has increased very dramatically! These results were from the same program as those of the previous figure.

this is an indication that gels may be a poor surrogate for behavior in tissue. In additon to this, which is well described by Biot's theories of linear poroelasticity, we also have the problem mentioned above, that the hydraulic conductivity changes very rapidly with the fluid fraction, so that the linear theories have extremely limited use in media such as the brain, in contrast to rocks where such expansions are quite small.

2.1.1. How Strongly Is the Hydraulic Conductivity Affected by Edema?

Expansion of the fractional extracellular volume ϕ can of course have dramatic effects on the flow of fluid in that space and the accompanying advection of particles carried by the fluid flow. The literature can report somewhat absurd levels of increase such as the almost 10^5 increase reported in (41). A commonly used expression for how the magnitude k of the hydraulic conductivity behaves with change of ϕ is the Kozeny–Carman relation from the early part of the twentieth century. This says

$$\frac{k}{k_0} = \left(\frac{\phi}{\phi_0}\right)^3 \left(\frac{1-\phi_0}{1-\phi}\right)^2 \qquad [3]$$

Thus if the extracellular space increases from the nominal 20% to, say, 60%, which likely occurs in the white matter tracts, this formula would say that the hydraulic conductivity increases by a factor of over a 100. However, it should be emphasized that this relation, even as a purely empirical one, is pertinent only to the case where the pore (= extracellular) space is filled with pure fluid, e.g., water. However, we know that the extracellular space is itself a porous medium, although more of a suspension than a rigid porous medium, so that our subsequent assumption will be incompatible with the Kozeny—Carman form. We therefore should not use this form in estimating the hydraulic conductivity after expansion of the interstitial space in a brain, and mention it here only as an indication of possible behaviors. In any case, such increases can be very dramatic, and quite variable in different parts of the brain, and so a purely linear theory is doomed if one demands some quantitative accuracy.

D'Arcy flow being linear is necessarily not turbulent, but it also says we have streamlines of flow. In an actual medium with many interstitial pathways that can merge or bifurcate, there is a somewhat chaotic nature to the actual flow pathways not captured in the above equation which refers to a homogenized medium. Thus our "marker" molecule (the solute we are interested in), even when we suppress any diffusive behavior, will not display the smooth profiles of concentration implied by pure advective transport with velocity described by D'Arcy's law; on the contrary, it will have quite irregular profiles. This, as mentioned above, is taken into account in the porous media literature by "hydrodynamic dispersion," which essentially posits a diffusion coefficient which is anisotropic with longitudinal and transverse components according to the local axis provided by the D'Arcy velocity, and with magnitudes that depend on these velocities. Forms for these are available in the literature (42).

We will further assume that the infusate has been treated to have the same effective viscosity as the interstitial fluid. If the infusate is in solution in a different (than saline solution) fluid which has significantly lower viscosity than interstitial fluid, there is the possibility of viscous fingering, which we shall ignore. There have been recent claims, from a group very active in intraparenchymal delivery, that infusates of significantly higher viscosity are beneficial for interstitial distribution of infusate ((43, 44) and further preprints from this group). If this is so, its mechanism is unclear and needs to be understood. (There is no evidence of reduction of backflow from using higher viscosity infusates in gel studies. Speculatively, if the effect reported is real, it could be due to other reasons, such as the osmotic effects of the infusate, for example.) The issue of non-Newtonian fluid flow, particularly in such highly polymeric environments in the narrow interstitial spaces may be raised. However, even in flow of macromolecular suspensions where the molecules are of the same (order of) width of the channels, it has been found that Newtonian fluid flow remains valid (45). So without compelling evidence to the contrary, we should retain that assumption.

Finally we should remark that in cases where the medium is subjected to rapidly varying pressures, such as when it is subjected to ultrasonic vibrations, it is no longer permissible to neglect accelerations. Newton returns into the picture, and the equations have to account for this. All this is treated in the literature, but is quite beyond the scope of this review. Again, there are really no results in the literature for the particular nonlinear phenomena we are concerned with in infusions, so this remains a subject we hope will be actively developed in the near future.

We trust this discussion highlights how complex it is to predict reliably what happens when we infuse fluids. We hope, however, that we have not also conveyed the impression that it is entirely hopeless! We now turn to an added complexity which further mandates a poroelastic description of the brain tissue.

2.2. Pathologies: BBB Disruption, Tumors, and Edema

In clinical care of patients with glioblastoma (GBM), it is customary to acquire T2-weighted images, since edema and its side effects is a frequent occurrence, and needs to be controlled, e.g., with steroid administration. Vasogenic edema associated with GBM is usually observed in the hemisphere containing the tumor. It tends to fill white matter structures fairly uniformly (for example, not dropping off with distance from tumor), right up to bordering grey matter.

It may fill only a portion of the hemisphere or nearly the entire hemisphere. It is not usually observed within the corpus callosum, or perhaps just advancing a short distance into it. It is not usually observed to cross the midline (though sometimes it does, while leaving the genu of the corpus callosum without noticeable edema). We can interpret all these observed effects in terms of the pore fraction expansion and the poroelastic properties of brain tissue discussed above. This amounts to a large change in the volume occupied by the white matter, and indeed a resulting deformation in the brain can often be observed, e.g., as midline shift, compression of the ventricles, etc. In these regions of course, the tissue cannot be assumed to be incompressible. Another argument in favor of this interpretation (of expansion of extracellular space) comes from diffusion measurements in magnetic resonance. It turns out that the apparent diffusion coefficient of water parallel to the fibers increases slightly by about a factor of 1.3; however, it is greatly increased in the transverse direction, by a factor of 3 or so. This is consistent with fibers spreading apart and increasing the extracellular space while lowering the fractional isotropy there. Similarly the more tightly packed regions in the genu or in gray matter, with perhaps tight cross links of extracellular matrix fibers as well, prevent these regions from expanding at the pressures encountered. We should mention that the initial expansion of the white matter fibers can and does occur with little or no increase of pressure. We can easily understand this by visualizing a suspension where the suspended particles are pushed apart when more fluid is introduced without the need for a pressure increase. Of course, in the brain, this cannot go on forever, and once some limits are reached, perhaps in the tautness of the extracellular matrix that helps tether the cells and their axons, a dangerous rise in pressure results.

This phenomenon consequent on edema is also observed during infusions. The issue of flow in a deformable, poroelastic, brain has been treated in analytical models quite extensively, particularly in the context of hydrocephalus (46–54). However, it has not been studied for infusion of therapies in normal or in diseased brains so extensively. Some good beginnings have been made in (50) and in his unpublished thesis. Infusions into tumors or in the presence of tumors is considerably complicated due to this edema, which also means that tumors have resting interstitial pressures which are elevated and strongly affect the distribution of infusates (55–16).

3. Devices

After this long tour (detour?) of the tissue characteristics which affect infusions, we finally turn to the devices. As stated in **Section 1**, devices for intraparenchymal delivery are, for the most part, tubes. We discuss these tubes and the behavior of infusions arising from these in the first subsection. We conclude with a brief look at the use of ultrasound in intraparenchymal delivery which is, to be optimistic, in its infancy.

3.1. Is a Tube a Tube a Tube?

Intraparenchymal delivery is effected essentially by controlled injection, or infusing fluid through a cannula or catheter inserted into tissue. Early on, catheters designed for other purposes, namely for ventricular access, were used. Such catheters used a number of ports on the walls of the tube, with the end closed. **Figure 5.11** illustrates the usual phenomenon that results: namely most or all the flow occurs out of one or two ports, frequently the most proximal.

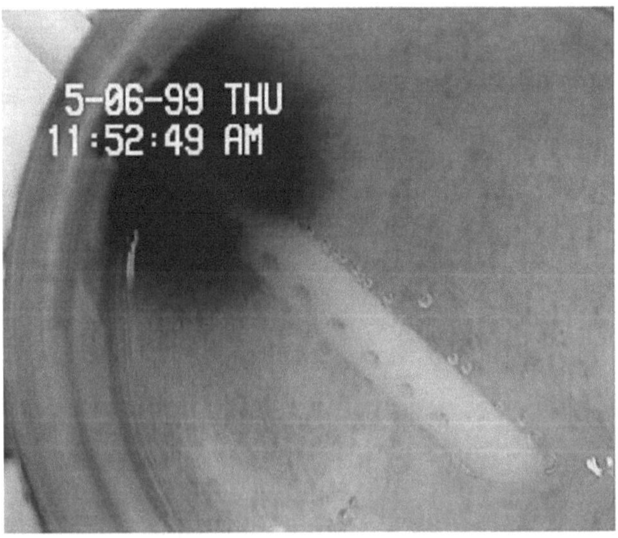

Fig 5.11. An older ventricular catheter which shows strong favoring of the proximal ports for efflux of infusate. (In this case, *bromophenol blue* was used as a marker, the infusion is into agar gel. See Ref [58].) Experiment conducted by Dr. Zhijian Chen of Virginia Commonwealth University.

Thus, while a sideport design is attractive from the point of view of avoiding otherwise unavoidable clogging of a port with tissue or the material forming the framework of the medium, it must overcome the problem illustrated. Lest we think that this problem may be particular to the medium of a gel as opposed to living tissue, we can illustrate the following occurrence in a clinical trial (**Fig 5.12** and **5.13**).

The usual explanation one hears, namely that "the flow follows the line of least resistance" is not particularly useful as an explanation. Using the corresponding electrical circuit analogy we know that in a linear resistive circuit, with several resistances in parallel, the current flow does *not* simply follow the path of least resistance. Rather, it distributes according to the (inverse of) the resistances, of course. Second, if all of the ports themselves offer the same resistance (as they should if they are all of the same size), however high this value might be, then the very slight resistance of the line (the flow along the length of the catheter between the two ports) is quite negligible. We can see this from some simple formulas from the elementary theory of viscous flow. The flow in question is called Poiseuille flow, which is the flow of a simple viscous fluid through the interior of a hollow cylinder. Suppose the cylinder is of length L, of internal radius R, and that the pressure across two of

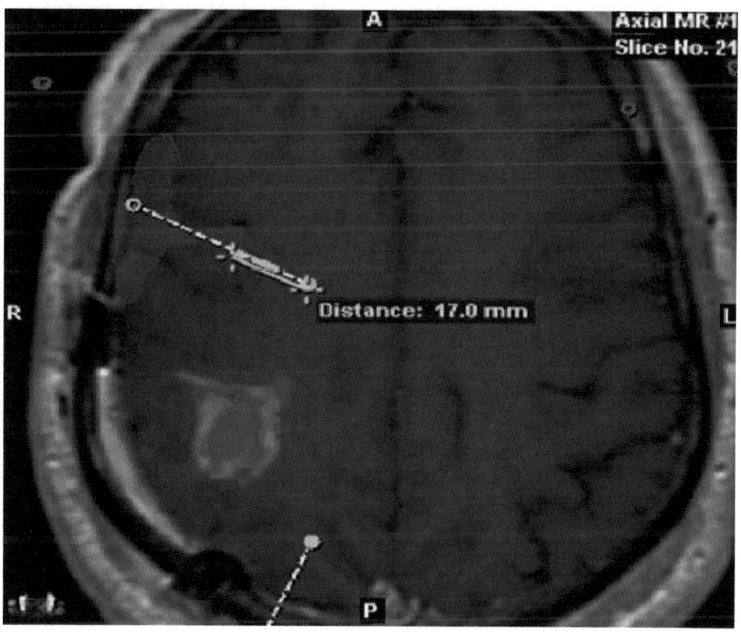

Fig. 5.12. (This and the next image have been produced by BrainLAB AG's iPlan™ software at Dr. Sampson's clinic at Duke University.) Magnetic resonance image of a section of a patient's brain undergoing intraparenchymal infusion from two catheters. The infusion from only one of the catheters which leaked via the Sylvian fissure into the sub-arachnoid space is shown. The tumor is the volume (blob) within the brain parenchyma.

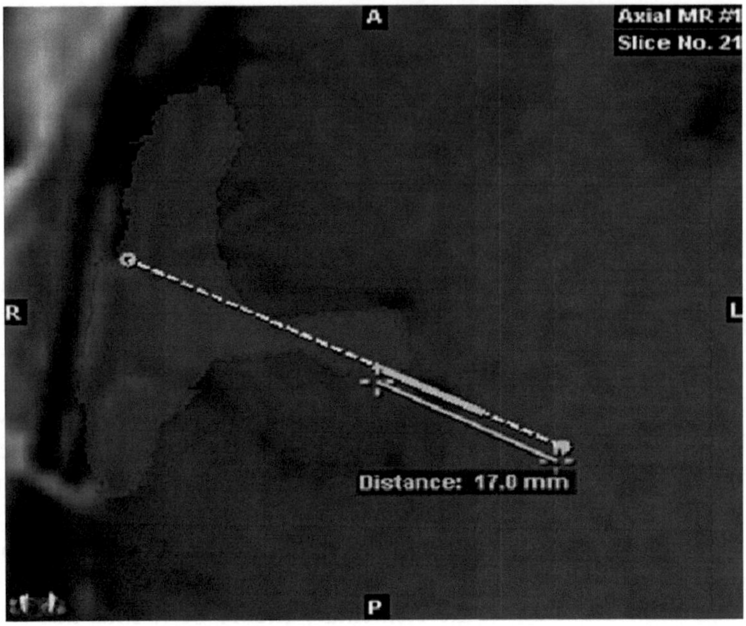

Fig. 5.13. A blow-up of the portion of the image showing the leak into the Sylvian fissure. By measuring the distance from the distal tip of the catheter, it may be confirmed that the entry of infusate into the tissue occurred predominantly from the proximal port alone.

its sections are are P_a and P_b respectively, with $P_a > P_b$ so that the fluid flows from *a* to *b*. The resistance to fluid flow — due to its viscosity — comes entirely from the walls of the cylinder, where the viscosity makes it stick. Then the pressure difference per unit length is related to the volume of fluid passing any cross section of this cylinder in unit time according to

$$\frac{P_a - P_b}{L} = \frac{8\mu}{\pi R^4} Q \qquad [4]$$

(This value is obtained by solving the Stokes equations for fluid mechanics directly, and the derivation is obtainable in any textbook and by simply searching for Poiseuille flow on the Web. Usually the formula is presented with Q being the left-hand side of the equation. We have rearranged it for our purposes.) Now, if the flow rate is as high as 5 μL/ min, the distance between ports as large as 5 mm, and the inner diameter 0.2 mm (these are fairly typical currently used values), then the pressure drop across that length is of the order of 20 Pa. This is of the order of 1% or so of the pressure required to drive the fluid through the tissue in intraparenchymal delivery, and such a small loss of pressure between ports cannot possibly account for the "all or nothing" phenomenon observed. In any case, while the most proximal port is usually the source of most of the fluid, it is not always, and so this "line resistance" is not the explanation. Nor can this result from the statistical variation of the resistance of the port depending on the tissue they are adjacent to. The effect is so pronounced that it is not explained from arguments based on *linear* resistive flow. The explanation, we feel, must involve the change of a port resistance from a high value to a low one. Perhaps there is either some blockage or surface tension effect that needs to be broken: once this is done, the pressure required to drive the flow drops substantially, the other ports with their still high pressure requirements fall out of contention. This explanation is of course purely speculative, and the phenomenon has still no satisfactory, validated, explanation. It is one of the many in this field that people have not, as we stated in **Section 1**, gotten around to solving. Meanwhile (usually incorrect) folklore gets people by.

We return to our theme: describing catheters used for intraparenchymal delivery and their behavior. As usual, there is more than one way to overcome the drawbacks of the multiple ports of the ventricular catheter. One is to abandon the multiport design and have a single port to the catheter, usually an open end. (This has the obvious drawback that clogging tends to occur which brings its own set of folklore — equally unsupported — on how to avoid or overcome it.) Another is to retain the sideport philosophy but design around the problem. One way to overcome the large ratios of the port resistances (after one port, say, is offering a low resistance) is to artificially raise the resistance of either the ports or the lines, so that such changes do not affect the ratios in any significant way. Both these solutions have been tried and both seem to work. We have ourselves filled the catheter itself with a porous material: in such a case the resistance per unit length *along* the length of the catheter is so high that any changes in port resistance should not affect the overall fluid resistances that determine the flow through the respective ports. Another solution is to have very small ports: Medtronic offers an intraparenchymal catheter with a large number of very small laser-drilled holes; Twin Star Medical offers another solution using microporous material. The idea is that these ports have now such large resistance from the geometry alone that any changes due to interactions with fluid or tissue does not significantly alter the ratios of their resistances. As stated, in both cases, it has been shown that the flow is not restricted to efflux from only one port but can be seen to be fairly uniformly distributed over all of them.

We now describe backflow which is important to understand and is always present, though different geometries may affect it differently.

3.1.1. Backflow

Backflow here is taken to mean a *reversible* flow of fluid along the outer shaft of the catheter, opening up a tissue-free region. (The term *tissue* can mean any porous medium.) Tissue tearing or other irreversible phenomena that can also result in reflux along the outer surface of the catheter are excluded from consideration, to restrict attention to the reversible effect. Obviously irreversible changes and coring of tissue can be very significant, and the catheter insertion procedure as well as catheter materials and tip geometry must reflect concern for this issue, but we shall not discuss it here, in part because so little is known with any certainty. So reversible backflow means that when the flow is reduced, so will the backflow. The backflow distance will be denoted Z here, while the *outer* radius of the catheter will be denoted r_c. The seminal paper on backflow is (40), and we have ourselves developed the theory considerably in papers to be published as well as in submission. We will quote some results from our work below.

An example of backflow is given in **Fig. 5.14**. This image is a T1-weighted MRI of an infusion into a live pig's brain. Two catheters were inserted: the one on the right side of the image shows a bright streak distal to a wider, brighter, streak. What was inserted was a host catheter almost 3 mm in diameter from which a smaller catheter just under 1 mm protruded a few millimeters. The infusate contained Gadodiamide which is a contrast reagent and shows up bright under this protocol. The image was taken just a few minutes into the infusion and clearly shows backflow along the catheter track which reaches the host catheter and continues to backflow around it.

Backflow is the result of a competition between two paths for the fluid flow: one through the resistive porous medium and the other along an annulus which is a tissue-free region between the outer surface of the catheter and the tissue. The flow through the tissue depends on its hydraulic conductivity K as well as the distance L to a sink or boundary which sets the scale over which the

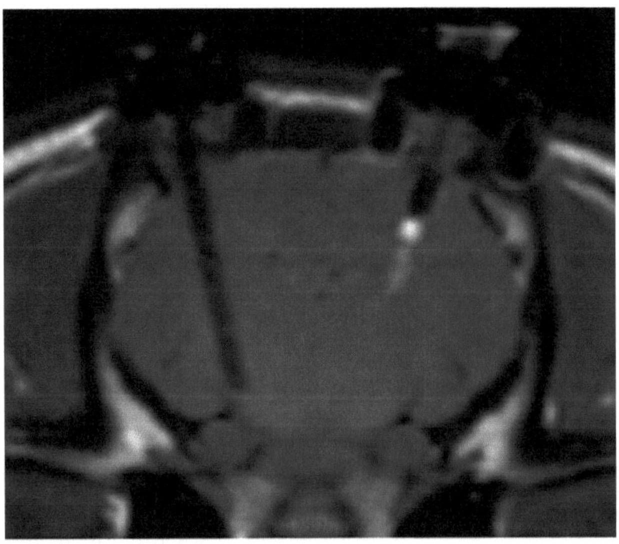

Fig 5.14. Backflow in an infusion into a pig's brain. See text for explanation. The experiment was performed by Dr. Michael Moseley of Stanford University in research directed by the author.

pressure of the fluid will fall to a background pressure, which we can take to be zero pressure. (Pressure, like electrical potential, is meaningful only upto an additive constant.) Instead of K, we use the hydraulic resistance R, which is proportional to $1/K$, with a slope that involves L and the catheter radius r_c in a dimensionless combination. The tissue-free annulus, on the other hand, is not in existence prior to its creation by backflow. It needs to be created by shearing the tissue adjacent to the catheter, and the stress that accomplishes this is the hydrostatic pressure of the fluid. The degree of resistance to shear is set by the shear modulus G. The more difficult it is to shear the tissue to create a fluid-filled region, i.e., the higher G is, the smaller Z will be. Further, the pressure available to shear the tissue is obviously higher with higher flow rate Q_0. So, Z will increase with Q_0. Finally, if R is very high, then the fluid will find it easier to backflow rather than make its way through the gel: Z will therefore be higher with higher R. Putting these three preceding statements together, we see that $Z \sim Q_0^a R^b / G^c$ where a, b, c are all positive exponents. At this point, if we remember that the other quantities in the problem are r_c, and the viscosity μ of the fluid, dimensional analysis alone will give us the correct exponents. [Q will henceforth denote the flow rate in the annular region at an arbitrary point up the catheter shaft while Q_0 denotes that quantity at the tip of the catheter: Q_0 will be close to the pump flow rate, with a bit taken away for the flow forward of the catheter port. To an approximation we will not examine here, it may be taken to be the pump flow rate.]

It is perhaps worth quoting the formulas for backflow distance and of pressure in the simplest theory to point out a feature. These formulas are as follows. In the simplest model, the resistance is given by the expression

$$R = \frac{\ln(L/r_c)}{2\pi K} \quad [5]$$

where the symbols have all been defined in the preceding paragraph. The backflow length is given by

$$Z = \frac{1}{3}\left(\frac{5000\pi \times Q_0^3 \times R^4 \times r_c^4}{192\mu G^3}\right)^{\frac{1}{5}} \quad [6]$$

where all the quantities have been defined in the answer to Question 2 above. As expected, backflow distance increases as R increases (it is harder for the fluid to flow into the tissue) and decreases as G increases (the tissue pushes back harder to close the gap). The corresponding pressure at the catheter tip becomes

$$p_0 = \left(\frac{120 R\mu G^3}{\pi r_c^4} Q_0^2\right)^{\frac{1}{5}} \quad [7]$$

Note that R and G now act in the *same* direction for determining pressure, with G being more important since it occurs in the third power compared to R. As the shear modulus of the medium increases, so does the pressure, which is what is observed.

$$p(z) = p_0\left(1 - \frac{z}{Z}\right)^{\frac{2}{3}} \quad [8]$$

Similar to equation [8], the profile of the flow rate becomes

$$Q(z) = Q_0\left(1 - \frac{z}{Z}\right)^{\frac{5}{3}} \quad [9]$$

We should point out that the simple theory which yields these formulas is ultimately unphysical. For example, if we look at the expression of pressure with distance, we see that the pressure gradient becomes infinite at the backflow distance Z, which is nonsense physically. These and other limitations, as well as a more elaborate theory which overcomes these, have been worked out by us.

The *type* of tissue and associated *edema*, as discussed in the previous section, significantly affects the fluid conductivity and hence the backflow. This is illustrated for a simple end port catheter in the accompanying cartoons. **Fig. 5.15** illustrates the beginning of an infusion into gray matter. The neuronal cell bodies are depicted surrounding the catheter or cannula. We depict neuronal cells in gray matter as having diameters of the order of 20 μm, while catheters typically have diameters that are at least one or more orders of magnitude larger. A streak of infusate has just entered the tissue region from the cannula. In **Fig. 5.16**, we see the backflow that, canonically at least, develops prior to significant penetration of infusate into the tissue itself, while the distribution of the neuronal cells remain as before. On the other hand, if we begin with infusion

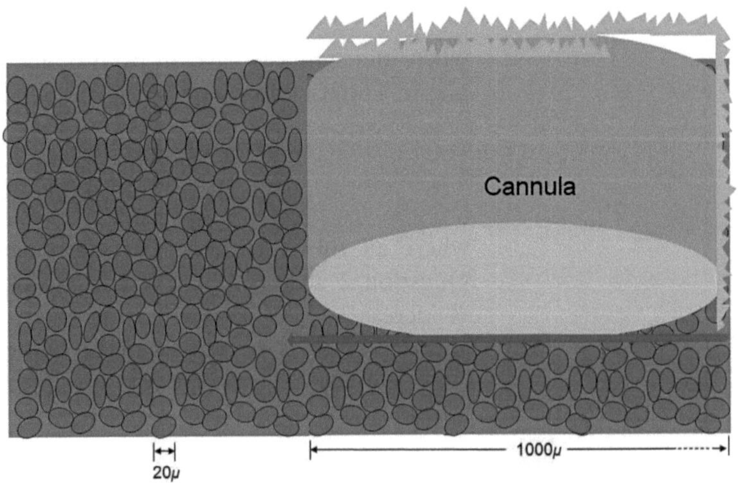

Fig 5.15. The beginning of an infusion into gray matter. See text for explanation.

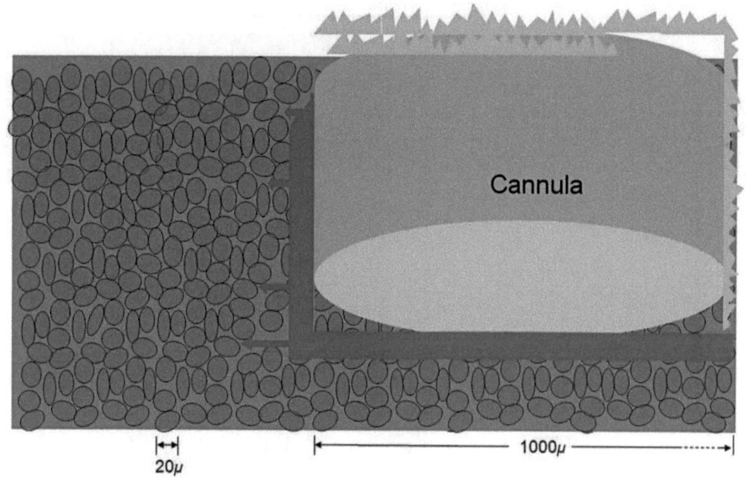

Fig 5.16. Development of backflow in gray matter. See text for explanation.

into white matter, as indicated by **Fig. 5.17**, we see, as illustrated by **Fig. 5.18**, that the white matter fibers are easily pushed apart, as already discussed in the subsection on pathological brain states in the previous section. The flow proceeds here more easily into the tissue, and the backflow is considerably lessened.

Also, if the tissue properties as well as the fluid in question are all fixed, then the formula says that

$$Z^5 \sim r_c^4 Q_0^3 \qquad [10]$$

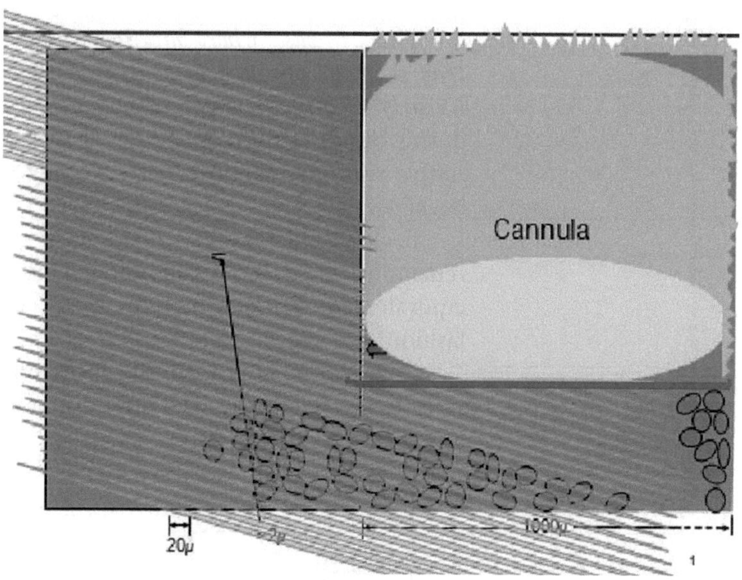

Fig. 5.17. The beginning of an infusion into white matter which is idealized as long myelinated axons. See text.

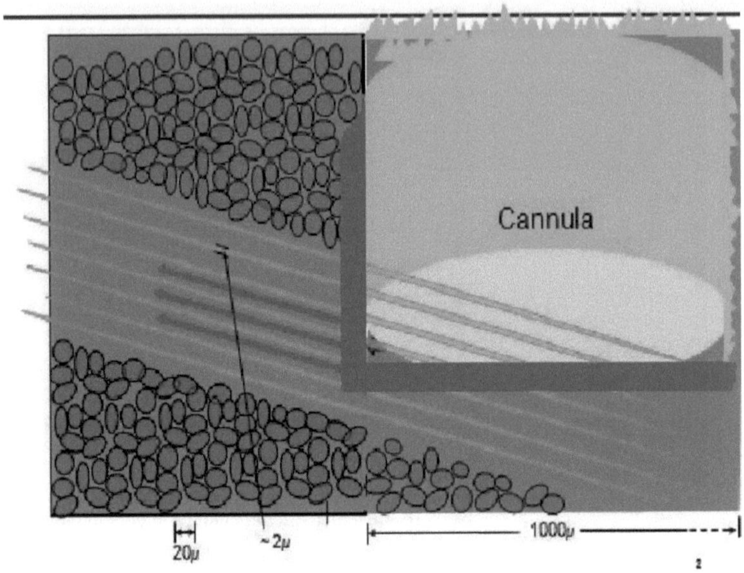

Fig 5.18. The development of backflow in white matter: the axons have been significantly pushed apart. See text.

so that the contours of constant backflow in a $r_c - Q_0$ plot will be hyperbolas which move away from the origin as Z increases, as shown in **Fig. 5.19**

We note that the theory of backflow says that the variation of pressure with the flow rate during infusion is *sub*linear:

$$p \sim Q^{\frac{2}{5}} \quad [11]$$

Thus, *in cases where backflow is important, it will no longer be true that the pressure measured at the catheter tip will increase linearly with the flow rate through the catheter,* as claimed above during our discussion of D'Arcy's law. We should mention that the simple scaling law above pertains only to the simplest theory of backflow.

Backflow increases with the *outer* radius r_c (other quantities remaining fixed). We can understand this by considering the equivalent question: why does $Z \to 0$ as $r_c \to 0$ in a monotonic fashion?

(First, let us note from the above equations, that other quantities being fixed, the backflow distance Z varies with the outer radius of the catheter as

$$Z \sim \left(r_c \ln \frac{L}{r_c} \right)^{\frac{4}{5}} \quad [12]$$

The distance L, it will be recalled from above, is the distance over which the pressure decays to its background value. This is just a boundary condition, so the backflow will vanish under the

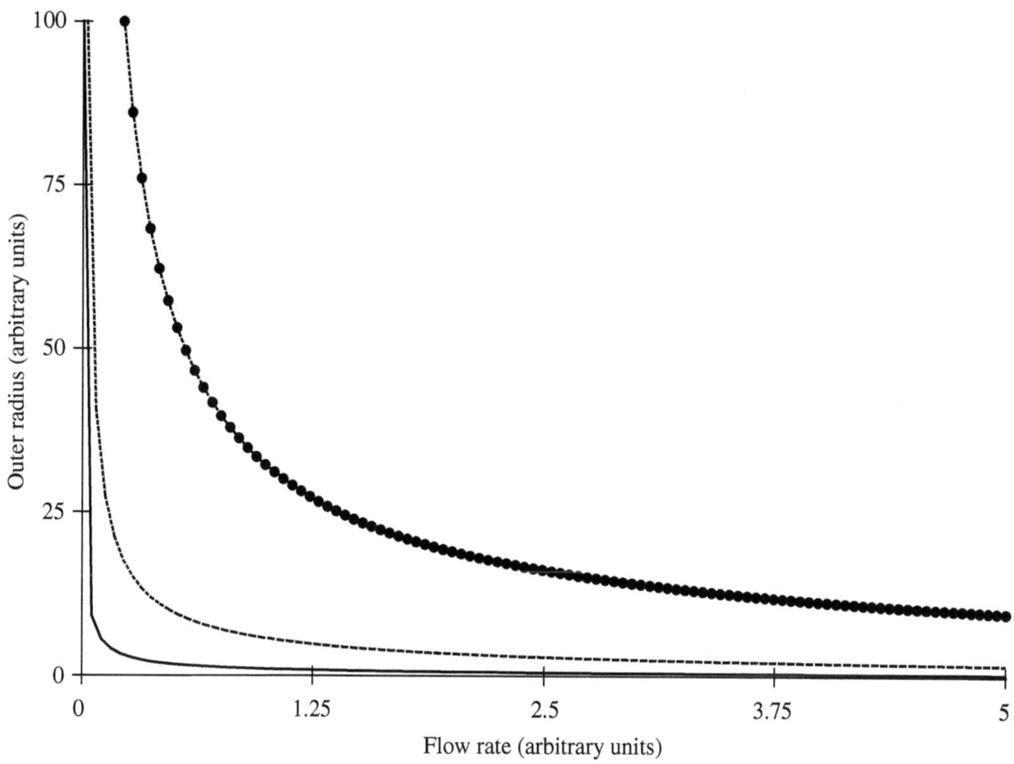

Fig 5.19. Contours of constant backflow in the $Q_0 - r_c$ plane: units of flow rate and radius are arbitrary. The hyperbolas are plotted for backflow Z increasing by a factor of 4, so that the hyperbola outermost from the origin has a backflow 16× that of the innermost hyperbola.

completely unphysical limit of $r_c \to L$. Obviously the catheter diameter must be much smaller than the extent of the gel or tissue. So we assume henceforth that $\frac{L}{r_c} >> 1$.)

So, to return to understanding why $Z \to 0$ as $r_c \to 0$, we can reason as follows. Consider Poiseuille flow in the annulus between the outside of the catheter and the tissue which is a distance h away. *We always assume* $h <_c$. Note that examining the problem of backflow means that we are given a certain flow Q, and this flow is going up the annulus. Poiseuille flow through an annulus has been mentioned above in the discussion surrounding equation [4]. From the equation for Poiseuille flow in an annulus, we expand the expression for the flow rate in powers of h and we find that for small h, when we keep only the lowest order term in h, the equation will read

$$Q \approx -\frac{\pi r_c}{6\mu} h^3 \frac{\Delta p}{l} \qquad [13]$$

where Δp is the pressure drop over a length l. Thus the flow rate goes down as $r_c \to 0$, since the pressure drop must be finite. However, *the flow rate from the pump* Q_0 *is fixed*; so the gradient of Q or the rate at which the flow rate changes as we move up the catheter (outside the catheter) must become very large to accommodate the vanishing of the flow rate at some finite distance Z away from the tip. As this gradient increases, it means also that the distance over which the flow rate vanishes becomes small. Thus as $r_c \to 0$, the backflow distance vanishes. This is a consequence of the Poiseuille flow alone, without reference to the tissue resistance. When that is taken into account, a more refined estimate follows, and it is seen that the backflow distance varies approximately as the 4/5 power of the catheter external diameter, for small diameters.

A formula that provides a backflow distance as a function of a steady flow rate is obviously a steady-state theory. As in any steady-state theory, *how* the pressure develops is outside the scope of such a picture. We have already explained above how the pressure creeps toward its final value as a result of a diffusive process under the slow infusion conditions we have been describing. **Figures 5.20** and **5.21** illustrate the behavior of backflow distances and pressure profiles as they behave over time, under conditions of flow rates that are maintained till the pressure reaches a plateau, and then increased to a next level.

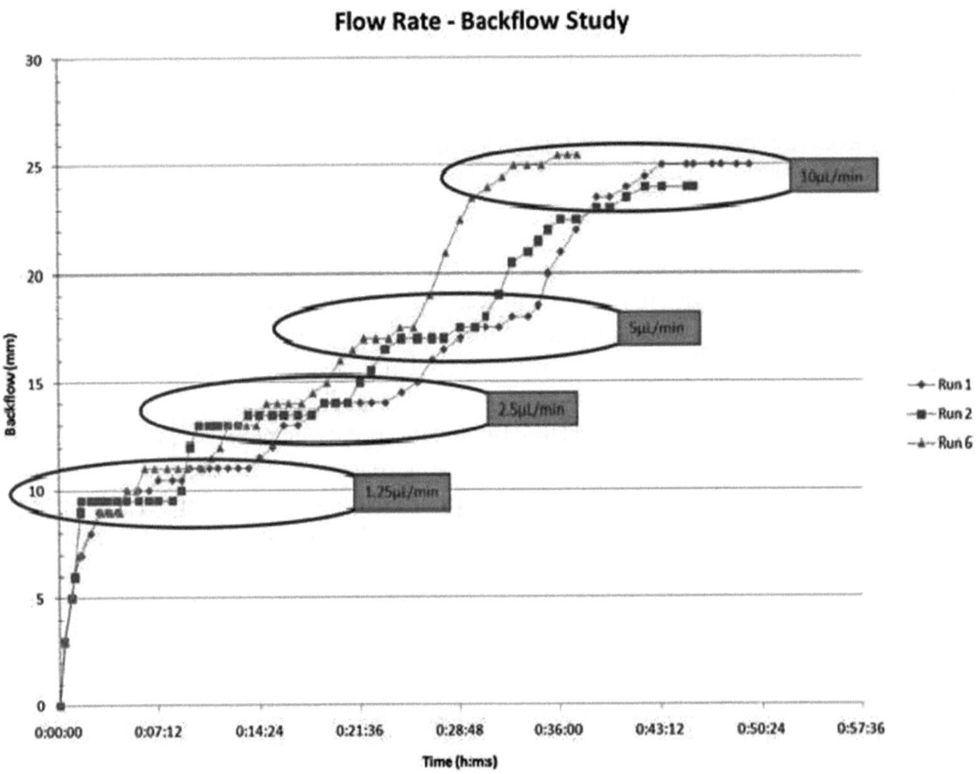

Fig 5.20. The figure illustrates backflow at different flow rates in a particular gel. The general characteristics of a sublinear increase of backflow with flow rate, and the variability between runs, can be seen. These results and the one in the next figure were obtained by Chris Ross of the Engineering Resources Group as part of experiments conducted on behalf of Therataxis with funding from the Kinetics foundation.

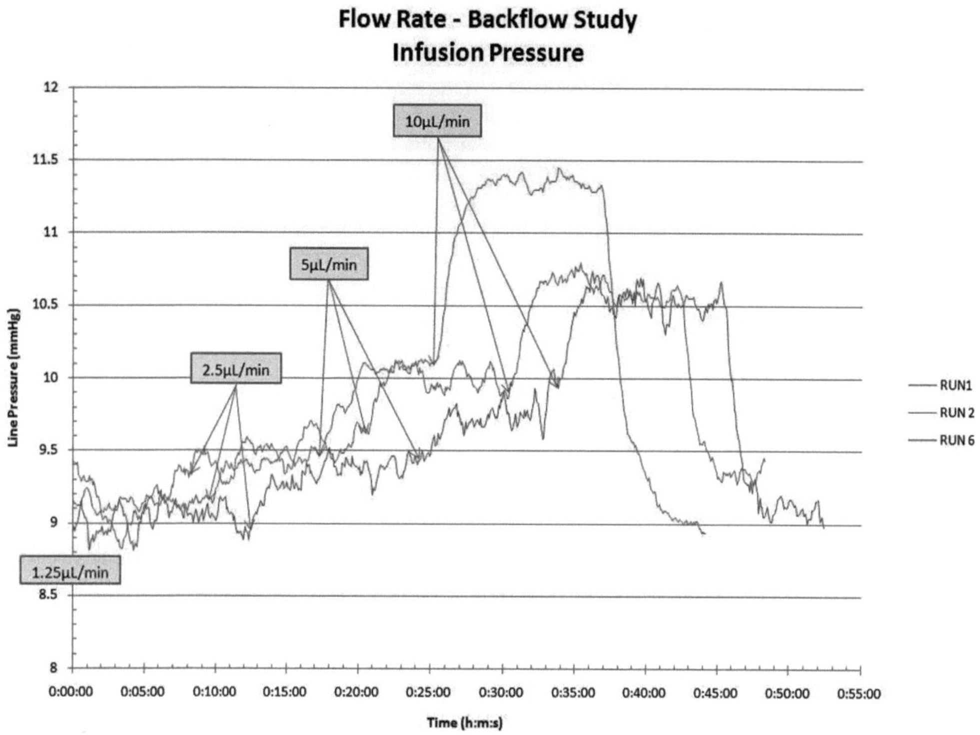

Fig 5.21. For the same runs as the preceding figure, the behavior of the pressure with time is indicated here.

Finally, we emphasize that the inner radius of the catheter and the velocity of efflux from the catheter is *completely irrelevant for questions relating to backflow, and indeed for the flow within a medium*. This is due to the nature of (linear) flow in a porous medium. The flow is dominated by viscosity, and not inertia (in other words, accelerations of the fluid particle are neglected). In that case, the problem of the steady velocity of the fluid flow is mathematically identical to that of the electric field of charges in a dielectric medium. For example, a purely spherical source of fluid in a homogeneous medium like a gel creates a velocity field in the medium that is independent of the size of the source, provided the total flow rate out of the sphere is fixed. A sideport breaks full spherical symmetry, by introducing the axis perpendicular to the sideport. However, examining the field of a charged disk will convince one that the point made remains. This is just due to the mathematics which is the same as that of Coulomb's law for the potential of charges in a uniform dielectric medium (of dielectric constant K, analoguous to the hydraulic conductivity, with the vacuum constant being unity).

3.1.2. Catheter Designs

We now describe several catheter designs in various stage of use in either animal or human clinical trials. Sideport catheters (**Fig. 5.22**) are in current use: it is hoped, as stated earlier, that

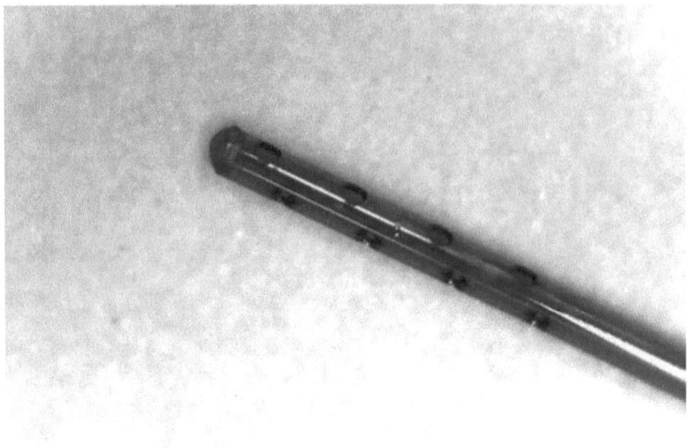

Fig 5.22. Catheter with side ports and closed end. This is more likely to avoid clogging by tissue during insertion, provided the ports are sufficiently small.

smaller ports with their higher resistance will overcome the problem of the "all or nothing" infusion through one port. This has indeed been shown to be the case. Besides, when a catheter has a large number of small ports, several of them are more likely to be open to infusion. The step catheter (58) in **Fig. 5.23** is a different design that addresses the need for a narrow diameter port to reduce backflow. It overcomes the lack of mechanical integrity of very narrow diameters by having a short length affixed to a wider, less compliant host. It also claims to significantly reduce backflow with this design. There are at least two reasons why one may argue that backflow may be reduced. One is that the flow redirection reduces the energy of the flow, and the second is that the insertion of such a catheter is more likely to accumulate pre-stress in the tissue just ahead of the host. However, there is no independent (e.g., from basic fluid mechanics) evidence that flow redirection on such small steps can reduce the energy of the flow very significantly. On the other hand, pre-stress might well increase the barrier to backflow, just as in the case of gels as we shall see below. The step requires further investigation to be of proven and well characterized efficacy. We also show a design of our own which is an even more preliminary stage of development compared with others. This channel catheter design (59) exploits a phenomenon which has been observed. Namely, if a catheter is inserted and then withdrawn, the coring of the tissue that it leaves behind can serve as an effective conduit for subsequent infusions (60). We therefore introduce a stylet to create such channels. The stylets can be quite narrow and with predetermined radii of

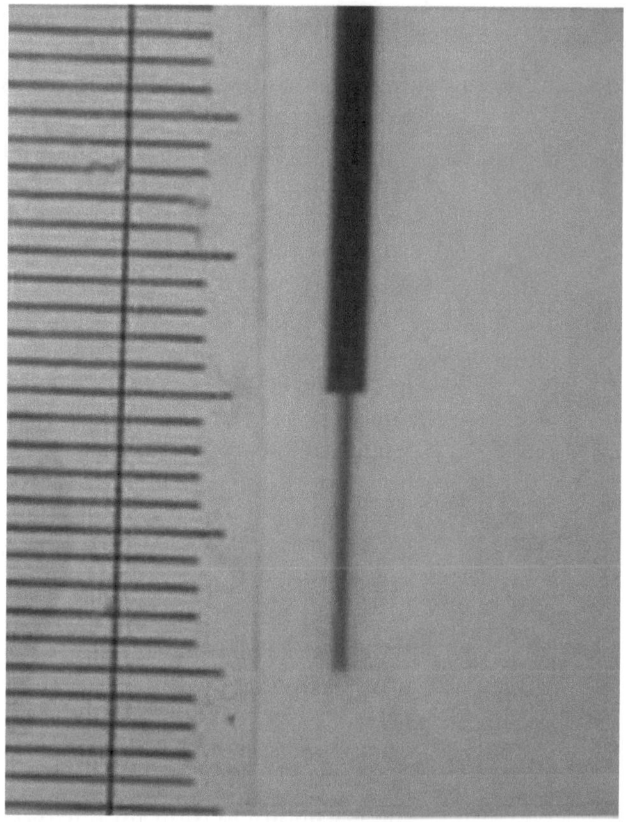

Fig 5.23. Catheter with step to smaller diameter catheter. One benefit is the progression to smaller sizes, which reduces backflow, while maintaining mechanical integrity. Another claimed is potential reduction of backflow due to flow redirection at the microcatheter/host interface.

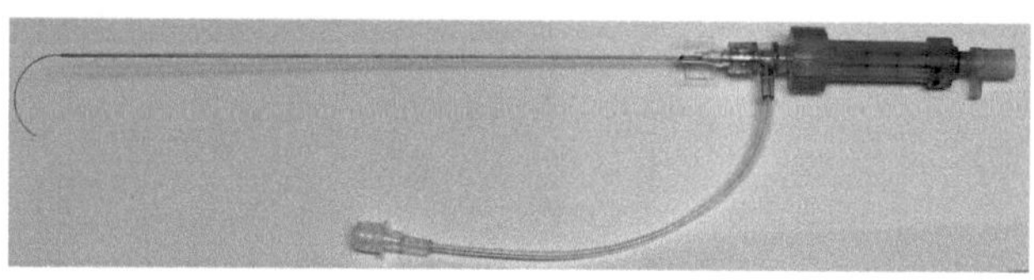

Fig. 5.24. A "channel" catheter. The idea is to simply core holes in tissue with shaped stylets, subsequently withdrawn. A dial at the proximal end allows rotational positioning of the stylet which then emerges in a pre-set direction.

curvature as they extend into tissue. **Figure 5.24** illustrates the device. Its principal advantage is the ability to create several paths of infusion from a single port of entry, in contrast to other catheters.

We can now begin to illustrate typical patterns of infusion, particularly in the homogeneous and isotropic gel medium. Our purpose is entirely qualitative: to illustrate the kinds of behaviors that can result. We therefore omit quantitative details on the infusions the particular flow rates and so on. The particular pattern obtained of course depends on these as well as the constitutive properties of the medium, as discussed in detail in the previous sections.

In **Fig. 5.25**, we see an infusion that is at a low enough rate, and perhaps with enough pre-stress in the gel, that backflow is invisible, and the infusion is an "ideal" ball.

However, if an infusion is at a high enough flow rate, the step will not present much of a challenge to overcome in backflow, as the following two figures illustrate. **Figure 5.26** shows an infusion within a couple of minutes, and an incipient backflow can be seen to be developing. **Figure 5.27** shows a late stage of this high flow rate infusion, with considerable backflow. We now compare different catheters under exactly the same conditions to see if the backflow is different for different geometries. In the next three figures, we compare the single end port catheter, a sideport catheter with 12 sideports, and a step catheter. The final step in the last catheter was 0.5 mm in outer diameter. Therefore for comparison, so was the outer diameter for the entire lengths of the other two catheters. The way in which the catheters were seated in gel was by first positioning the catheter in the sol phase of the gel, and so the gel cooled and solidified with the catheter in place. Thus there was no pre-stress in gel, in contrast to when we insert a catheter into pre solidified gel. In fact, one can visibly observe the stretching of the gel upon insertion until it reaches breaking point and snaps back. So the level of pre-stress is hard to control when inserting,

Fig. 5.25. An ideal ball of infusion in gel.

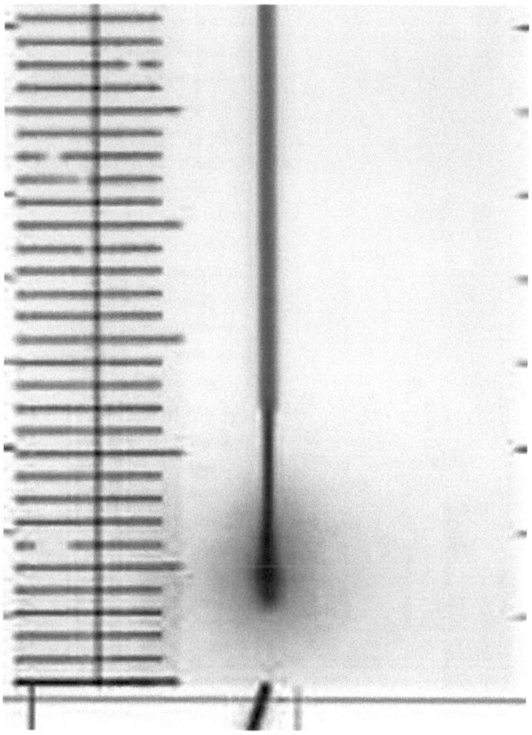

Fig. 5.26. The very early stage of an infusion with a high flow rate. The backflow is beginning to develop.

and depends on the depth of the insertion as well as its dynamics. Furthermore, the gel behaves quite differently from tissue in this regard. To remove all such complications, we proceeded with the method mentioned which avoids pre-stress. However, at this point, the gel adheres to the catheter so that this too is a form of pre-stress at least as far as the backflow is concerned. We therefore performed one final step, namely to lift up the catheter a couple of millimeters to break the adhesion, and then to reseat it. Under these conditions, the backflow is under conditions that confirm closest to the theory summarized above. The gel used had 0.2% agar and was at "room temperature" — close to 20°C. It is clear from these figures (**Fig. 5.28**, and **5.29**: see captions) that the total length of backflow is not different between the single end port and sideport case: as expected the sideport is more symmetrical since flow occurs out of the entire length over which the sideports are distributed. In repeated runs, this behavior was reproducible. There are no surprises here.

However, the backflow does indeed seem to be stemmed by the step in **Fig. 5.30**. However, this particularly nice demonstration was not faithfully reproducible, and some runs showed some asymmetrical backflow. We believe that these were due to

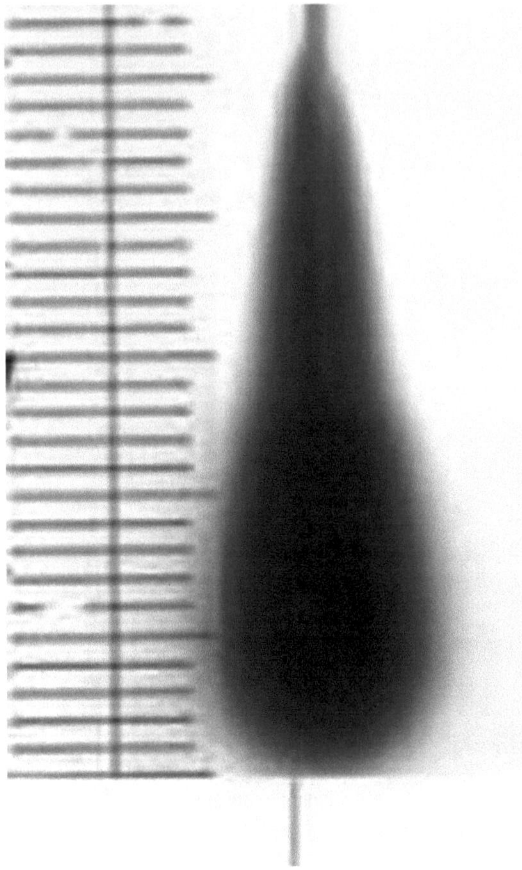

Fig. 5.27. A high flow rate infusion which has considerable backflow. The infusion spreads out into the gel (or tissue) from this backflow region which extends considerably up the catheter.

fissures in the gel perhaps. We do *not* show the behavior of backfow from a single channel of the channel catheter: as expected there was no particular improvement over the single end port or step catheter. However, the following pictures *which are taken from an entirely different experimental configuration of gel* show some methods of using the channel catheter. **Figures 5.31** and **5.32** are simply frontal and top views, respectively, of the same infusion. In this case, the catheter was simply inserted into the gel. However, the gel used has a concentration of 0.6% agar, it is maintained at "body temperature," 37°C, and is also unconstrained, i.e., surrounded by water, instead of the walls of a container. This allows the gel to expand upon infusion (see the discussions of poroelasticity above). This and the pre-stress on insertion all serve to reduce backflow, so these figures cannot be compared to the previous ones. However, they serve to show that the channels created can serve as conduits for infusion and that any number of channels (in this

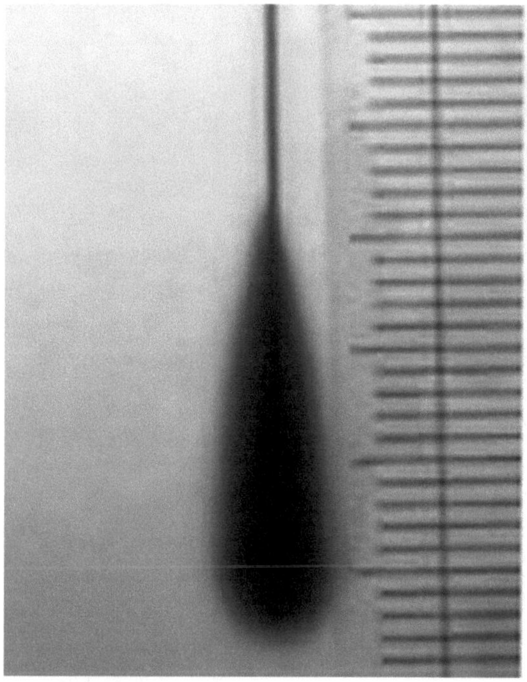

Fig 5.28. This and the next two figures show a comparison among a single end port catheter with the sideport catheter and a step catheter. This figure shows the backflow from an endport catheter. All data were collected with a 5 μL/ min infusion rate, and a special method of creating a gel/catheter configuration that avoids pre-stressing the gel that would otherwise occur if the catheter is simply inserted, as it would have to be in clinical use.

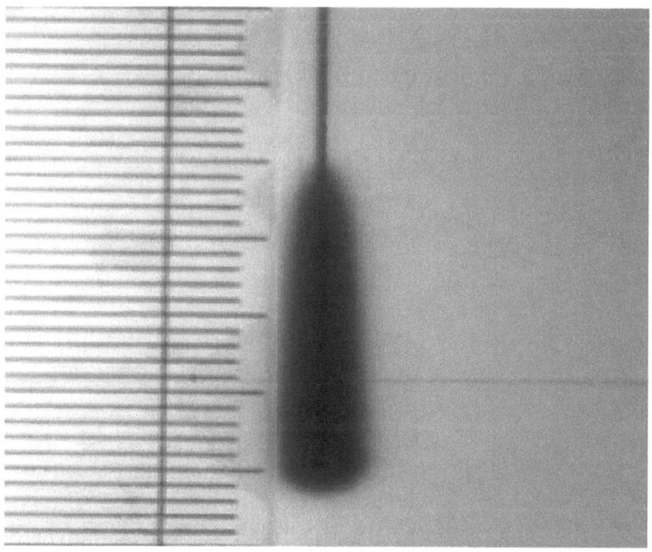

Fig 5.29. Backflow from a catheter with 12 sideports, analogous to that shown in figure (22). While the total length of backflow is not altered in any significant way, the flow is more even leaving the side of the catheter compared to the end-port case, as would be expected.

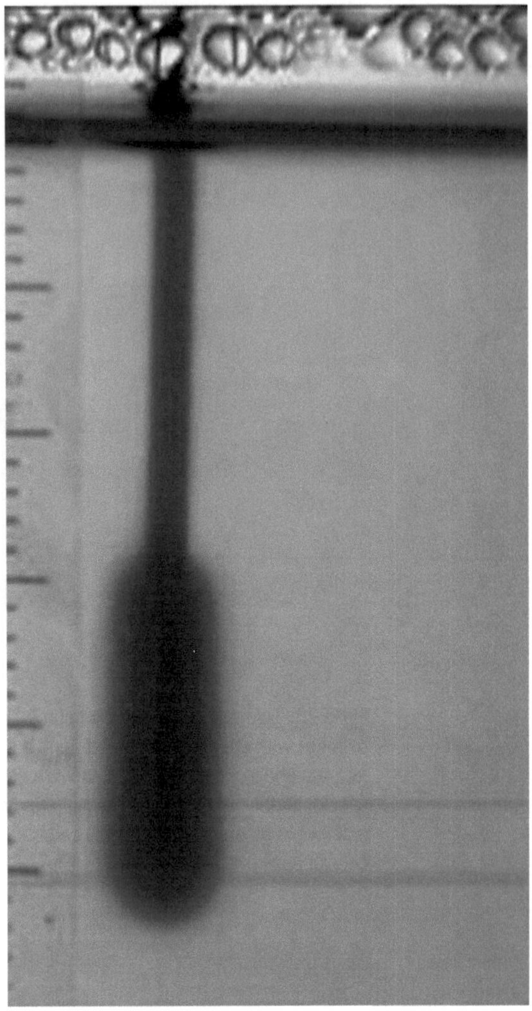

Fig 5.30. This is flow from the step catheter shown in figure (23). The final small catheter has the same outer diameter as the other two catheters shown in this set of figures. In this instance, it seems clear that the step seems to have an effect in curtailing backflow, compared to the configuration without the step.

case three) can be created to spread the infusion into the medium from a larger initial area than either a single sideport or endport catheter.

We summarize by a resistive circuit analogy to a setup for infusion. Like all analogies, this one too must not be carried too far, as our treatment above of the complexities of the behavior of the fluid in tissue indicates. Nevertheless it is illustrative to consider the analogy shown in **Fig. 5.33**.

An entirely different form of infusion source is the GliaSite balloon, which was invented for brachytherapy of recurrent gliomas (61). As its name implies, it is a balloon which is connected to

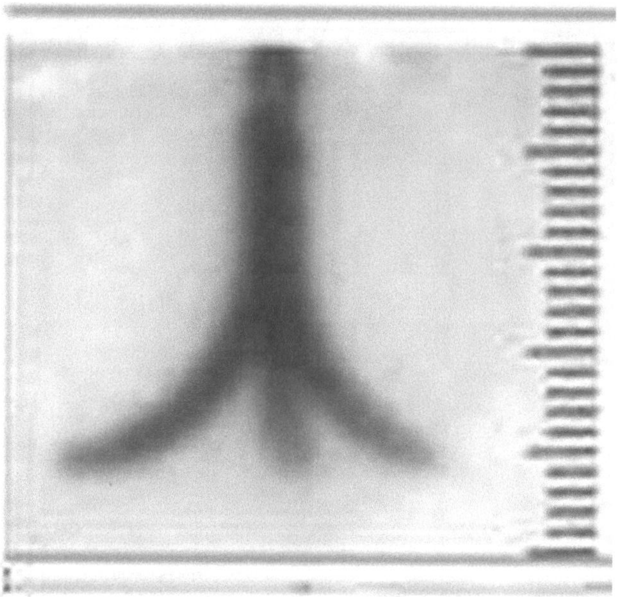

Fig 5.31. Frontal view of an infusion from a channel catheter with three channels created 120° apart.

Fig 5.32. Top view of the same infusion as the previous one with frontal view.

a pump: the entire apparatus is shown in **Fig. 5.34**, with a closeup of the balloon shown in **Fig. 5.35**. When inflated, its configuration is shown in **Fig. 5.36**. Let us call the distal end of the balloon with its port the "south pole." Then its use is suggested for intraparenchymal delivery by insertion into a resection cavity,

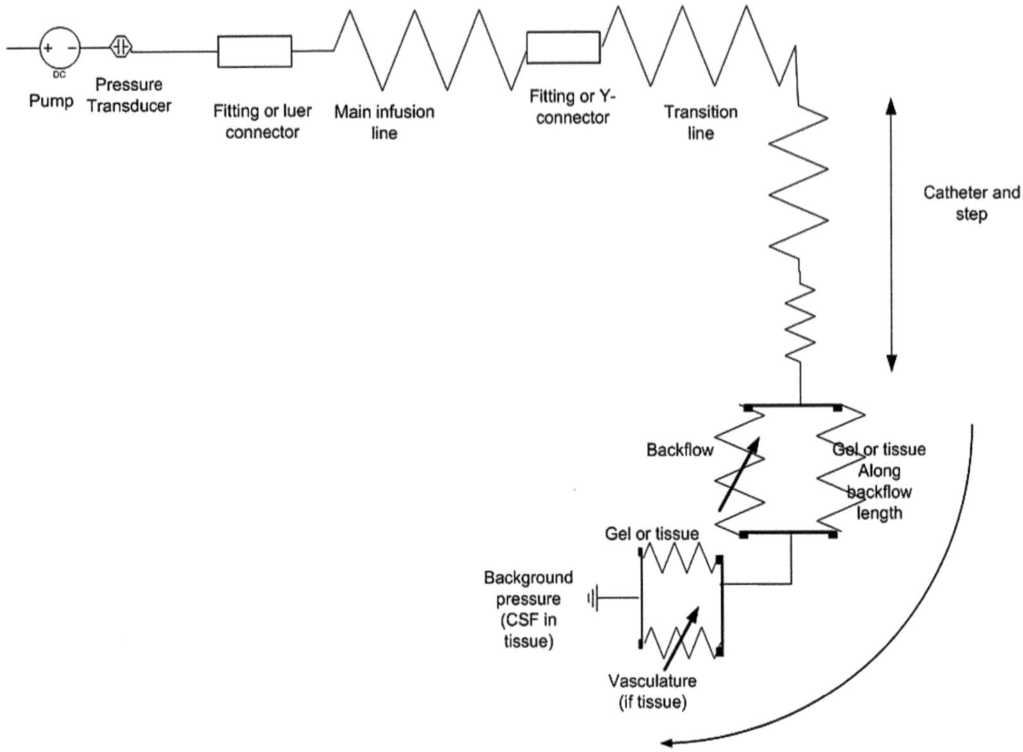

Fig 5.33. An electrical circuit analogy for accounting for the steady-state pressure that drives flow through a medium including all the intervening apparatus such as infusion lines and catheter. Note that we have used a "parallel" resistor analogy used above to indicate the paths of backflow and of flow in the tissue. This would imply, for example, that the pressure will be lower in cases where backflow is allowed to develop versus cases where the only flow is through the porous medium (all other parameters such as flow rate being fixed). This is indeed so, and has been mentioned further in the text.

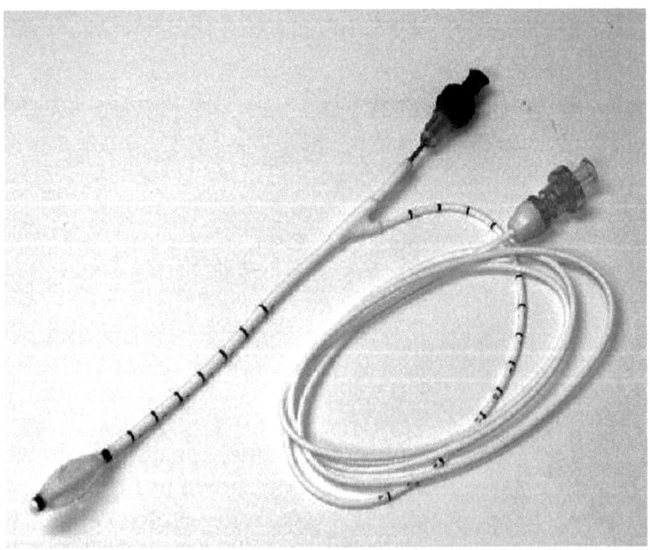

Fig 5.34. The Gliasite balloon apparatus.

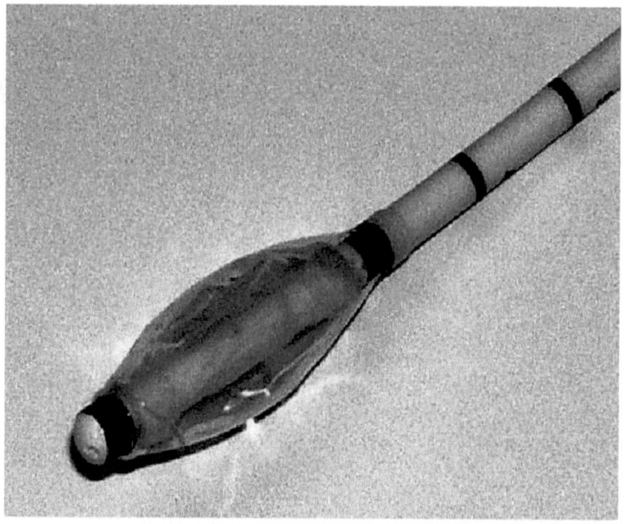

Fig. 5.35. A closeup of the balloon in its uninflated state.

Fig. 5.36. The GliaSite balloon inflated.

inflation, and subsequent infusion of the drug across the "south pole." Its effectiveness was favorably reviewed in (62), but we would like to point out one drawback apparent in **Fig. 5.37**. Namely, the protrusion or "knob" in the south pole may result in a pooling as shown (in part because the balloon is significantly stiffer than the gel, or indeed tissue). It would also be better to have an areal infusion (over the whole or part of the surface area of the balloon) rather than through a single port. Such redesigns are easy to envisage, but their implementation remains for the future.

Fig. 5.37. Infusion from the south pole of the balloon pools in the cavity caused by the protrusion.

3.2. Infusion Protocols

Finally, in the discussion on catheters, we mention a word on infusion protocols. We have so far been considering a steady or particular flow rate. It is true that one often changes the flow rate from one value to another, but a different method of infusion is a periodic one: a flow rate is maintained for a short time, then turned off, and the cycle repeats. If we were to compare the performance of such an infusion with the corresponding steady infusion at a rate equal to the average flux over a period, some differences may obtain. We recall the discussion on poroelasticity: the time dependence of the pressure, for example, *may* make the backflow characteristics different. Further, since the peak flow rate is higher, there may be an additional "Lagrangian" velocity that aids convection over the usual "Eulerian" velocity. However, we have completed a theoretical study of this question and have concluded that there is no such effect, in contrast to a weak effect we would observe in a traveling wave. We do not have space to describe this distinction: we simply refer to (63) for a lucid discussion of this point. Finally, if we were really optimistic, we would also note that the differences in concentration profiles in the two cases may affect the diffusivity of the therapeutic molecule differently, though this effect too is likely negligible. Meanwhile, we do have preliminary results that indicate that, *in gels*, the pulsed infusion seldom performs worse than the corresponding steady-state infusion. In a fraction of the cases, it does have better backflow characteristics. We display the typical of these cases where the backflow shows up in a steady infusion, but not in a pulsed. **Figure 5.38** shows a steady infusion at the low flow rate of 1.36 μL/min. Correspondingly, **Fig. 5.39**

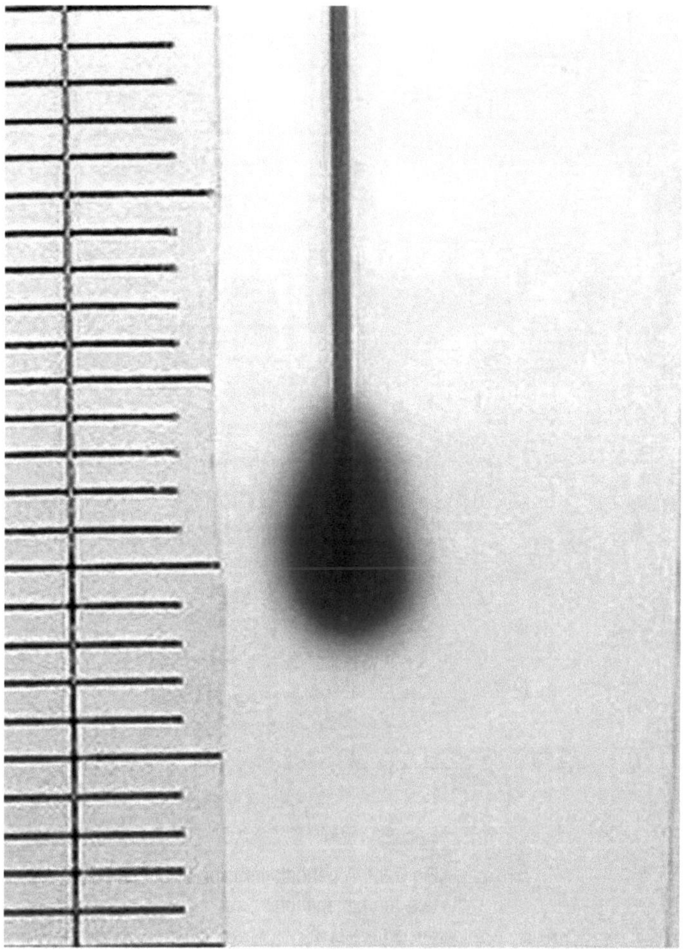

Fig 5.38. An infusion at a steady flow rate for comparison with a periodic infusion at the same average flow rate, shown in the next figure.

shows an infusion which had a period of 220, with the pump being on for only 10. Thus to correspond to the average rate mentioned, the flow rate is 30 μL/min. (In fact, the average flow rate was decided *after* the pulsed parameters were chosen.) It is of interest to compare how the pressure (at the pump, it must be remembered) behaves in these two protocols. This is displayed in **Fig. 5.40**. The reasons for this difference in behavior between pulsed and steady infusion in gels remains to be elucidated.

3.3. Ultrasound for Intraparenchymal Delivery

In the last several years, people have proposed the use of ultrasound to assist in intraparenchymal delivery (64, 65). As we have seen above, when fluid in a porous medium is subjected to a non-zero pressure gradient, it is pushed by that pressure gradient. The velocity of the fluid is linear in the pressure gradient in accord with D'Arcy's law, when the forces are not too excessive. Such a pressure gradient

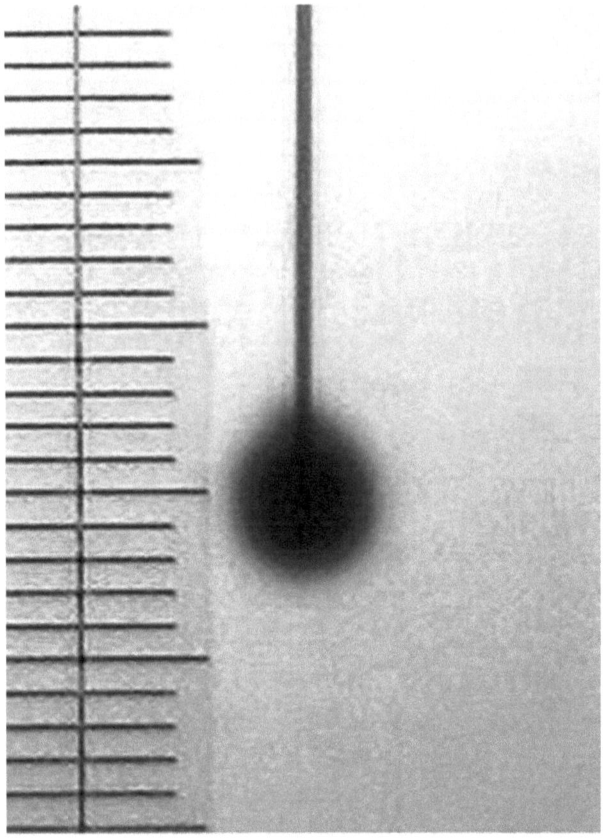

Fig 5.39. A periodic infusion at the same average flow rate as the previous figure. See text for explanation.

may be produced from an oscillating sound wave in two ways. The usual way is to induce a momentum transfer through the absorption of acoustic energy. The magnitude of this force, called the radiation force, depends on characteristics of the acoustic field and the medium. Fluid flow produced by the acoustic radiation force is called acoustic streaming: see (63) for an authoritative review.

To date, acoustic radiation force and streaming have been used for a range of biomedical applications including, but not limited to, calibration of therapeutic ultrasonic transducers; inducing motion in tissue for elastography imaging of cysts, lesions, and tumors; and manipulating contrast-enhancing micro-bubbles injected into the blood stream. (We omit references since our focus is on intraparenchymal delivery of therapeutic agents.) In recent years a growing interest has developed in using ultrasound to facilitate drug delivery in brain. Researchers in the field of chemical engineering and oil exploration have demonstrated that fluid flow through porous solids can be enhanced using sound waves (66). Studies also have been conducted to investigate the feasibility of using unfocused as well as focused ultrasound at low

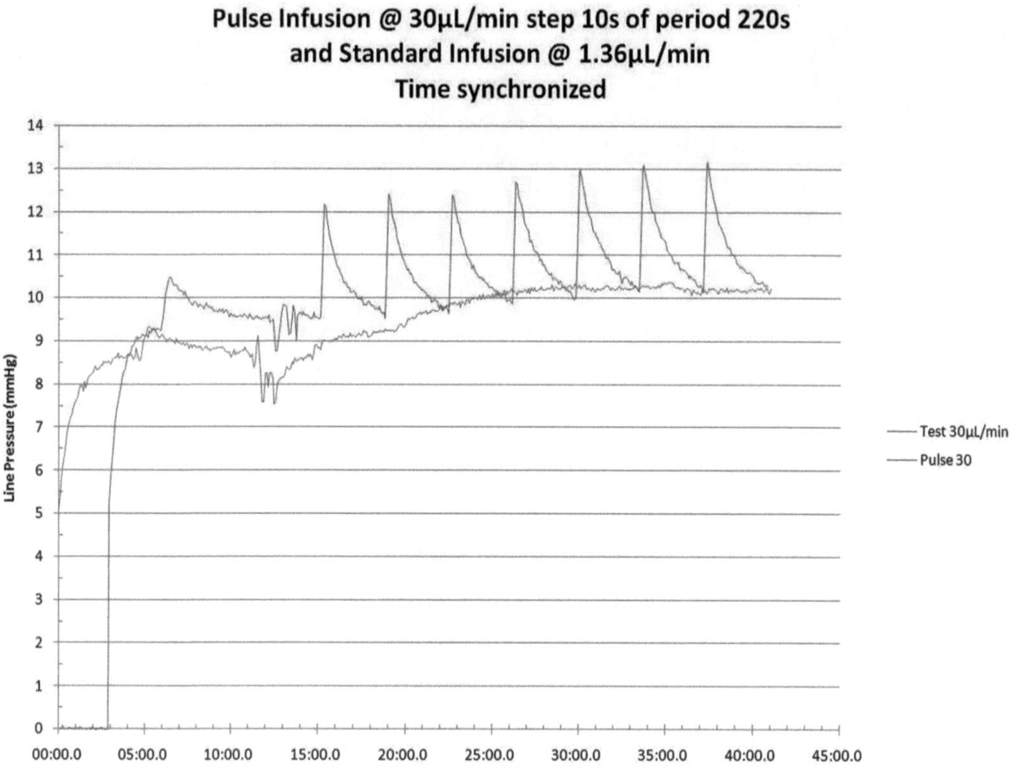

Fig 5.40. The behavior of the pressures for the periodic and the steady infusion protocols. The average pressures must coincide for the two protocols, so we attribute the increased average of the pulsed infusion to experimental error.

frequencies (100 kHz–1 MHz) to enhance drug transport within the brain (67, 68). Although the *ex vivo* results of the study (67) indicated that ultrasound application led to a 24-fold increase in permeation of tritium labeled molecules (70 kDa) in the excised brain, ultrasound application did not lead to a significant enhancement in drug transport in the in vivo experiments. We show some results from their papers below. Some of the Cytodome and UCSF researchers' results are shown in **Fig. 5.41** and **5.42**, while the Cornell results are shown in **Fig. 5.43**. It should be noted that the Cytodome results, while showing enhanced spread, were due to two hours of sonication which is very considerable.

Part of the problem with the Cytodome results, we believe, is due to the beam being unfocused. **Figure 5.44** shows, on the left, the streaming that results when the sound wave is not focused in the plane of the dye, while the figure on the right shows the effect of focusing accurately. The pictures were taken upon *only* 20 seconds of sonication and hence show a very strong effect (about 1 cm along the acoustic axis). In clinical trials, acceptable rates of infusion from a catheter are restricted to less than 5 μL/min. In such a case, assuming no backflow, it would take over a day to

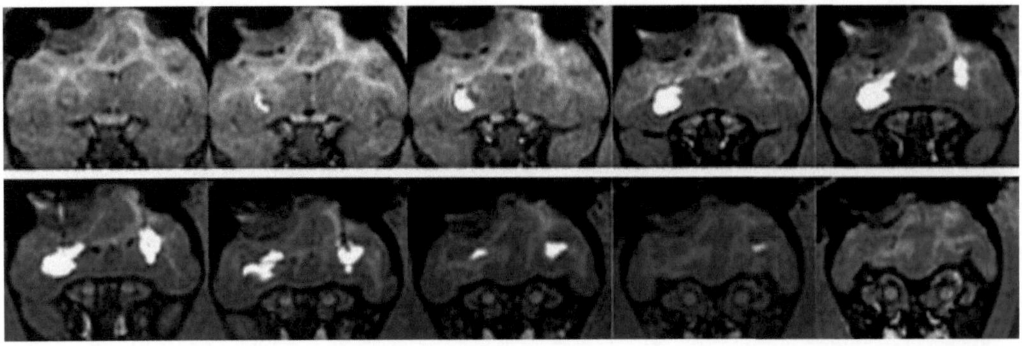

Fig 5.41. (This and the next image are from the paper referred to in the text: the ultrasound technology is due to Cytodome, Inc.) Cynomolgous monkeys where simultaneous infusions were conducted bilaterally (in each hemisphere). One hemisphere (the left one in the image) was assisted by ultrasound, the other was not. See the cited paper for details on the ultrasound and infusion parameters.

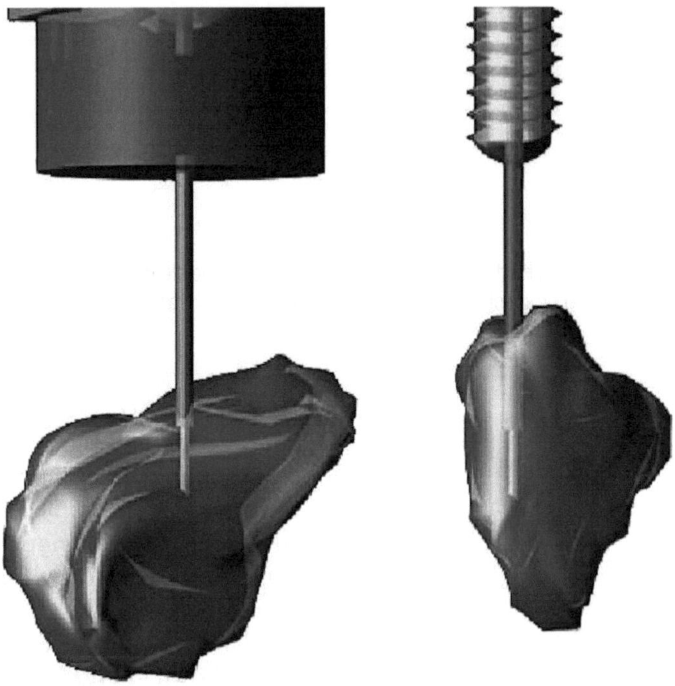

Fig 5.42. A reconstruction of the measured volumes of the contrast agent. The *left image* shows a clear, but not spectacular, enhancement of the volume of distribution due to two hours of sonication.

reach a lateral extent of a centimeter! Of course, the flow would fill a ball, or interior of a sphere, in contrast to just a line as in this case. Nevertheless, there is a significant difference between the two cases which deserves to be explored in greater depth. The field of ultrasound assisted transport of particles is very much in the research phase.

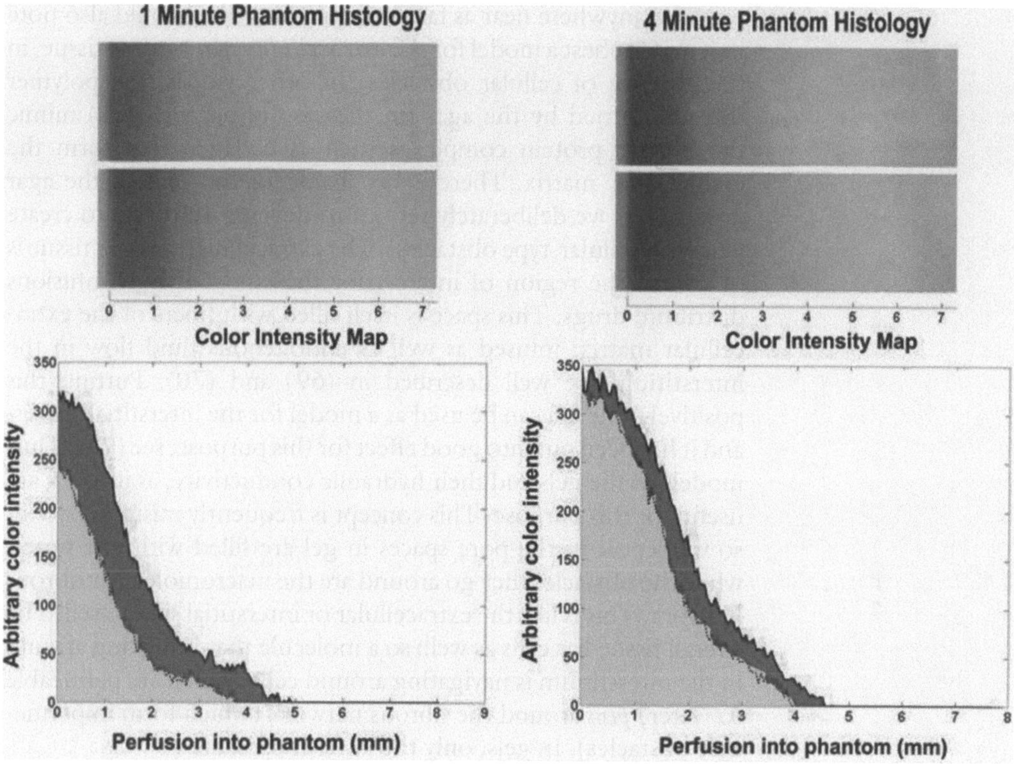

Fig 5.43. In this study at Cornell, focused ultrasound was used in a gel. The enhancement visible in the images of the distribution seems less significant than in the tissue example used by Cytodome and the UCSF collaborators of the previous images.

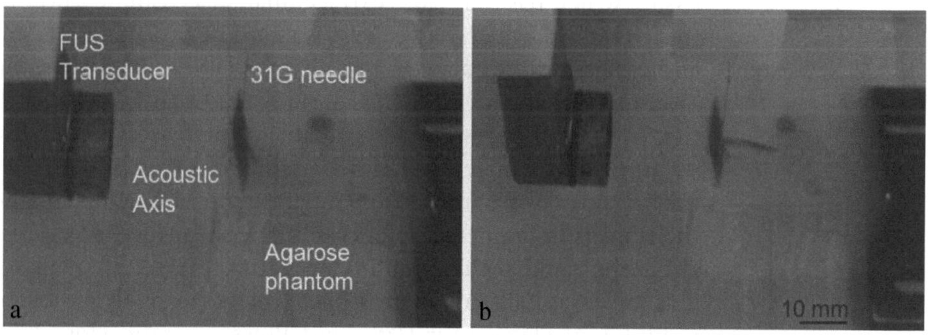

Fig 5.44. Unconstrained gel also shows strong streaming effects. Focusing is critical in obtaining strong results and is the difference between the left figure, where the focus was not in the plane of the initial dye deposition, and the right, where it was either on or closer to that plane. There is about 1 cm of displacement along the axis. This research is part of a collaboration between Therataxis and the Riverside Research Institute. The experiment was conducted by Dr. Parag Chitnis of the Institute.

3.4. Gel Phantoms: Their Uses and Abuses

In the above discussion, we have quoted extensively from results from gel studies. Gels indeed are frequently studied as surrogates for tissue. We have already mentioned one problem with this hypothesis, namely as a poroelastic medium, it does not propagate pressure anywhere near as fast as tissue does. We should also note that gel is at best a model for the *extracellular space only* of tissue, in the absence of cellular obstacles. In other words, the polymer threads formed by the agar (in the case of agarose gels) mimic the fibrous protein complexes such as collagen that form the extracellular matrix. There is no analog of the cells in the agar gels (unless we deliberately set out to develop a process to create gels with cellular-type obstacles). The extracellular space in tissue is of course the region of interest for the study of how infusions distribute drugs. This space is itself filled with fibers of the extracellular matrix: infused as well as endogenous fluid flow in the interstitium are well described in (69) and (70). Putting this positively, the gel can be used as a model for the interstitial spaces, and it has been put into good effect for this purpose, see (71). Thus models of the gels and their hydraulic conductivity, as in (72), are useful for this purpose. This concept is frequently misunderstood, so we repeat it: the pore spaces in gel are filled with just water, while the obstacles they go around are the macromolecular fibrous network. This is like the extracellular or interstitial space itself. The overall tissue has cells as well: so a molecule that is moving around in the interstitium is navigating around cells (which are permeable to water) *plus* around the fibrous network (which form impermeable obstacles). In gels, only the latter is present.

Generally speaking, the hydraulic resistance of tissue is actually lower or of the same order of magnitude, and the shear modulus higher, than that of gel, depending on the agar concentrations. It is indeed surprising that the hydraulic resistance of tissue, which has only 20% extracellular space can be lower than that of the gel, which is over 99% water. The explanation is subtle and is outside the scope of this chapter. The fact that the shear modulus of, say, 0.2 – 0.6% gel at body temperature is lower than that of brain tissue is literally palpable. One reason for the lower shear resistance in gels is that body temperature is close to that of the gel to sol phase transition temperature in the agar gels. Above the transition temperature of course the shear modulus vanishes altogether, and at these high temperatures, it is very small. (The gels are essentially incompressible, which means that their bulk modulus of elasticity is high, and Poisson's ratio close to one half.)

An excellent paper on the theory of poroelasticity with applications to gels is (39). In that reference the pore size of "gels" is reported as being only about 100 Å; however, the gels he examines there are rather dense gels of a different material and host than the agar gels, which have been used in the figures shown here. There is an atomic force microscopy study of agarose gels (73) which reports pore sizes of almost 400 nm, which is 40 times as large as reported by Johnson. An NMR study (74) reports sizes closer to 100 nm. However, these numbers pertain to a concentration of

2% agarose, which is the the lowest concentration of agar studied in (73), while we have been working with only a tenth of that concentration of agar. If we use the sublinear behavior of pore size versus agar concentration reported by them, then we may expect the pore sizes in our gels to be over 1 μm. *These pore sizes are larger than that reported in tissue* (24). Other references which picture the microstructure of gels by various techniques include (75) and (76) on other related gel systems.

We already know that gels are homogeneous whereas tissue is not, and further that gels do not have blood vessels which are a sink for fluid under excess pressure: the gel medium conserves water or infused fluid while the tissue does not. Gels are also very brittle, and it takes great care to not have infusions dominated by cracks or fissures in gels, which result in behaviors seldom seen in live tissue. These cautions must be borne in mind when examining gel data. However, a *relative* comparison of backflow amongst different catheters is reliably obtained from the proper gel studies.

4. Conclusions

We have described in some detail the difficulties involved with intraparenchymal deliveries, and some of the devices used in such deliveries. We have described the extreme variability of brain tissue. While we have not described tumors in any depth, they add considerably more heterogeneity and other complications to the process. People are of course trying hard to circumvent the invasiveness of this technique, which itself was born out of the difficulties of systemic delivery techniques. At the very beginning of **Section 2**, we have mentioned some of the other methods that are being attempted. In addition, there is an intriguing suggestion from the Mayo clinic, stemming from the invention disclosed in (77) on some catheterization of cerebral veins. This would be intermediate in invasiveness between the direct delivery into tissue and systemic delivery. Finally, there are a large number of patents filed on "smart catheters" that allow real time monitoring of various properties of tissue or drug action during delivery. Such devices will of course allow better feedback control of delivery, though space as well as focus on delivery rather than monitoring has prevented us from discussing these here. (We have, however, briefly discussed the behavior of pressure during delivery, and characteristics of these when there is occluding material.) Time will tell if the difficulties in intraparenchymal delivery will be overcome to have successful and reproducible infusions, or if these other technologies will become competitive with, and obviate the need for, intraparenchymal delivery. However, the intraparenchymal route is at the moment, the only delivery method that can successfully transport certain therapies across the blood—brain barrier and into the tissue where they need to go.

Acknowledgement

The research described in this paper was supported in part by the Kinetics Foundation. Most of the figures have been obtained during collaborative research with a number of collaborators. These collaborators have been mentioned in the figure captions and I owe them a debt of gratitude in permitting use of the figures in this chapter. The title of course is borrowed from a well-known source (78).

References

1. W. Pardridge. This is the adjective used by William Pardridge of the University of California, Los Angeles to describe intraparenchymal infusions.
2. R. Dermietzel (Ed.) *Blood-Brain Barriers: From Ontogeny to Artificial Interfaces.* John Wiley and Sons, New York, 2006. In two volumes.
3. W. M. Pardridge. The blood-brain barrier: bottleneck in brain drug development. *NeuroRx: The Journal of the American Society for Experimental Neurotherapeutics*, 2:3–14, 2005.
4. J. F. Deeken and W. Loscher. The blood-brain barrier and cancer: transporters, treatment and Trojan Horses. *Clinical Cancer Research*, 13:1663–1674, 2007.
5. D. R. Groothuis. The blood-brain and blood-tumor barriers: a review of strategies for increasing drug delivery. *Neuro-Oncology*, 2:45–59, 2000.
6. K. Hynynen, N. McDannold, N. Sheikov, F. Jolesz, and N. Vikhodtseva. Local and reversible blood-brain barrier disruption by non-invasive focused ultrasound at frequencies suitable for trans-skull sonications. *NeuroImage*, 24:12–20, 2005.
7. J. J. Choi, M. Pernot, S. A. Small, and E. E. Konofagou. Non-invasive, transcranial and localized opening of the blood-brain barrier using focused ultrasound in mice. *Ultrasound in Medicine and Biology*, 33:95–104, 2007.
8. J. J. Choi, M. Pernot, T. R. Brown, S. A. Small, and E. E. Konofagou. Spatiotemporal analysis of molecular delivery through the blood-brain barrier using focused ultrasound. *Physics in Medicine and Biology*, 52:5509–5530, 2007.
9. P. Kumar, H. Wu, J. I. McBride, K. eun Jum, M. H. Kim, B. L. Davidson, S. K. Lee, P. Shankar, and N. Manjunath. Transvascular delivery of small interfering RNA to the central nervous systems. *Nature*, 448: 39–43, 2007.
10. M. Hammarlund-Udenaes, M. Friden, S. Syvanen, and A. Gupta. On the rate and extent of drug delivery to the brain. *Pharmaceutical Research*, 25:1737–1750, 2008.
11. X. Liu, C. Chen, and B. J. Smith. Progress in brain penetration evaluation in drug discovery and development. *Journal of Pharmacology and Experimental Therapeutics*, 325:349–356, 2008.
12. J. Burns and D. F. Weaver. A mathematical model for prediction of drug molecule diffusion across the blood brain barrier. *Canadian Journal of Neurological Sciences*, 31:520–527, 2004.
13. R. Cecchelli, V. Berezowski, S. Lundquist, M. Culot, M. Renftel, M. P. Dehouck, and L. Fenart. Modelling of the blood-brain barrier in drug discovery and development. *Nature Reviews Drug Discovery*, 6:650–661, 2007.
14. K. Rose, L. H. Hall, M. Hall, and L. B. Kier. Modeling blood-brain barrier partitioning using topological structure descriptors. *MDL Information Systems White Paper*, 2003.
15. L. Rosso, C. S. Brock, J. M. Gallo, A. Saleem, P. M. Price, F. E. Turkheimer, and E. O.

Aboagye. A new model for prediction of drug distribution in tumor and normal tissues: pharmacokinetics of Temozolomide in glioma patients. *Cancer Research*, 69: 120–127, 2009.

16. P. F. Morrison, D. W. Laske, H. Bobo, E. H. Oldfield, and R. L. Dedrick. High flow microinfusion: tissue penetration and pharmacodynamics. *American Journal of Physiology*, 266:R292–R305, 1994.
17. R. R. Lonser, N. Gogate, J. D. Wood, P. F. Morrison, and E. H. Oldfield. Direct convective delivery of macromolecules to the peripheral nerves. *Journal of Neurosurgery*, 89:610–615, 1998.
18. R. R. Lonser, R. J. Weil, P. F. Morrison, L. S. Governale, and E. H. Oldfield. Direct convective delivery of macromolecules to the spinal cord. *Journal of Neurosurgery*, 89:616–622, 1998.
19. M. Y. Chen, R. R. Lonser, P. F. Morrison, L. S. Governale, and E. H. Oldfield. Variables affecting convection-enhanced delivery in the striatum: a systematic examination of rate of infusion, cannula size, infusate concentration, and tissue-cannula sealing time. *Journal of Neurosurgery*, 90: 315–320, 1999.
20. R. R. Lonser, S. Walbridge, K. Garmestani, J. A. Butman, H. A. Walters, A. O. Vortmeyer, M. W. Brechbiel, P. F. Morrison, and E. H. Oldfield. Successful and safe perfusion of the primate brainstem with a macromolecule: in vivo magnetic resonance imaging of macromolecular distribution during infusion. *Journal of Neurosurgery*, 97:905–913, 2002.
21. D. Croteau, S. Walbridge, P. F. Morrison, J. A. Butman, A. O. Vormeyer, D. Johnson, E. H. Oldfield, and R. R. Lonser. Real time, in vivo imaging of convective distribution of a low molecular weight tracer. *Journal of Neurosurgery*, 102:90–107, 2005.
22. R. H. Bobo, D. W. Laske, A. Akbasak, P. F. Morrison, R. L. Dedrick, and E. H. Oldfield. Convection-enhanced delivery of macromolecules in the brain. *Proceedings of the National Academy of Sciences U. S. A.*, 91:2076–2080, 1994.
23. E. R. Kandel, J. Schwartz, and T. Jessell (Eds.). *Principles of Neural Science*, 4th edition. McGraw Hill, New York, 2000.
24. R. G. Thorne and C. Nicholson. In vivo diffusion analysis with quantum dots and dextrans predicts the width of brain extracellular space. *Proceedings of the National Academy of Sciences*, 103:5567–5572, 2006.
25. H. Davson and M. B. Segal (Eds.). *Physiology of the Cerebrospinal Fluids and Blood-brain Barriers*. CRC Press, Boca Raton, FL 1996.
26. N. J. Abbott. Dynamics of CNS barriers: evolution, differentiation, and modulation. *Cellular and Molecular Neurobiology*, 25:5–23, 2005.
27. R. O. Weller. Pathology of cerebrospinal fluid and interstitial fluid of the CNS: significance for Alzheimer disease, prion disorders and multiple sclerosis. *Journal of Neuropathology and Experimental Neurology*, 57:885–894, 1998.
28. M. E. Jonhagen, A. Nordberg, K. Amberia, L. Backman, T. Ebendal, B. Meyerson, L. Olson, A. Seigar, M. Shigeta, E. Theodorsson, M. Viitanen, B. Winblad, and L. O. Wahlund. Intracerebroventricular infusion of nerve growth factor in three patients with Alzheimer's disease. *Dementia and Geriatric Cognitive Disorders*, 9:246–257, 1998.
29. C. E. Johanson, J. A. Duncan, E. G. Stopa, and A. Baird. Enhanced prospects for drug delivery and brain targeting by the Choroid Plexus – CSF route. *Pharmaceutical Research*, 22:1011–1037, 2005.
30. K. Sawamoto, H. Wichterle, O. Gonzalez-Perez, J. A. Cholfin, M. Yamada, N. Spassky, N. S. Murcia, J. M. Garcia-Verdugo, O. Marin, and J. L. Rubenstein et al. New neurons follow the flow of cerebrospinal fluid in the adult brain. *Science*, 311:629–632, 2006.
31. G. Craciun, A. Brown, and A. Friedman. A dynamical system model of neurofilament transport in axons. *Journal of Theoretical Biology*, 237:316–322, 2005.
32. A. Friedman and G. Craciun. A model of intracellular transport of particles in an axon. *Journal of Mathematical Biology*, 51:217–246, 2005.
33. L. W. Bruch, E. Zaremba, and M. W. Cole. *Physical Adsorption: Forces and Phenomena*. Dover Mineola, NY, 2007 edition, 1997.
34. U. Hornung (Ed.). *Homogenization and Porous Media*. Springer, New York, 1997.
35. H. Wang. *Linear Theory of Poroelasticity*. Princeton University Press, Princeton, NJ, 2000.
36. O. Coussy. *Mechanics of Porous Continua*. John Wiley and Sons, New York, 2004.
37. O. Coussy. *Poromechanics*. John Wiley and Sons, New York, 2004.
38. D. L. Johnson, J. Koplik, and R. Dashen. Theory of dynamic permeability and tortuosity in fluid-saturated porous media. *Journal of Fluid Mechanics*, 176:379–402, 1987.

39. D. L. Johnson. Elastodynamics of Gels. *Journal of Chemical Physics*, 77:1531–1539, 1982.
40. P. F. Morrison, M. Y. Chen, R. S. Chadwick, R. R. Lonser, and E. H. Oldfield. Focal delivery during direct infusion to brain: role of flow rate, catheter diameter, and tissue mechanics. *American Journal of Physiology*, 277:R1218–R1229, 1999.
41. X. Y. Zhang, J. Luck, M. W. Dewhirst, and F. Yuan. Interstitial hydraulic conductivity in a fibrosarcoma. *American Journal of Physiology Heart and Circulatory Physiology*, 279:H2726–H2734, 2000.
42. J. Bear. *Dynamics of Fluids in Porous Media*. Dover, New York 1988.
43. Y. Mardor, O. Rahav, Y. Zauberman, Z. Lidar, A. Ocherashvilli, D. Daniels, Y. Roth, S. E. Maier, and Z. Ram. Convection-enhanced drug delivery: increased efficacy and magnetic resonance image monitoring. *Cancer Research*, 65:6858–6863, 2005.
44. B. Perlstein, Z. Ram, D. Daniels, A. Ocherashvilli, Y. Roth, S. Margel, and Y. Mardor. Convection enhanced delivery of maghemite nanoparticles: increased efficacy and MRI monitoring. *Neuro-Oncology*, 10:153–161, 2008.
45. X. Fan, N. Phan-Thien, N. T. Yong, X. Wu, and D. Xu. Microchannel flow of a macromolecular suspension. *Physics of Fluids*, 15:11–21, 2003.
46. G. Powathil, M. Kohandel, S. Sivaloganathan, A. Oza, and M. Milosevic. Mathematical modeling of brain tumor: effects of surgery, radiotherapy and chemotherapy. *Physics in Medicine and Biology*, 52:3291–3306, 2007.
47. S. Sivaloganathan, M. Stastna, G. Tenti, and J. M. Drake. A viscoelastic approach to the modelling of hydrocephalus. *Applied Mathematics and Computation*, 163:1097–1107, 2005.
48. S. Sivaloganathan, M. Stastna, G. Tenti, and J. M. Drake. A viscoelastic model of the brain parenchyma with pulsatile ventricular pressure. *Applied Mathematics and Computation*, 165:687–698, 2005.
49. S. Sivaloganathan, M. Stastna, G. Tenti, and J. M. Drake. Biomechanics of the brain: A theoretical and numerical study of Biot's equations of consolidation theory with deformation-dependent permeability. *International Journal of Non-Linear Mechanics*, 40:1149–1159, 2005.
50. J. H. Smith and J. A. C. Humphrey. Interstitial transport and transvascular fluid exchange during infusion into brain and brain tissue. *Journal of Microvascular Research*, 2:871–882, 2006.
51. M. Stastna, G. Tenti, J. M. Drake, and S. Sivaloganathan. Brain biomechanics: an exact solution and asymptotics of slow variations, for the time dependent equations of consolidation theory. *Dynamics of Continuous, Discrete and Impulsive Systems*, 5:415–425, 1999.
52. G. Tenti, J. M. Drake, and S. Sivaloganathan. Mathematical pressure-volume models of the cerebrospinal fluid. *Applied Mathematics and Computation*, 94:243–266, 1998.
53. G. Tenti and K. Rohlf. The role of the Womersley number in pulsatile blood flow: a theoretical study of the Casson model. *Journal of Biomechanics*, 34:141–148, 2001.
54. G. Tenti, S. Sivaloganathan, and J. M. Drake. The synchrony of arterial and CSF pulsations is not due to resonance. *Pediatric Neurosurgery*, 37:221–222, 2002.
55. R. K. Jain and L. E. Gerlowski. Extravascular transport in normal and tumor tissues. *Critical Review of Oncology and Hematology*, 5:115–170, 1986.
56. R. K. Jain and T. Baxter. Mechanisms of heterogeneous distribution of monoclonal antibodies and other macromolecules in tumors: significance of interstitial pressure. *Cancer Research*, 48:7022–7032, 1988.
57. R. K. Jain. Vascular and interstitial barriers to delivery of therapeutic agents in tumors. *Cancer Metastasis Reviews*, 9:253–266, 1990.
58. M. T. Krauze, R. Saito, C. Noble, M. Tamas, J. Bringas, J. W. Park, M. S. Berger, and K. Bankiewicz. Reflux-free cannula for convection-enhanced high-speed delivery of therapeutic agents. *Journal of Neurosurgery*, 103:923–929, 2007.
59. M. Solar, C. Ross, R. Raghavan, and M. Brady. Device and method for delivering therapeutic agents to an area of the body. *United States Patent Application 12/152,280*, filed May 13, 2008.
60. J. H. Sampson, M. L. Brady, N. A. Petry, D. Croteau, A. H. Friedman, H. S. Friedman, T. Wong, D. D. Bigner, I. Pastan, and C. Pedain. Intracerebral infusate distribution by convection-enhanced delivery in humans with malignant gliomas: descriptive effects of target anatomy and catheter positioning. *Neurosurgery Supplement*, 1:89–99, 2007.
61. J. F. Dempsey, J. A. Williams, and J. B. Stubbs et al. Dosimetric properties of a novel brachytherapy balloon applicator for the treatment of malignant brain tumor resection cavity margins. *International Journal of Radiation Oncology and Biological Physics*, 42:421–429, 1998.

62. J. J. Olson, Z. Zhang, D. Dillehay, and J. Stubbs. Assessment of a balloon-tipped catheter modified for intracerebral convection-enhanced delivery. *Journal of Neurooncology*, 89:159–168, 2008.
63. J. Lighthill. Acoustic Streaming. *Journal of Sound and Vibration*, 61:391–418, 1978.
64. G. Heart, G. Tolkowsky, A. Brisken, and J. Karratt. Implantable ultrasound systems and methods for enhancing localized delivery of therapeutic substances. *U. S. Patent Application 2004/0267234*, 2004.
65. R. Raghavan. Device, methods, and control of sonic guidance of molecules and other materials using time-reversal acoustics. *United States Patent Application 12/319, 311*, 2009.
66. P. Poesio, G. Ooms, A. Schraven, and F. van der Bas. Theoretical and experimental investigation of acoustic streaming in a porous material. *Physical Review* E, 66:016309-1–016309-9, 2002.
67. Y. Liu, S. Paliwal, K. Bankiewicz, J. Bringas, G. Heart, S. Mitragotri, and M. Prausnitz. Ultrasoundassisted drug delivery in the brain. *preprint*, 2009.
68. G. Lewis and W. Olbricht. Acoustic targeted drug delivery in neurological tissue. *BMES Annual Conference*, 2007.
69. J. R. Levick. Flow through interstitium and other fibrous matrices. *Quarterly Journal of Experimental Physiology*, 72:409–438, 1987.
70. M. A. Swartz and M. E Fleury. Interstitial flow and its effects in soft tissues. *Annual Reviews of Biomedical Engineering*, 9:229–256, 2007.
71. S. Ramanujan, A. Pluen, T. D. McKee, E. B. Brown, Y. Boucher, and R. K. Jain. Diffusion and convection in collagen gels: implications for transport in the tumor interstitium. *Biophysical Journal*, 83:1650–1660, 2002.
72. T. Stylianopoulos, A. Yeckel, J. J. Derby, X. J. Luo, M. S. Shephard, E. A. Sander, and V. H. Barocas. Permeability calculations in three-dimensional isotropic and oriented fiber networks. *Physics of Fluids*, 20: 123601-1-123601-10, 2008.
73. N. Pernodet, M. Maloum, and B. Tinland. Pore size of agarose gels by atomic force microscopy. *Electrophoresis*, 18:55–58, 1997.
74. M. M. Chui, R. J. Phillips, and M. J. McCarthy. Measurement of the porous microstructure of hydrogels by nuclear magnetic resonance. *Journal of Colloid and Interface Science*, 174:336–344, 1995.
75. T. K. Attwood, B. J. Nelmes, and D. B. Sellen. Electron microscopy of beaded agarose gels. *Biopolymers*, 27:201–212, 1988.
76. R. Nisslert, M. Kvarnstrom, N. Loren, M. Nyden, and M. Rudemo. Identification of the three-dimensional gel microstructure from transmission electron micrographs. *Journal of Microscopy*, 225:10–21, 2007.
77. S. J. Asirvatham, P. A. Friedman, C. J. Bruce, and D. R. Holmes. Detecting and treating nervous system disorders. *U. S. Patent Application 60/738, 718*, 2005.
78. S. Freud. *Civilization and Its Discontents*. W. W. Norton and Company, New York reissue edition, 1989.

Chapter 6

Intrathecal Drug Delivery by Implanted Pumps

Shawn Belverud, Alon Mogilner, and Michael Schulder

Abstract

Direct central neural axis neuromodulation has become a viable means to treat chronic neurologic disease and injury. Although first described in 1898, technology has only recently allowed for strict modulation and adjustments of drug delivery into the centeral nervous system. Evolving congruently with understanding of the altered neurophysiology and an expanding pharmacologic armamentarium, intrathecal pump systems augment treatment in an ever expanding number of disorders. In this review, we present a general historical overview of direct neuroaxis delivery and discuss current indications for intrathecal delivery with considerations in patient and drug selection. Complications, both surgical and medical, are discussed in detail with focus on avoidance and management.

Key words: Intrathecal baclofen, intrathecal narcotics, intraventricular drug delivery, neuropathic pain, nociceptive pain, spasticity.

1. Introduction

Direct pharmacologic delivery to the central neural axis is an effective means to treat diseases involving the central nervous system (CNS) when other methods have failed to relieve symptoms or have resulted in significant side effects (1). While advances are being made in the engineering of new pharmaceuticals to penetrate the central nervous system (2), treatment of acute and chronic pain, spasticity, and a growing number of other neurologic disorders are effectively treated with this form of neuromodulation. This chapter offers a review of intrathecal (IT) drug delivery and its indications with considerations in patient selection. Also discussed will be surgical technique, medical and surgical complications as well as the pump systems used for the drug delivery.

K.K. Jain (ed.), *Drug Delivery to the Central Nervous System*, Neuromethods 45,
DOI 10.1007/978-1-60761-529-3_6, © Humana Press, a part of Springer Science+Business Media, LLC 2010

1.1. Intrathecal Analgesia

"The greatest evil is physical pain," stated by St. Augustine, speaks to an emotional and physical milieu that nearly all people have experienced. Due to the complexity of pain, patients respond in vastly disparate ways, and the means to treat their plight have evolved to allow for greater patient specificity (3–5). Furthermore, nociceptive circuitry is altered as acute sensation evolves to chronic; these changes require a dynamic pharmaceutical regimen to meet a patient's changing needs (6). On a range from NSAIDs to both long and short acting narcotics, when traditional means of enteral analgesics fail to relieve pain symptoms or cause deleterious symptoms that preclude their use, alternative means of delivery are warranted.

Direct CNS drug delivery for pain is a logical evolution of the usual systematic administration for it provides two advantages: avoidance of the blood–brain barrier and selective permeability to the structures that possess the receptors upon which the drug is acting. These benefits, coupled with mechanical advancements in pump technology, allow a constant level of analgesia to be maintained. Intrathecal delivery allows for fractional amounts of the drug being administered, though many authors urge caution in transitioning from oral to IT morphine. The commonly used ratio is 300:1 oral to IT, respectively (7, 8). Furthermore, IT analgesics bypass hepatic metabolites, minimizing systematic side effects.

The first report of direct CNS drug delivery, "cocainization of the spinal cord," was described by August Bier in 1898. This was followed later by numerous reports using IT phenol for malignant pain in the 1960s (9). These reports preceded the finding of opioid receptors within the CNS, which did not take place until the early 1970s (10, 11). This discovery, coupled with an increased understanding of pharmacology and pain physiology, led to IT morphine injections in animal models (12) followed by the first human trials in cancer patients (13), both of which reported efficacious analgesia. Since the original trials, IT morphine and other opioids have become routine in treating malignant pain (14–17), but the use of IT analgesics have now evolved to be the most commonly used for the management of non-oncologic pain.

Soon after IT opioids proved efficacious for the treatment of oncologic pain, a number of small studies began to identify a role for IT analgesics in non-oncologic analgesia (18–21). In 1996, Winkelmuller and Winkelmuller reported 92% satisfaction and 81% with improved quality of life in a heterogeneous group of 120 non-malignant somatic and neuropathic pain patients (22). That same year, Paice et al. published a separate multicenter retrospective study of 429 patients reporting improved analgesia in those suffering from somatic pain compared to neuropathic (23). The most common cause of the somatic pain in these studies, which continues to be the most common indication of IT analgesics currently, was failed back syndrome (FBS).

Since inception, IT morphine has been the first line pharmaceutical used for direct CNS analgesic delivery. However, attempts to maximize individual patient satisfaction has led to the use of a range of IT opioids alone, or admixed with different classes of drugs to include local anesthetics and alpha-2 agonists (24, 17). Ziconotide, an N-type calcium channel blocker and the only other IT analgesic FDA approved for chronic pain (both cancer and non-malignant), can also be efficacious in the 20% of patients that do not respond to IT morphine (25–27). These combinations of IT drugs have broadened the scope of non-malignant pain treatment to now include FBS, complex regional pain syndrome, post-herpatic neuralgia, and peripheral neuralgias (28).

To maintain prolonged, constant, and stable analgesia, pump technology has evolved to become both queriable and adjustable by both the physician and the patient. These features have now given practitioners a system that has high fidelity and greatly limits the potential for side effects and inconsistent drug delivery. From the initial external devices that functioned via time-separated boluses, recent pumps now have internal reservoirs of both 20 and 40 ml and offer patient controlled administration (PCA) for breakthrough pain. (Specifics regarding surgical technique and pump complications will be discussed later in this chapter.)

1.2. Patient Selection in Intrathecal Analgesia

A strict criterion governing the selection of surgical candidates and input from multiple specialties helps to ensure both surgical success and minimal operative morbidity. As a group, chronic pain patients generally need large doses of drugs to treat their severe pain and thus are at greater risk to develop tolerance, addiction, and suffer from the many systematic side effects. Approximately 2.5–5% of these patients have pain that is resistant to conventional oral narcotics, anticonvulsants, and antidepressants or have side effects that preclude their usage (29). It is often difficult to determine how long to attempt different enteral drug combinations before turning to direct neural axis administration and, furthermore, what constitutes an unacceptable side-effect. Even subtle side effects associated with oral narcotics can hinder mood, cognitive ability, and alertness and have a harmful effect on the patient's daily lives. When transitioned to IT analgesic delivery, doses of oral medications can usually be reduced and the side effects cease, resulting in improved mental health (30). Other selection criteria for IT delivery include status an adequate social support system, an organic etiology for the pain that cannot be corrected by surgery, absence of significant psychological dysfunction, a patent spinal subarachnoid space, and lack of issues related to secondary gain (31).

Given the cost of the surgical implantation and hardware, life expectancy becomes an issue of important consideration, particularly in oncology patients. Cost analysis indicates that IT systems become a cost-effective mode of analgesic delivery for cancer and non-cancer patients with at least a 3–6 and 11–22 months life expectancy, respectively (32). If the life expectancy likely is under 3 months, an external IT catheter system may be considered with the explanation that risk of infection, cerebrospinal fluid (CSF) leaks, and catheter displacement is increased.

Once a patient is deemed a suitable candidate for IT analgesia, a screening trial is necessary to determine the patient's response. This is most commonly accomplished with a single bolus of preservative-free medication via lumbar puncture, but can also be tested by an external delivery system that allows a constant, titratable infusion over a predetermined time period. Continuous use of the base-line oral analgesic regime should continue through this period, for possibility that the assessment of IT response may be confounded by symptoms of narcotic withdrawal.

After the trial, the patient should be sent to a monitored setting with appropriate nursing care, pulse oximetry, and hourly neurological checks and vigilance toward respiratory depression. Naloxone should also be readily available at the bedside. Given the hydrophilic pharmacokinetics of morphine, analgesic effect should commence in approximately 30 min to 1 h, with effects lasting 18–24 h (33). If a patient does not attain adequate pain relief from an appropriate trial, it is unlikely that the pain will be adequately treated by chronic therapy. However, effective analgesia from the trial does not necessarily predict long-term success. Depending on the results, up to one-third of trial responders fail to have adequate pain relief after 12–18 months (22, 34). Biologic tolerance is certainly a possible mechanism for this phenomenon and changing the IT opioid or adding an alpha-2 agonist or local anesthetic such as bupivicaine can often work to regain analgesia. If these changes fail to meet the pain control needs, ziconotide should be considered.

Once the pump system has been placed, vigilance should be maintained for other pharmaceutical side effects that can develop in both the short and the long term. Constipation is quite common along with urinary retention. The latter usually resolves over time or can be relieved with decreased IT dosing (35, 36). Pruritis is frequently encountered but most usually resolves over days. Myoclonus can develop months after the commencement of treatment and may necessitate dose reduction. A switch to IT hydromorphone should be considered for patients with resilient pruritis and/or myoclonus if either should become intolerant to the patient (37). Sex hormones may become suppressed but are responsive to supplementation (38). Persistent dependent edema can also arise that is not readily amenable to diuretics. Certain authors consider pre-existing venous insufficiency to be a relative contraindication to IT opioid administration (39). For further reading related to specific IT drugs, the reader is referred to the Polyanalgesic Consensus Conference statement, which outlines drug selection algorithms and reviews the side-effect profile of these various individual agents (40).

1.3. Intrathecal Baclofen

Baclofen, a synthetic pre- and postsynaptic receptor type B ($GABA_B$) agonist (41), was developed in the 1920s to be used as an antiepileptic. Perceptive clinicians noted that despite its poor ability to control seizures, it significantly diminished muscle tonicity in those with concomitant spasticity. These findings led to the transformation of baclofen from a seizure treatment to its primary use today as an anti-spasmodic. Despite oral Baclofen having excellent absorption through the intestinal system, its hydrophilicity results in attenuated CNS penetration Moreover, raising oral levels of Baclofen ultimately leads to side effects consisting of sedation, confusion, urinary frequency, and insomnia. These limiting systemic side effects and pharmacodynamics make IT Baclofen (ITB) an ideal treatment of debilitating spasticity.

Penn et al. reported the initial use of ITB in 1984 (42). This was followed by a double-blinded study 5 years later involving adults with spinal spasticity secondary to multiple sclerosis or spinal injury (43). In these early series, authors described diminished clinical spasticity, as measured by the Ashworth score, of up to 95% in all patients treated and a concomitant decrease in the side effects associated with oral baclofen. Reports soon followed involving pediatric patients, describing similar results in the reduction of not only spasticity but also dystonias (44–47). ITB, now considered the gold standard for the treatment of spasticity, has also an added analgesic effect both centrally and periphally that further helps to assuage patient discomfort (48–50). As with IT treatment for chronic pain, ITB has been shown to be cost effective when used for more than 5 years and when taking into account improvement in the recipient's quality of life (51).

1.4. Patient Selection for Intrathecal Baclofen

As with IT pain management, stringent guidelines and input from multiple specialties are the foundation of sound patient selection and these minimizes complications. Commonly used selection criteria (52) is illustrated in **Table 6.1**.

Table 6.1
Indications for baclofen pump

Severe, chronic spasticity ($\geq$ 12 months) secondary to CNS lesion
Refractory to oral medication and/or severe side effects
Patent sub-arachnoid space to place catheter
Positive response to IT baclofen trial
No allergy to baclofen

Table 6.2
Ashworth spasticity grading scale

Ashworth score	Degree of muscle tone
1	No increase in tone
2	Slight increase in tone
3	More marked increase in tone, passive movement performed easily
4	Considerable increase in tone, passive movement are difficult
5	Affected limb is rigid in flexion or extension

Physical therapy evaluation is required for formal Ashworth scale (**Table 6.2**) assessment, which will provide a baseline exam for Baclofen trial response. ITB offers greater benefit to those with an Ashworth score of 3–5 in the upper and lower extremities, which indicates a diminished ability for both the care and function of the patient. Prior to any surgical procedure, expectations of treatment should be clearly explained to the patient's family to include increased flexibility, functionality, and greater ability for the care of the patient. Furthermore, it is important to explain the need and process of the reservoir refills and the considerations in the size of the pump to be placed. Improper expectations or implications of diminished compliance should be considered contraindications to pump placement.

Age ranges for ITB pump placement have expanded with time to include patients ever younger and older. Nevertheless, the use of baclofen pumps in children less than 4–5 years old is rarely warranted secondary to an adequate response to oral Baclofen. As the age range of patients treated with ITB has increased, so have the diseases and injuries treated. Spinal cord injury resulting in spastic diplegia or familial spastic paraparesis patients often benefit from ITB (53, 54). Interestingly, ITB is also effective in patients with severe spastic hemiplegia such that iatrogenic contralateral hypotonia is not induced.

While ITB has been an efficacious means of treating spasticity secondary to spinal cord injury, the most common indication has now evolved to spasticity of cerebral origin. In the treatment of di- or tetraplegic spasticity secondary to cerebral palsy or following cerebral infarction, traumatic brain injury (TBI) or multiple sclerosis (MS), ITB has proven to be equally effective (45–58). Furthermore, children with dystonia and concomitant spasticity of cerebral origin have displayed improvement with ITB as well. Albright et al. in 2001 described enhanced ease of care and quality of life improvement in 86%, and speech improvement in 33% of 77 patients who had undergone pump placement (59). It is imperative to differentiate dystonias resulting from brain injury versus primary dystonias, for the latter are best treated with deep brain stimulation (60). Patients with hemidystonia respond in a similar fashion to their spastic counterparts, displaying good outcomes. Expanding on the indications, recent reports describe effective treatment of post spinal cord injury priapism (61) as well as dysautonomias (62) with ITB.

Once the IT patient has been determined to be a good surgical candidate and sound expectations for the treatment have been accepted by the family and caregivers, the patient should undergo an ITB trial. This can be accomplished either by giving a single bolus via a lumbar puncture or by placing an external pump and catheter system, with infusion and monitoring over a matter of days. The trial ensues with sequential incremental test doses of 50, 75, and up to 100 μg as necessary followed by a physical therapy assessment within 2–4 h of placement. A bolus provides maximal benefit to the lower extremities at 1 h, the upper extremities at 2 h, and has maximal effects at 4 h with anti-spasmodic effects waning at 8–10 h (63, 45). An improvement in clinical spasticity following the trial, corresponding to a 2-point reduction in the Ashworth score (*see* **Table 6.2**), without any significant side effects, serves as adequate confirmation of ITB response. An internal pump and catheter system should then be placed. If lethargy is seen at the time of the diagnosis with the initial test bolus, the concentration of Baclofen solution infused into the pump should be diluted and the patients should be watched with vigilance for appropriate titration of ITB.

Pharmacologic side effects from ITB most often consist of lethargy with bradycardia and respiratory depression occurring less often. These symptoms are generally mild and can often be limited by appropriate titration of the ITB dosing. New onset seizures have been reported (64); however, most of these cases occurred after traumatic head injury and a report put forth by Buonaguro et al. described 13% reduction in 150 patients (65). Symptoms related to impotence and loss of libido, as well as descriptions of scoliosis exacerbation, are no longer believed to be significantly related to ITB (66–68).

1.5. Surgical Techniques

1.5.1. Presurgical Planning

Once the decision has been made to proceed to surgery, it is important to select the appropriate hardware. There are numerous factors that play a role in choosing an ever-growing number of pump options to include the size, weight, age as well as the expected drug dosing. A large majority of patients will receive a programmable pump system; non-programmable types are cheaper but should be only used in patients undergoing pump replacement after adequate pain control on a static medication dose. Recently, the Synchromed II has become available in a 40-ml pump (**Fig. 6.1**), but placement is limited in younger

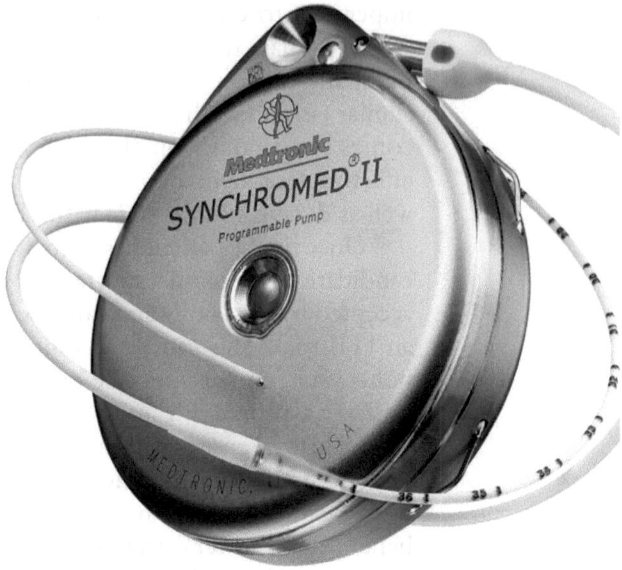

Fig. 6.1. Medtronic Synchromed II pump and one-piece sutureless catheter (Courtesy of Medtronic, Minneapolis, MN).

patients and those with minimal abdominal wall thickness. In most patients, the pump is generally placed in the abdominal subcutaneous adipose tissue or infrequently in the gluteal region. In thin patients who lack adequate space for pump placement, sub-fascial placement may be required for improved cosmesis as well as diminished incidence of wound and skin breakdown (69).

Once the pump type and size are selected, it is then appropriate to lend attention to the one- or two-piece catheter systems available (**Fig. 6.2**). Each system has its respective advantages: one-piece catheter is technically easier to implant for it requires fewer connections while the two-piece catheter offers a larger diameter, which is thought by some authors to be preventative against kinking, obstruction, and migration out of the IT space. Additionally, repair of IT catheter malfunction is easier with a two-piece system. Considering these factors, our preference is to use the two-piece catheter for physically active and ambulatory patients, with the belief that one-piece placement in active patients may be more likely to malfunction.

When planning the surgical implantation, it is also important to consider the location of the catheter tip. The location of the catheter largely depends on the disease state and the medication(s) infused (70). McCall and MacDonald reported that upper extremity spasticity scores improved and lower extremity remained unchanged when a catheter is positioned in the cervical region (71). Regarding pain patients, hydrophilic medications such as morphine diffuse throughout the spinal thecal space; however,

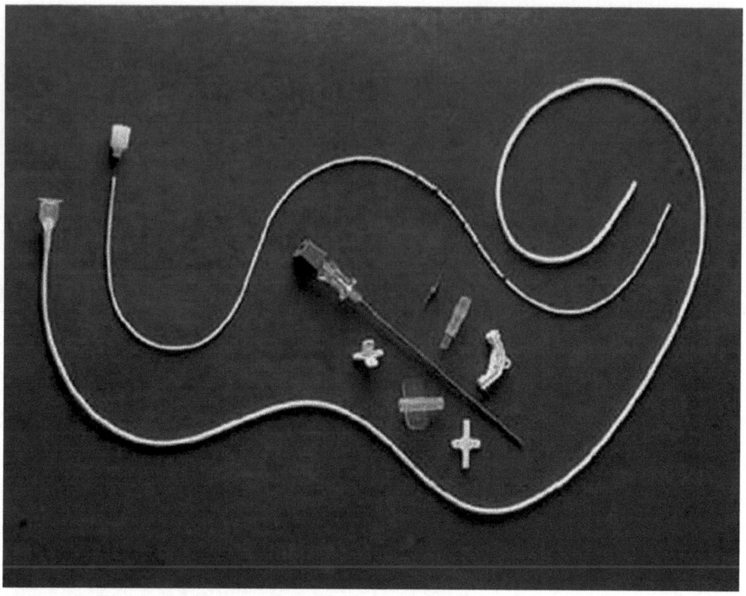

Fig. 6.2. Medtronic two-piece catheter system and accessories including InDura free-flow intrathecal catheter model 8711 (Courtesy of Medtronic, Minneapolis, MN).

lipophilic analgesics (fentanyl) rapidly penetrate spinal tissue near adjacent the catheter tip, thereby making tip placement adjacent to the pain level is paramount.

1.5.2. Implant Technique

Surgical implantation can be performed under local or general anesthesia, in the lateral decubitus position or prone position depending on where the pump will be placed (**Fig. 6.3**). We recommend fluoroscopic guidance for all catheter implants, for this goes to assure appropriate placement of the catheter tip without looping or kinking. The Tuohy needle supplied with the catheter is used to enter the lumbar IT space via an oblique paramedian approach under fluoroscopic guidance, which can facilitate catheter placement and minimize the incidence of catheter shear compared to placement at the midline through the interspinous ligament. Once CSF flow is noted, the catheter is then advanced to the appropriate level. Subsequently, a linear cut-down is made above and below the needle down to the lumbodorsal fascia. A purse-string stitch is then placed, using a non-absorbable suture, around the needle entry point to minimize catheter movement and CSF leakage. After removing the needle and stylet, CSF flow is confirmed at the proximal catheter and the purse-string suture is secured. Once again, CSF flow is confirmed, ensuring that the stitch did not obstruct the catheter. A silicone elastomer anchor is then used to affix the catheter to the fascia.

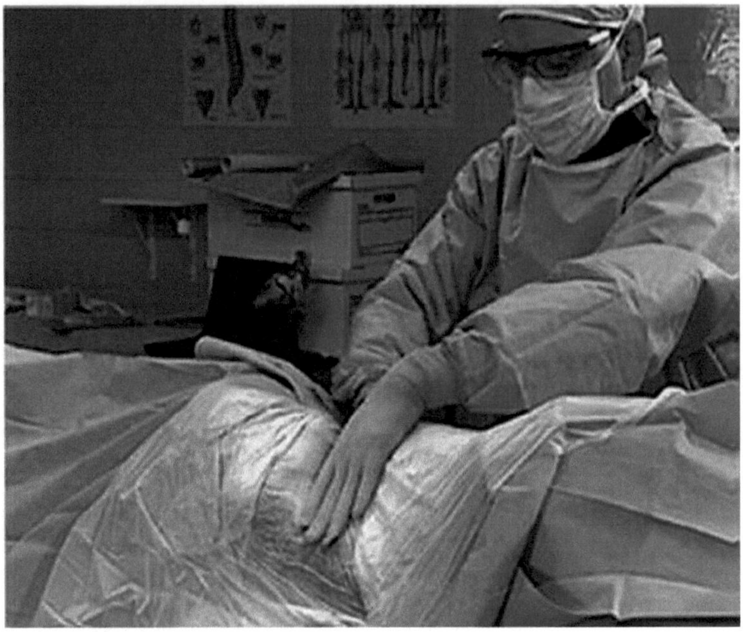

Fig. 6.3. Patient in lateral position, prepared and draped for placement of internal intrathecal system with placement of pump in abdominal wall (Courtesy of Medtronic, Minneapolis, MN).

After the catheter is anchored and CSF flow and a patent catheter verified, the pump pocket is made in the pre-determined location. We make certain that we mark the incision pre-operatively for there often is significant shifting of the soft tissue, especially in larger patients, after placing the patient in the lateral decubitus position. Furthermore, in obese individuals one should avoid pump placement too deep within the abdominal wall, as it may result in difficulty refilling. For subfascial placement, a plane is developed between the rectus fascia and muscle using blunt dissection. The semicircular line will need to be divided to ensure an adequately sized pocket.

Once the pocket has been formed, a catheter passer is then used to navigate the catheter from one incision to the other. When a two-piece catheter is used, the pieces are connected, and a plastic strain-relief sleeve is placed over the connection, which usually remains in the lumbar wound and is anchored to the fascia. The pump is then prepared appropriately, and may be placed in a Dacron pouch according to the surgeon's preference (although it is important to note that these will likely calcify, which can later impede refilling). The proximal catheter is then trimmed appropriately, leaving sufficient catheter coiled behind the pump as a strain relief to allow for normal movement. It is then connected to the pump with the appropriate connector, which may require one or more ties, depending on the particular connector used.

Dickerman and Schneider reported an increase in catheter fracture when the placement of the catheter connection site of the catheter and pump was placed near the iliac crest and recommend placement such that the neck of the pump is in the superiomedial position (72, 73). After a layered closure, the pump is programmed to deliver the appropriate priming dose and pre-determined daily dose (**Fig. 6.4**).

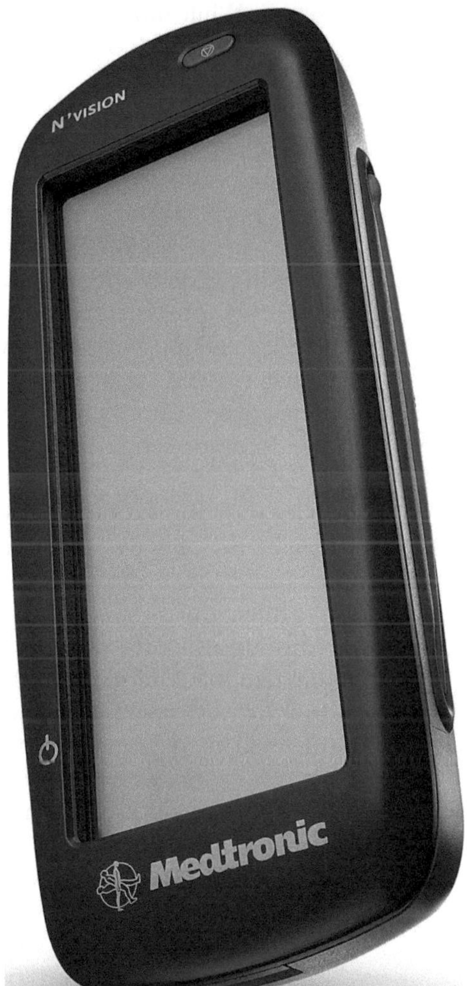

Fig. 6.4. Medtronic programmer for synchromed and synchromed II pumps (Courtesy Medtronic Corporation Minneapolis, MN).

Post-operative management is largely surgeon dependent; some authors support maintaining the patient in a flat position for 24 h as others view this operation as a same day procedure. The length of stay is largely dependent on the familiarity of the surgeon with that specific patient. Therefore, for all patients who are undergoing initial pump system placement, we feel it is best to monitor the patient in a hospital setting at least overnight before discharge. Pump rate is titrated to the patient's subjective relief of symptoms and corresponding medicinal side effects heretofore discussed.

1.5.3. Complications

Apart from incisional pain, which can often be reduced by local injection after closing the wound and followed by oral analgesics, patients may report positional headaches consistent with CSF leakage. While most often these symptoms are transient, occasionally an epidural blood patch, performed under fluoroscopy to avoid inadvertent catheter damage, may be indicated. Leakage of CSF through the lumbar wound indicates the need for urgent surgical exploration. The onset of acute non-incisional type pain may suggest compression of nerve roots, spinal cord, or both (74). Pump pocket seromas may arise and are usually treated conservatively with an abdominal binder. CSF collections can also occur following surgery at the catheter entry site and may track forward to the pump pocket. Again, conservative management is recommended and most often resolve over a period of weeks to months. Persistent CSF collections should require investigation into undiagnosed causes of increased CSF pressure (hydrocephalus or pseudotumor cerebri). Furthermore, one must always rule out a catheter leak or disconnect as a source of a postoperative fluid collection. Replacement of the IT catheter at a different level may be the only way to eliminate a persistent CSF fistula around the catheter exit site.

Infection of IT pump system will ordinarily require device removal, although certain authors have reported successful in situ treatment of infected pumps (75, 76). Malfunction of the drug delivery system is a clinical diagnosis. Confirmation can be obtained via radiography demonstrating a catheter disconnect or fracture. Aspiration of CSF via the catheter access port suggests that the catheter is patent and injection of myelographic contrast media can be performed to confirm delivery to the IT space. Aspiration of the pump reservoir can result in a larger quantity of drug than is predicted, indicating a blockage to flow and malfunction of the pump alarm. Although these, and other diagnostic studies, can be used as confirmation of a malfunction, unequivocal clinical symptoms of medication underdose or withdrawal, even in the presence of normal radiographic studies, warrants surgical exploration of the system. Abrupt cessation of IT baclofen therapy can result in life-threatening baclofen withdrawal, with symptoms of hyperthermia, tachycardia, hypertension, seizures, altered mental status, and psychomotor agitation (77, 78). Oral baclofen replacement is usually not successful, and the time to ITB pump revision or external catheter placement should be bridged with intravenous valium.

The use of concentrated analgesics, both narcotic and non-narcotic, is associated with the formation of catheter granulomas, inflammatory masses at the distal tip of the IT catheter that may cause pain or neurologic deficit secondary to spinal cord compression (79, 80). Clinicians should be vigilant for new onset radicular pain with or without sensory and/or motor deficits in these cases. Magnetic resonance imaging will display a mass at the catheter tip. If the patient has no significant motor deficit, the granuloma may resolve on its own if the therapy is discontinued or if the analgesic is diluted. Therefore, surgical exploration indicated only in acute neurologic deficit secondary to neural compression. Medication overdose can occur with either opioids or baclofen, due to injection next to the pump, into the side port when trying to refill it, due to changes in programming or malfunctions of the pump. Suspected drug overdose should be treated with immediate deactivation of the pump and administration of naloxone for narcotic agents and physostigmine for baclofen overdose, as well as airway protection with intubation if necessary. A high-volume lumbar puncture will facilitate rapid removal of remaining drug in the CSF space. Inadvertent subdural (as opposed to subarachnoid) placement of the catheter and drug delivery, followed by delayed communication between the two spaces, has also been reported as cause of overdose (81).

2. Intraventricular Pumps

In cases of diffuse (often in cancer patients) or specifically upper body pain, greater efficacy may be achieved by targeting opioid receptors in the brain (82). This is most commonly done with placement of a catheter into the frontal horn of the lateral ventricle (83). As previously described with IT catheter systems, medication(s) may be administered via bolus injections into a subcutaneous reservoir, or with the aid of an implanted programmable pump system (84). Surgical technique of an intraventricular pump system (vis-à-vis the device itself) is equivalent to that of an intraspinal device, save the supine operative position. Surgical complications likewise are in general the same, especially the risk of infection and the need for device explantation; additionally, one should be vigilant for intracerebral or intraventricular hemorrhage from catheter passage. Side effects of drug delivery are again equivalent, with possibly a higher incidence of nausea and respiratory depression (85). The technical ease of the procedure and the severity of disparate neurological disorders have led to protean human trials and animal studies involving intraventricular delivery. Patients with Alzheimer's disease and animal models of storage diseases and spongiform encephalopathies (such as Creutzfeldt–Jakob disease) (86–88) have been included in this. From these preliminary efforts one can see the possibility of neurosurgical therapies alleviating diseases now deemed untreatable and terminal.

3. Conclusion

Neuromodulation in the form of direct neuraxial drug delivery offers excellent therapeutic benefit and established cost efficiency to properly selected patients. Patients with pain of malignant or other origin, chronic neurologic disease or acute injury leading to spasticity, and other disease states noted above may realize significant benefit from direct CNS drug delivery. Using sound surgical techniques, the complications of implanting these devices are minimal and well tolerated, resulting in dramatic improvement in quality of life and overall function.

References

1. Belverud S, Mogilner A, and Schulder M. Intrathecal therapies. Neurotherapeutics Jan 2008;5(1):114–122.
2. Patel MM, Goyal BR, Bhadada SV, Bhatt JS et al. Getting into the brain: approaches to enhance brain drug delivery. CNS Drugs 2009;23(1):35–58.
3. Hazard RG, Haugh LD, Green PA et al. Chronic low back pain: the relationship between patient satisfaction and pain, impairment and disability outcomes. Spine 1994;19:81–87.
4. Arnstein P, Caudill M, Mandle CL et al. Self efficacy as a mediator of the relationship between pain intensity, disability and depression in chronic pain patients. Pain 1999;80:483–491.
5. Schmitz U, Saile H, and Nilges P. Coping with chronic pain: flexible goal adjustment as an interactive buffer against pain-related distress. Pain 1996;67:41–51.
6. Zhang X and Bao L. The development and modulation of nociceptive circuitry. Curr Opin Neruobiol 2006 Aug;16(4): 460–466.
7. Pereira J, Lawlor P, Vigano A et al. Equianalgesic dose ratios for opioids. A critical review and proposals for long-term dosing. J Pain Symptom Manage 2001 Aug;22(2): 672–687.
8. Lindsay SM. The conversion challenge: from intrathecal to oral morphine. Am J Hosp Palliat Care 2004;21:143–147.
9. Brill S, Gurman GM, and Fisher A. A history of neuraxial administrationof local analgesics and opioids. Eur J Anaesthesiol 2003;20: 682–689.
10. Pert CB and Snyder S. Opiate receptor: demonstration in nervous tissue. Science 1973;179(77):1011–1014.
11. Terenius L. Characteristics of the “receptor” for narcotic analgesics in synaptic plasma membrane fractions from rat brain. Acta Pharmacol Toxicol (Copenh) 1973;33(5): 377–384.
12. Yaksh TL and Rudy TA. Studies on the direct spinal action of narcotics in the production of analgesia in the rat. J Pharmacol Exp Ther 1977;202(2): 411–428.
13. Wang JK, Nauss LA, and Thomas JE. Pain relief by intrathecally applied morphine in man. Anesthesiology 1979;50(2): 149–151.
14. Rico RC, Hobika GH, Avellanosa AM, Trudnowski RJ, Rempel J, and West CR. Use of intrathecal and epidural morphine for pain relief in patients with malignant diseases: a preliminary report. J Med 1982;13: 223–231.

15. Leavens ME, Hill CS, Jr, Cech DA, Weyland JB, and Weston JS. Intrathecal and intraventricular morphine for pain in cancer patients: initial study. J Neurosurg 1982; 56:241–245.
16. Foley KM. Treatment of cancer pain. N Engl J Med 1985;313(2):84–95.
17. Brogan SE. Intrathecal therapy for the management of cancer pain. Curr Pain Headache Rep 2006;10:254–259.
18. Auld AW, Maki-Jokela A, and Murdoch DM. Intraspinal narcotic analgesia in the treatment of chronic pain. Spine 1985; 10:777–781.
19. Krames ES. Intrathecal infusional therapies for intractable pain: patient management guidelines. J Pain Symptom Manage 1993; 8(1):36–46.
20. Kanoff RB. Intraspinal delivery of opiates by an implantable, programmable pump in patients with chronic, intractable pain of nonmalignant origin. J Am Osteopath Assoc 1994;94(6):487–493.
21. Hassenbusch SJ, Stanton-Hicks M, Covington EC et al. Long-term intraspinal infusions of opiods in the treatment of neuropathic pain. J Pain Symptom Manage 1995;10(7):527–543.
22. Winkelmuller M and Winkelmuller W. Long-term effects of continuous intrathecal opioid treatment in chronic pain of nonmalignant etiology. J Neurosurg 1996;85(3): 458–467.
23. Paice JA, Penn RD, and Shott S. Intraspinal morphine for chronic pain: a retrospective, multicenter study. J Pain Symptom Manage 1996;11:71–80.
24. Nitescu P, Dahm P, Appelgren L, and Curelaru I. Continuous infusion of opioid and bupivacaine by externalized intrathecal catheters in long-term treatment of "refractory" nonmalignant pain. Clin J Pain 1998; 14:17–28.
25. Staats PS, Yearwood T, Charapata SG et al. Intrathecal ziconotide in the treatment of refractory pain in patients with cancer or AIDS: a randomized trial. J Am Med Assoc 2004 Jan 7;291(1):63–70.
26. Jain KK. An evaluation of intrathecal ziconotide for the treatment of chronic pain. Expert Opin Invastig Drugs 2000 Oct;9(10): 2403–2410.
27. Webster LR, Fisher R, Charapata S, and Wallace MS. Long-term intrathecal ziconotide for chronic pain: an open label study. J Pain Symptom Manage 2008; Aug 19.
28. Kaplan KM and Brose WG. Intrathecal methods. Neurosurg Clin N Am 2004;15: 289–296.
29. Meuser T, Pietruck C, Radbruch L et al. Symptoms during cancer pain treatment following WHO guidelines: a longitudinal follow-up study of symptoms prevalence, severity, and etiology. Pain 2001;93: 247–257.
30. Penn RD. Neuraxial analgesic administration. In: Neurosurgical pain manage. Elsevier Saunders, Philadelphia, PA, 2004, 150–155.
31. Follett KA and Doleys DM. Selection or candidates for intrathecal drug administration to treat chronic pain: considerations in pre-implantation trials. Medronic, Inc., Minneapolis, MN, 2002.
32. Hassenbusch SJ, Paice JA, Patt RB et al. Clinical realities and economic considerations: economics of intrathecal therapy. J Pain Symptom Manage 1997;14(3 Suppl):S36–S48.
33. France JC, Jorgenson SS, Lowe TG, and Dwyer AP. The use or intrathecal morphine for analgesia after posterolateral lumbar fusion: a prospective, double-blind, randomized study. Spine 1997;22:2272–2277.
34. Penn RD and Paice JA. Chronic intrathecal morphine for intractable pain. J Neurosurg 1997;67:182–186.
35. Jacobson L, Chabal C, and Brody MC. A dose – response study of intrathecal morphine: efficacy, duration, optimal dose and side-effects. Anesth Analg 1988;67: 1082–1088.
36. Uppal GS, Haider TT, Dwyer A, and Uppal JA. Reversible urinary retention secondary to excessive morphine delivered by an intrathecal morphine pump. Spine 1994;19: 719–720.
37. Anderson VC, Cooke B, and Burchiel KJ. Intrathecal hydromorphone for chronic nonmalignant pain: a retrospective study. Pain Med 2001;2:287–297.
38. Abs R, Verhelst J, Maeyaert J et al. Endocrine consequences of long-term intrathecal administration of opioids. J Clin Endocrinol Metab 2000;85:2215–2222.
39. Aldrete JA and Couto da Silva JM. Leg edema from intrathecal opiate infusions. Eur J Pain 2000;4:361–365.
40. Hassenbusch SJ, Portenoy RK, Cousins M et al. Polyanalgesic Consensus Conference 2003: an update on the management of pain by intraspinal drug delivery – report of an expert panel. J Pain Symptom Manage 2004;27:540–563.

41. Zieglgänsberger W, Howe JF, and Sutor B. The neuropharmacology of baclofen. In: Müller H, Zierski J, Penn RD, editors. Local-spinal therapy of spasticity. Springer-Verlag, Berlin, 1988, 37–49.
42. Penn RD and Kroin JS. Intrathecal baclofen alleviates spinal cord spasticity. Lancet 1984;1(8385):1078.
43. Penn RD, Savoy SM, Corcos D et al. Intrathecal baclofen for severe spinal spasticity. N Engl J Med 1989;320:1517–1521.
44. Müller H. Treatment of severe spasticity: results of multi-center trial conducted in Germany involving the intrathecal infusion of by an implantable drug delivery system. Dev Med Child Neurol 1992;34:739–745.
45. Albright AL, Cervi A, and Singletary J. Intrathecal baclofen for spasticity in cerebral palsy. J Am Med Assoc 1991;265(11): 1418–1422.
46. van Hilten BJ, van de Beek WT, Hoff JI et al. Intrathecal baclofen for the treatment of dystonia in patients with reflex sympathetic dystrophy. N Engl J Med 2000;343(9): 625–630.
47. Taricco M, Adone R, Pagliacci C et al. Pharmacological interventions for spasticity following spinal cord injury. Cochrane Database Syst Rev 2000;2:CD00131.
48. Herman RM, D'Luzansky JC, and Ippolito R. Intrathecal baclofen suppresses central pain in patients with spinal lesions. Clin J Pain 1992;8(4):338–345.
49. Loubser PG and Akman NM. Effects of intrathecal baclofen on chronic spinal cord injury pain. J Pain Symptom Manage 1996;12(4):241–247.
50. Young RR and Delwaide PJ. Drug therapy: spasticity (second of two parts). N Engl J Med 1981;304:96–99.
51. de Lissovoy G, Matza LS, Green H, Werner M, and Edgar T. Cost-effectiveness of intrathecal baclofen therapy for the treatment of severe spasticity associated with cerebral palsy. J Child Neurol 2007 Jan;22(1): 49–59.
52. Coffey RJ, Cahill D, Steers W et al. Intrathecal baclofen for intractable spasticity of spinal origin: results of a long-term multicenter study. J Neurosurg 1993;78: 226–232.
53. Penn RD. Intrathecal baclofen for spasticity of spinal origin: seven years of experience. J Neurosurg 1992 Aug;77(2):236–240.
54. Taricco M, Adone R, Pagliacci C, and Telaro E. Pharmacological interventions for spasticity following spinal cord injury. Cochrane Database Syst Rev 2000;2: CD001131.
55. Armstrong RW, Steinbok P, Cochrane DD et al. Intrathecally administered baclofen for the treatment of children with spasticity of cerebral origin. J Neurosurg 1997;87: 409–414.
56. Ivanhoe CB, Francisco GE, McGuire JR et al. Intrathecal baclofen management of poststroke spastic hypertonia: implications for function and quality of life. Arch Phys Med Rehabil 2006;87:1509–1515.
57. Ben Smail D, Peskine A, Roche N et al. Intrathecal baclofen for treatment of spasticity of multiple sclerosis patients. Mult Scler 2006;12:101–103.
58. Francisco GE, Hu MM, Boake C, and Ivanhoe CB. Efficacy of early use of intrathecal baclofen therapy for treating spastic hypertonia due to acquired brain injury. Brain Inj 2005 May;19(5):359–364.
59. Albright AL, Barry MJ, Shafton DH, and Ferson SS. Intrathecal baclofen for generalized dystonia. Dev Med Child Neurol 2001;43:652–657.
60. Albright AL. Neurosurgical treatment of spasticity and other pediatric movement disorders. J Child Neurol 2003;18 (Suppl 1): S67–S78.
61. D'Aleo G, Rifici C, Kofler M et al. Favorable response to intrathecal, but not oral, baclofen of priapism in a patient with spinal cord injury. Spine 2009 Feb 1;34(3): E127-E129.
62. Cuny E, Richer E, and Castel JP. Dysautonomia syndrome in the acute recovery phase after traumatic brain injury: relief with intrathecal baclofen therapy. Brain Inj 2001 Oct;15(10):917–925.
63. Muller H, Zierski J, Dralle D et al. Pharmacokinetics of intrathecal baclofen. In: Muller H, Zierski J, Penn RD, editors. Local-spinal therapy of spasticity. Springer-Verlag, Berlin, 1988, 155–214.
64. Kofler M, Kronenberg MF, Rifici C et al. Epileptic seizures associated with intrathecal baclofen application. Neurology 1994;44: 25–27.
65. Buonaguro V, Scelsa B, Curci D et al . S. Epilepsy and intrathecal baclofen therapy in children with cerebral palsy. Pediatr Neurol 2005;33:110–113.
66. Jones ML, Leslie DP, Bilsky G, and Bowman B. Effects of intrathecal baclofen on perceived sexual functioning in men with spinal cord injury. J Spinal Cord Med 2008;31(1):97–102.

67. Senaran H, Shah SA, Presedo A et al. The risk of progression of scoliosis in cerebral palsy patients after intrathecal baclofen therapy. Spine 2007 Oct 1;32(21):2348–2354.
68. Shilt JS, Lai LP, Cabrera MN et al. The impact of intrathecal baclofen on the natural history of scoliosis in cerebral palsy. J Pediatr Orthop 2008 Sep;28(6):684–687.
69. Kopell BH, Sala D, Doyle WK et al. Subfascial implantation of intrathecal baclofen pumps in children: technical note. Neurosurgery 2001;49:753–756.
70. Albright AL, Turner M, and Pattisapu JV. Best-practice surgical techniques for intrathecal baclofen therapy. J Neurosurg 2006; 104:233–239.
71. McCall TD and MacDonald JD. Cervical catheter tip placement for intrathecal baclofen administration. Neurosurgery 2006;59: 634–640.
72. Dickerman RD and Schneider SJ. Recurrent intrathecal baclofen pump catheter leakage: A surgical observation with recommendations. J Pediatr Surg 2002 Jun;37(6):E17.
73. Dickerman RD, Stevens QE, and Schneider SJ. The role of surgical placement and pump orientation in intrathecal pump system failure: A technical report. Pediatr Neurosurg 2003 Feb;38(2):107–109.
74. Huntoon MA, Hurdle MF, Marsh RW, and Reeves RK. Intrinsic spinal cord catheter placement: implications of new intractable pain in a patient with a spinal cord injury. Anesth Analg 2004;99:1763–1765.
75. Atiyeh BS, Hayek SN, Skaf GS, Al Araj A, and Chamoun RB. Baclofen pump pocket infection: a case report of successful salvage with muscle flap. Int Wound J 2006;3: 23–28.
76. Boviatsis EJ, Kouyialis AT, Boutsikakis I et al. Infected CNS infusion pumps. Is there a chance for treatment without removal? Acta Neurochir (Wien) 2004;146:463–467.
77. Shirley KW, Kothare S, Piatt JH, Jr, and Adirim TA. Intrathecal baclofen overdose and withdrawal. Pediatr Emerg Care 2006;22:258–261.
78. Greenberg MI and Hendrickson RG. Baclofen withdrawal following removal of an intrathecal baclofen pump despite oral baclofen replacement. J Toxicol Clin Toxicol 2003;41:83–85.
79. Wadhwa RK, Shaya MR, and Nanda A. Spinal cord compression in a patient with a pain pump for failed back syndrome: a chalk-like precipitate mimicking a spinal cord neoplasm: case report. Neurosurgery 2006; 58:E387.
80. Bejjani GK, Karim NO, and Tzortzidis F. Intrathecal granuloma after implantation of a morphine pump: case report and review of the literature. Surg Neurol 1997;48: 288–291.
81. Lew SM, Psaty EL, and Abbott R. An unusual cause of overdose after baclofen pump implantation: case report. Neurosurgery 2005;56:E624.
82. Lazorthes Y, Verdié JC, Bastide R et al. Spinal versus intraventricular chronic opiate administration with implantable drug delivery devices for cancer pain. Appl Neurophysiol 1985;48:234–241.
83. Lazorthes YR, Sallerin BAM, and Verdié JCP. Intracerebroventricular administration of morphine for control of irreducible cancer pain. Neurosurgery 1995;37: 422–429.
84. Dennis GC and DeWitty RL. Long-term intraventricular infusion of morphine for intractable pain in cancer of the head and neck. Neurosurgery 1990;26:404–407.
85. Karavelis A, Foroglou G, Selviaridis P, and Fountzilas G. Intraventricular administration of morphine for control of intractable cancer pain in 90 patients. Neurosurgery 1996;39:57–61.
86. Baskin DS, Browning JL, Pirozzolo FJ et al. Brain choline acetyltransferase and mental function in Alzheimer disease. Arch Neurol 1999;56:1121–1123.
87. Passini MA, Watson DJ, Vite CH et al. Intraventricular brain injection of adeno-associated virus type 1 (AAV1) in neonatal mice results in complementary patterns of neuronal transduction to AAV2 and total long-term correction of storage lesions in the brain of ß-glucuronidase-deficient mice. J Virol 2003;77:7034–7040.
88. Dohura K, Ishikawa K, Murakami-Kubo I et al. Treatment of transmissible spongiform encephalopathy by intraventricular drug infusion in animal models. J Virol 2004;78:4999–5006.

Chapter 7

Intracarotid Drug Delivery: Revisiting the Past or Looking into the Future

Shailendra Joshi

Abstract

Intracarotid drug delivery was extensively investigated in the past; however, the approach was largely abandoned due to inconsistent benefits. Yet, intracarotid drug delivery is anecdotally used for treating a variety of brain diseases. The rapid advances in endovascular techniques now require a renewed evaluation of this therapeutic approach. This review describes the pharmacological principles, applications, and pitfalls of intraarterial drug delivery to brain tissue.

Key words: Blood–brain barrier, drug delivery to brain, endovascular procedures, intraarterial drugs, intracarotid drugs.

1. Introduction

The rationale behind intracarotid drug delivery is fairly straightforward. Intracarotid drug delivery uses the body's physiological network of capillaries to deliver drugs regionally to brain tissue. The adaptive architecture of the capillary network in any tissue is finely tuned to ensure nutrient delivery to each cell. Thus, the distance through which drugs have to diffuse after an intracarotid injection is fairly constant. However, this is where the simplicity ends. The injection of drug into an artery is a complex biomechanical phenomenon. The drug rapidly transits through the capillary network. Streaming of drug can generate unpredictable concentrations within the arterial distribution. The ultra-fast first pass kinetics of intracarotid drug delivery is hard to define by conventional pharmacokinetic models. Rapid changes in drug concentrations are beyond the temporal resolution of

K.K. Jain (ed.), *Drug Delivery to the Central Nervous System*, Neuromethods 45,
DOI 10.1007/978-1-60761-529-3_7, © Humana Press, a part of Springer Science+Business Media, LLC 2010

conventional tissue drug sampling techniques. Not surprisingly therefore in the past attempts understand intracarotid drug delivery were largely unsuccessful. Like many treatments for brain diseases that were promising in pre-clinical studies, intracarotid treatments too failed to make a significant impact in clinical trials (1). Unexplained neurological complications and relatively little improvement in outcome limited the applications of intracarotid drugs and the technique was virtually abandoned by the early 1990s.

The major hurdles in selectively delivery drugs to the brain have been the presence of a blood–brain barrier (BBB), the vulnerability of brain tissue to transient ischemia, and the devastating effects of even limited local brain injury. In contrast to other methods (**Fig. 7.1**), intracarotid delivery has a broad appeal. Intracarotid drug delivery could increase the effectiveness of other methods of selective drug delivery to the brain that are based on targeting specific molecules or transport mechanisms, by increasing their concentrations at the treatment site (**Table 7.1**).

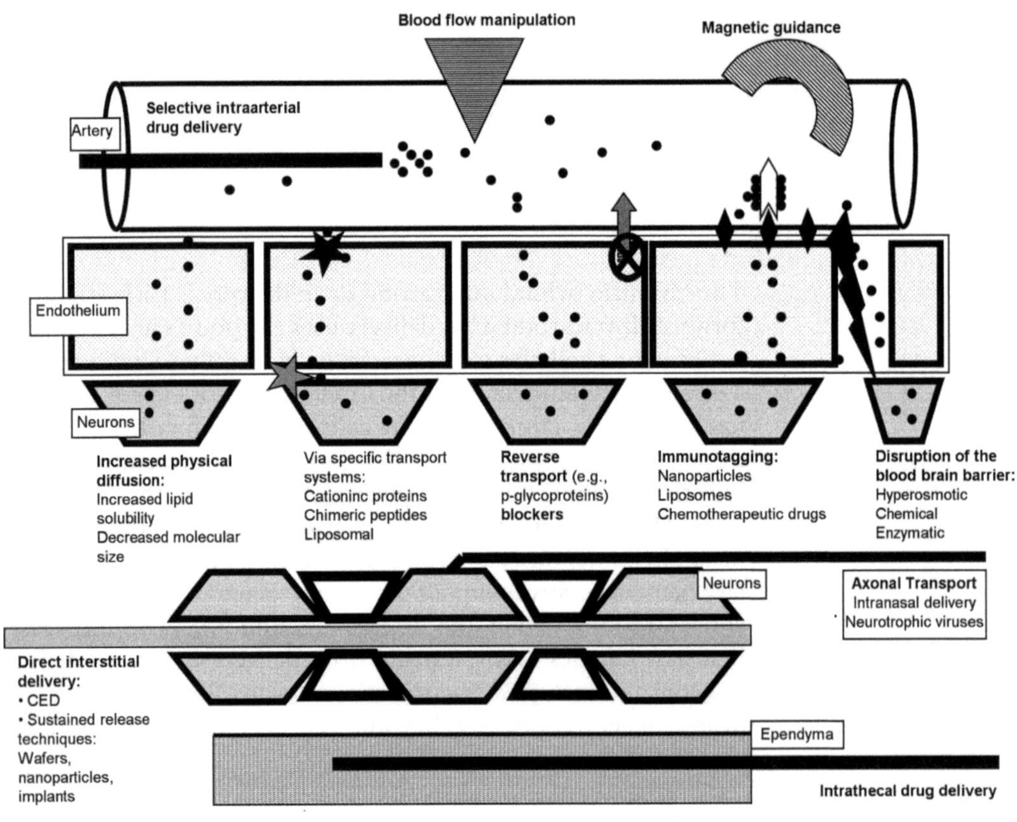

Fig. 7.1. Novel methods of brain selective drug delivery.

Table 7.1
A comparison of emerging methods for selective drug delivery to the brain

	Intravenous	Nanoparticles (75–81)	Liposomes (78, 80, 81, 82, 83, 84, 85, 86, 87)	P-gap modulators (86, 88, 89, 90)	Peptides-based delivery (76, 80, 91, 92)	Conventional intracarotid infusions (23, 93, 94, 95, 96, 97)
Route of delivery	Intravenous	Intravenous	Intravenous	Intravenous	Intravenous	Intracarotid
Method of brain selectivity	Drug design, increasing lipid solubility	Immuno-tagging; magnetic guidance	Immuno-tagging; BBB transport	BBB transport	BBB transport	± Enhanced systemic elimination
Onset of action	Moderate	Slow	Slow	Slow	Slow	Instantaneous
Maximum tissue concentrations	Moderate	Moderate-high	Moderate	Moderate	Moderate	High
Systemic neutralization	No	No	No	No	No	Yes
Limitations	Systemic side effects	Embolism; insufficient magnetic field may limit use	Embolism; may require immuno-targeting	Limited to certain drugs	Limited to certain drugs	Unexplained white matter lesions
Clinical experience	Extensive	None	Minimal	Minimal	Minimal	Extensive

2. Theoretical Effectiveness of Intracarotid Drugs

Figure 7.2 shows the simple kinetic framework described by Dedrick et al. (2) that can determine the advantage of intracarotid drug delivery. If C_1 and C_2 are regional and systemic drug concentrations, then the pharmacokinetic advantage of intracarotid (ic) over intravenous (iv) infusion (R) can be expressed as $R = (C_1/C_2)_{ic}/(C_2/C_1)_{iv}$. In this model $R = 1 +$ [Clearance total body/(regional blood flow (Q))*(1 – regional drug extraction (E))]. The analysis shows that the maximum advantage of intracarotid drug infusion is achieved by the following:

(i) drugs with high systemic clearance,

(ii) drugs with high regional drug extraction, and/or

(iii) low regional arterial blood flow.

There are several assumptions in this kinetic model proposed by Dedrick et al. These include uniform mixing of drugs in the blood, an assumption of free drug concentration that is based on steady-state drug–protein interactions, constant drug clearances over time, constant regional blood flows, and homogenous distribution within the arterial irrigation and tissue compartments. **Figure 7.3** shows the complexity of drug delivery in the brain tissue and the potential barriers to drug delivery. Considering the potential steps at each stage of drug transport to the target site, and time constants at each step, the current pharmacokinetic models

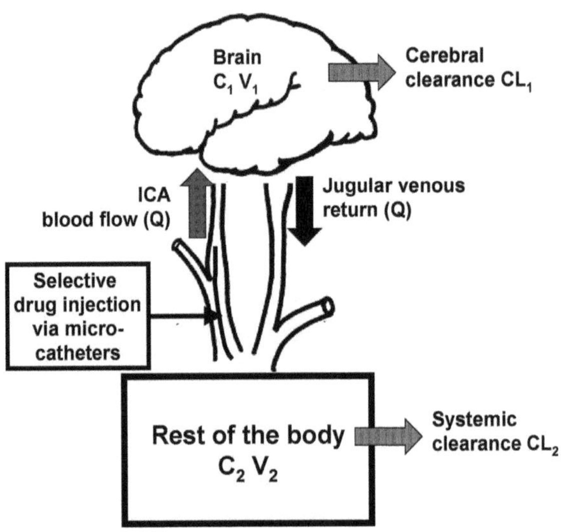

Fig. 7.2. Model of intracarotid drug delivery, C_1, C_2 and V_1, V_2 are concentrations and volumes in the brain (1) and remainder of the body (2), respectively. Q is the regional blood flow and CL_1 and CL_2 are cerebral and remaining body clearances. Modern microcatheters can restrict intraarterial interventions to tumor tissue.

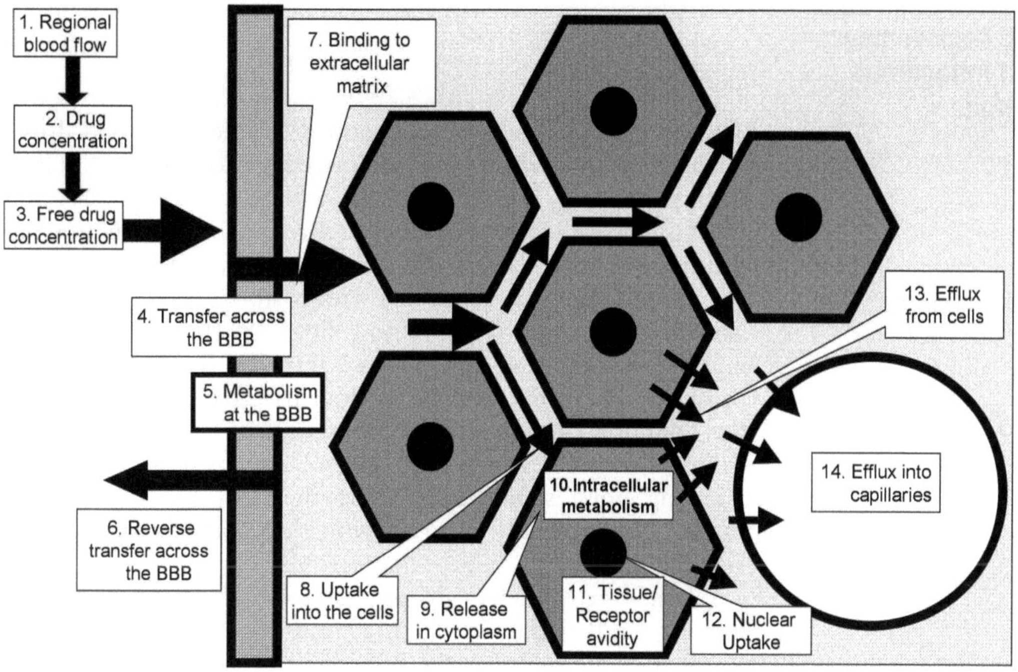

Fig. 7.3. Micropharmacokinetic factors in intracarotid drug delivery.

can at best describe the broad correlations but they fail to predict drug concentrations. Jones et al. compared the uptake of different benzodiazepines after intracarotid bolus injections in rats. They found that in vitro modeling *grossly underestimated* the in vivo uptake of drugs. They observed that free drug concentrations after intracarotid bolus injections were 3- to 60-fold greater in vivo than those predicted by in vitro parameters (3).

3. Advantages of Intracarotid Drugs

Intracarotid drug delivery has several unique advantages including the generation of high tissue concentrations at a specific site with a small total dose of the drug (2, 4). Intracarotid injections can instantaneously achieve therapeutic concentrations at the target site. The side effects of intracarotid drugs can be decreased by their physical removal from the jugular venous return (5). They can also be minimized by the administration of an intravenous antidote concurrent with intracarotid drug delivery. Intracarotid drugs are relatively safe evidenced by over 50 years of experience with the intracarotid administration of anesthetic, amytal, for neurological testing without any significant complications (6, 7).

4. Disadvantages of Intracarotid Drugs

The dose advantage of intracarotid drugs decline when drugs need to be infused over a prolonged period of time. High resting cerebral blood flow (CBF) considerably decreases the effectiveness of intracarotid drug delivery because it dilutes the drug and predisposes to drug maldistribution due to streaming. Streaming typically occurs with a slow rate of drug infusion relative to regional blood flow (8–10). Streaming is increased when drugs are injected into the distal branches of the internal carotid artery compared to proximal injections in the common carotid artery. Streaming can be decreased with drug infusion rates exceeding 20% of arterial flow volume, by injecting boluses of drugs during the diastolic phase of the cardiac cycle or with catheters with side ports for drug injection (9, 11).

Although a primary driving force for the development of intracarotid therapies was for the treatment of brain neoplasms, it was soon realized that the BBB prevented effective transfer of most chemotherapeutic agents (12). To overcome this limitation intracarotid mannitol has been used for over two decades to disrupt the BBB. Several alternatives to mannitol including cereport, leukotrienes, and acyl-glycerols are currently under development that may prove to be more selective for disruption of the blood–tumor barrier (13–16).

5. Recent Technical Improvements That Facilitate Intracarotid Drug Delivery

Multiple recent advances offer specific promise to overcome previous limitations realized with intracarotid drug delivery. These include the following: (i) development of microcatheters and small balloon-directed occluding catheters which target small regions of the brain compared to previous larger catheters that targeted the entire brain (17); (ii) development of flow directed catheters that are made of less thrombogenic materials (18); (iii) development of drugs that prevent platelet activation (e.g., aspirin, ticlopidine, clopidogrel) and strategies that minimize the risk of thromboembolism (19–21) with newer anticoagulant drugs including hiuridin; (iv) recognition and prevention of air embolism which is directly related to the rate of injection and inversely to the viscosity of the injected fluid by careful handling of injected drugs (22).

6. Strategies to Improve Intracarotid Drug Delivery

Several strategies have been developed to improve intracarotid drug delivery. (i) As the equation in **Fig. 7.2** shows the first attempts to improve intracarotid drug delivery used removal of chemotherapeutic drugs from the systemic blood by extracorporeal hemoperfusion. However, the method was invasive and although pharmacokinetically beneficial, it did not find clinical application. (ii) To circumvent the problem of streaming an effort was made to minimize it by injecting the drug as boluses during the diastolic phase of the cardiac cycle. However, bolus injections of drug rapidly transit the cerebral circulation and provide very minimal time for the dugs to diffuse into the brain parenchyma, thereby limiting the usefulness of the method. (iii) More recently a spatial dose-fractionating algorithm was proposed. The method calculates intracarotid drug dose based on the volume of tissue perfused which improves safety by minimizing the side effects (23). (iv) Our attempt at improving intracarotid drug delivery relies on bolus injection of drugs during states of carefully reduced blood flow.

In classical pharmacology blood flow is an independent parameter; however, in the clinical realm, blood flow to the brain is often transiently interrupted during high-risk surgery (24). Considering the fatal outcome of brain tumors, we believe that risks of manipulating blood flow to the brain are justified if there are tangible benefits to improve chemotherapy of brain tumors. To maximize drug delivery to the brain we have proposed transient reductions in blood flow and modulating bolus characteristic (25). Preliminary data suggest considerable increase the tissue concentrations or the pharmacological effects of the lipid soluble drugs when they are injected in states of reduced blood flow (26, 27). (v) In addition, to understand the complex kinetics of intracarotid drugs we are relying of novel high speed optical tissue drug concentration measurements using diffuse reflectance spectroscopy. The technique can track tissue concentration of drugs externally, without compromising tissue integrity (28). Non-invasive measurements are particularly important while investigating drug delivery to the brain because of changes in tissue diffusion characteristic by injury or inflammation. Although our optical technique is limited to drugs with an absorption spectrum distinct from hemoglobin and oxhemoglobin, it can be applied to several anticancer drugs (29, 30). Diffuse reflectance spectroscopy is capable of measuring drug concentration in less than 50 ms; such high-speed measurements will greatly advance the pharmacokinetics of intracarotid drug delivery (31).

7. Applications of Intracarotid Drugs

7.1. Functional Brain Mapping with Intracarotid Anesthetics

The most extensive experience with intracarotid drugs has been the injection of anesthetic drugs for the localization of brain functions, i.e., the Wada test. Anesthetic drugs that have been used for Wada test include amytal, methohexitone, and, recently, propofol (32–34). Compared to amytal, methohexitone has a shorter duration of effect, making it possible to test multiple arterial territories during the same procedure (35). The use of propofol has resulted in transient neurological complications in elderly patients with doses exceeding 20 mg (36). However, there is limited clinical justification for the selection of various intracarotid doses of anesthetic drugs for the Wada test. Typically a graded dose of amytal (75–125 mg) mixed with contrast is selectively administered until there is an upper limb drift. Speech, language, and memory functions are assessed as the deficit resolves. Excessively high doses of 250–300 mg of amytal may be required for neurological testing. Such a dose is sufficient to affect the contralateral hemisphere on recirculation therefore prevents proper localization of neurological functions (37).

7.2. Intracarotid Chemotherapy

The clinical indications for intracarotid chemotherapy are presently limited to the treatment of high-grade gliomas, CNS lymphomas, germ cell tumors, and metastatic brain tumors (38, 39). Nitrosourea derivatives including BCNU, and ACNU and serCNU which have high lipid solubility are better suited for intracarotid delivery due to the high first pass extraction, short elimination half life, and rapid onset of drug effects (40, 41). Other chemotherapeutic drugs including methotrexate and carboplatin require disruption of the BBB (38, 42). Removal of chemotherapeutic drugs using extracorporeal hemoperfusion of the jugular venous return can improve selective CNS drug delivery (5, 43, 44). Systemic rescue by intravenous antidotes, e.g., thiosulphate, is used to minimize the side effects of intracarotid cisplatin, such as nephrotoxity (45, 46). However, despite initial response to treatment the impact on long-term survival of patients with intracarotid chemotherapy has been disappointing (47). Intracarotid chemotherapy has been associated with unexplained white matter lesion (48). The modest impact on long-term survival, the invasive nature of the therapy, and potential for stroke have significantly limited the use of intracarotid chemotherapy.

7.3. Intracarotid Vasodilator Therapy for Cerebral Vasospasm

Intracarotid *papaverine* has been the mainstay of such treatment, but other drugs including nimodipine, nicardipine, amrinone, milrinone, mannitol, verapamil, and prostaglandin E have also been used intraarterially for the treatment of vasospasm (**Table 7.2**).

Table 7.2
Intracarotid vasodilators for the treatment of cerebral vasospasm

Author	Diagnosis subjects (n)	Intracarotid drug	Significant complication	Outcome	References
Fraticelli et al. (2008)	SAH vasospasm (24)	Milrinone		53±37% inprovement in angiographic diameter	(98)
Keuskamp et al. (2008)	SAH vasospasm (10)	Verapamil (20 mg+)	No adverse hemodynamic effects	Improved	(99)
Suzuki et al. (2006)	SAH vasospasm (2)	Cloforsin		Improved	(51)
Feng et al. (2002)	SAH vasospasm (29)	Verapamil	No hemodynamic complications	Improved neurological outcome in 5/29	(100)
Fortin et al. (2000)	Catheter induced spasm (1)	Mannitol		Resolved	(101)
Yoshida et al. (1997)	SAH vasospasm (2)	Amrinone		Reversal of spasm	(102)
Kaku et al. (1992)	SAH vasospasm (10)	Papaverine with nicardipine		8/10 patients improved	(103)
Kassell et al. (1992)	SAH vasospasm (10)	Papaverine	Two had transient neurological deficits	8/12 anigiographically improved 4/12 clinically improved	(104)
Boker et al. (1985)	SAH+ vasospasm: (3)	Nimodipine	None	Reversal of spasm	(105)

Intracarotid vasodilator therapy is generally used in conjunction with the triple-H (hypertensive, hypervolemic, and hemodilution) therapy of cerebral vasospasm (49). Adenosine has been proposed for the treatment of vasospasm, and despite its significant vasodilator effects, it has not yet been used in clinical settings. The exceedingly short half-life of adenosine and its poor transfer across the blood–brain barrier limit its potential usefulness in settings of cerebral vasospasm that last several days (50). Anecdotally, intracarotid colforsin, an adenylyl cyclase stimulator, which causes vascular smooth muscle relaxation by increasing intracellular cyclic adenosine monophosphate has been used in treating cerebral vasospasm after subarachnoid hemorrhage (51). Intracarotid vasodilator therapy includes at least two significant risks: (a) cerebral steal and (b) an increase in intracranial pressure (ICP). Unless vasodilator therapy is restricted to the arteries afflicted by cerebral vasospasm, vasodilator therapy can result in cerebral steal, i.e., redistribution of blood flow away from the focal ischemic region due to vasodilation and enhanced perfusion in healthy vascular beds. Due to the potential increase in ICP it is desirable to monitor ICP during intracarotid vasodilator therapy (52).

7.4. Intracarotid Thrombolysis

Intravenous thrombolysis can achieve a significant degree of recanalization within 3 h of the onset of neurological symptoms (53). However, it is often difficult to complete thrombolytic therapy within such a narrow time window. Intracarotid delivery of thrombolytic drugs can extend the time window for successful intervention up to 6 h (54). Intracarotid urokinase, streptokinase and more recently recombinant tissue plasminogen activator have been used for such therapy (**Table 7.3**) (55). With the current treatment strategy neurological outcome is improved when primary recanalization can be achieved within 3 h of injury (56). The development of endovascular clot retrieving devices, laser, or ultrasonic clot lysis techniques, and the placement endovascular stents has supplemented or provided alternatives to intracarotid thrombolytic therapy that can further extend the time window for successful intervention up to 8 h (57–62). The primary advantage of mechanical disruption of clot is the rapid restoration of cerebral reperfusion compared to the slow dissolution of the clot by pharmacological means that could take an hour or more. The principle concern with intracarotid thrombolysis is the conversion of an ischemic infarction to a hemorrhagic infarction.

7.5. Treatment of Infections

Medical treatment of purulent meningitis and encephalitis is sometimes complicated by the inability to deliver sufficient antibiotics to the infected site (63). In experimental animals intracarotid hyperosmolar disruption of BBB has been shown to enhance the delivery of tobramycin and vancomycin to the brain (64). Even

Table 7.3
Intracarotid thrombolytic therapies for ischemic stroke

	Number of patients/ procedures	Time window	Drug and dose	Complications	Outcome	References
Nagel et al. (2009)	106 patients	Acute basilar artery stroke	IV abciximab and IA rt-PA		Improved outcome with bridging therapy	(106)
Khatri et al. (2005)	61 angiograms	Acute stroke	IV and IA rTPA		49–54% improved. Clinical outcome was related to degree of revascularization	(107)
Abou-Chebl et al. (2005)	12	3.8±2.2 h	IA rTPA, IA or IV abciximab, heparin, and stent	1/12 hemorrhagic infarction	11/12 good recovery	(108)
Sorimachi et al. (2005)	25		IA rTPA		21/25 recanalized. Clinical outcome depended on the distal pressure >30 mm Hg	(109)
Zaidat et al. (2005)	96	Acute stroke	IA or combined IV and IA rTPA	ICH 7.6–13.3%	24–69% recanalization which was a function of time to treatment. Clinical outcome was related to the improvement in perfusion after thrombolysis	(110)
Bourekas et al. (2004)	36	<3 h	IA	11% ICH	75% recanalization	(56)

(continued)

Table 7.3 (continued)

	Number of patients/procedures	Time window	Drug and dose	Complications	Outcome	References
Li et al. (2003)	19	Acute stroke	IA thrombolysis+ stent		18/19 No stenosis	(60)
Lee et al. (2002)	26	Acute stroke	IA urokinase alone (16), IA Urokinase with IV abciximab		Abciximab increased the recanalization rate and decreased the urokinase dose	(111)
Rabinstein et al. (2002)	1	<3hr	IA-rTPA		Complete recovery	(112)
Keris et al. (2001)	12 Treaterd 33 controls	<6 h	rTPA (actilyse) 50 mg IA initially and IV subsequently	ICH	Good outcome in 66% in treatment group vs 21% in control	(54)
Fukuda et al. (2001)	5	< 6 h	IA thrombolysis in 3/5		Recanalization achieved in all 3	(113)
Kase et al. (2001)	120 of the 180 patients	<6 h	IA pro-urokinase	Symptomatic ICH in 10.9% with 83% mortality		(114)
Moazami et al. (2001)	13	<6 h			Canalization artial to completer in 6/13 and neurological improvement in 5/13	(115)
Sylaja et al. (2001)	5		IA-urokinase		Two patients treated within 3 h had complete recovery	(116)

Suwanwela et al. (2001)	2	<6 h	IA-rTPA		2/2 Angiographic but not clinical recovery	(117)
Endo et al. (1998)	33	< 6 h	IA thrombolysis		Good prognosis with treatment <3 h followed by stent or angioplasty	(118)
Jahan et al. (1999)	26	<6 h	IA TPA	Hemorrhage in 10/26	11/26 Partial or complete recanalization, 9 had good outcome. 15/26 little or no recanalization	(119)
Barr et al. (1994)	12	<8 h	Urokinase (160–500 KUnits)	3/12 Failed	9/12 recanalized	(120)
Casto et al. (1992)	5	<5 h	Urokinase 560 K units		4/5 Recanalized	(121)

without hypertonic disruption, intracarotid antibiotics have yielded beneficial results in treating intracranial infections (63, 65–67).

7.6. Treatment of Raised ICP

Yokota et al. used low-dose bilateral intracarotid infusion of mannitol to treat severely raised intracranial pressure in 18 human subjects with head trauma. They observed that intracarotid mannitol significantly decreased ICP and caused no hemodynamic side effects or disturbances in the electrolyte status (68).

8. Future Applications

Intracarotid gene therapy: Perhaps the most potent application of intracarotid drugs in the future is to deliver genes that could alter the course of a disease or help us understand disease processes. Intracarotid delivery of viral vectors, DNA liposomes, and stem cells has all been successfully demonstrated (69, 70). Selective delivery of Herpes viral vectors to tumor region by concurrent use of cereport (RMP-7) is one such example (71). Bone marrow stem cells have been successfully delivered to traumatized regions of the brain after intracarotid injection (72). Although the field of intracarotid gene therapy is in its infancy, it could play a key role in the future treatment of brain diseases due to degeneration, ischemia, trauma, or neoplasia. Intracarotid drug delivery promises to play a critical role in molecular neurosurgery and restorative neurosurgery (73, 74).

9. Conclusions

The rapid technical developments in the field of endovascular surgery can greatly improve intracarotid drug delivery to the brain and call for a re-evaluation of previously abandoned therapeutic approaches. Improvements in intracarotid drug delivery and emerging optical methods to monitor drug concentrations could greatly improve the safety and efficacy of the technique. Intracarotid drug delivery could be a force-multiplier for other brain selective methods of drug delivery such as liposomes, nanoparticles, and chimeric peptides. Combining the potential for gene and stem cell therapies with enhanced CNS selective drug design, and novel intracarotid drug delivery protocols, could greatly advance conventional and future therapies for brain diseases.

Acknowledgments

Acknowledgments are due to Ms. Mei Wang, MPH, Staff Researcher in the Department of Anesthesiology for help in preparing this manuscript, and to the funding support of National Institute of Heath NCI R01 127500 Grant, from the National Institute of Health, Bethesda, MD.

References

1. Del Zoppo GJ: Why do all drugs work in animals but none in stroke patients? 1. Drugs promoting cerebral blood flow [Review]. J Intern Med 1995; 237: 79–88.
2. Dedrick RL: Arterial drug infusion: pharmacokinetic problems and pitfalls. J Natl Cancer Inst 1988; 80: 84–9.
3. Jones DR, Hall SD, Jackson EK, Branch RA, and Wilkinson GR: Brain uptake of benzodiazepines: effects of lipophilicity and plasma protein binding. J Pharmacol Exp Ther 1988; 245: 816–22.
4. Fenstermacher J and Gazendam J: Intra-arterial infusions of drugs and hyperosmotic solutions as ways of enhancing CNS chemotherapy. Cancer Treat Rep 1981; 65: 27–37.
5. Oldfield EH, Clark WC, Dedrick RL, Egorin MJ, Austin HA, DeVroom HD, Joyce KM, and Doppman JL: Reduced systemic drug exposure by combining intraarterial cis-diamminedichloroplatinum(II) with hemodialysis of regional venous drainage. Cancer Res 1987; 47: 1962–7.
6. Wada DR, Harashima H, Ebling WF, Osaki EW, and Stanski DR: Effects of thiopental on regional blood flows in the rat. Anesthesiology 1996; 84: 596–604.
7. Fitzsimmons BF, Marshall RS, Pile-Spellman J, and Lazar RM: Neurobehavioral differences in superselective Wada testing with amobarbital versus lidocaine. Am J Neuroradiol 2003; 24: 1456–60.
8. Saris SC, Wright DC, Oldfield EH, and Blasberg RG: Intravascular streaming and variable delivery to brain following carotid artery infusions in the Sprague-Dawley rat. J Cereb Blood Flow Metab 1988; 8: 116–20.
9. Saris SC, Blasberg RG, Carson RE, deVroom HL, Lutz R, Dedrick RL, Pettigrew K, Chang R, Doppman J, Wright DC et al.: Intravascular streaming during carotid artery infusions. Demonstration in humans and reduction using diastole-phased pulsatile administration. J Neurosurg 1991; 74: 763–72.
10. Blacklock JB, Wright DC, Dedrick RL, Blasberg RG, Lutz RJ, Doppman JL, and Oldfield EH: Drug streaming during intra-arterial chemotherapy. J Neurosurg 1986; 64: 284–91.
11. Lutz RJ, Dedrick RL, Boretos JW, Oldfield EH, Blacklock JB, and Doppman JL: Mixing studies during intracarotid artery infusions in an in vitro model. J Neurosurg 1986; 64: 277–83.
12. Neuwelt EA, Glasberg M, Frenkel E, and Barnett P: Neurotoxicity of chemotherapeutic agents after blood-brain barrier modification: neuropathological studies. Ann Neurol 1983; 14: 316–24.
13. Black KL, Cloughesy T, Huang SC, Gobin YP, Zhou Y, Grous J, Nelson G, Farahani K, Hoh CK, and Phelps M: Intracarotid infusion of RMP-7, a bradykinin analog, and transport of gallium-68 ethylenediamine tetraacetic acid into human gliomas. J Neurosurg 1997; 86: 603–9.
14. Bhattacharjee AK, Nagashima T, Kondoh T, and Tamaki N: The effects of the Na(+)/Ca(++) exchange blocker on osmotic blood-brain barrier disruption. Brain Res 2001; 900: 157–62.
15. Cloughesy TF, Black KL, Gobin YP, Farahani K, Nelson G, Villablanca P, Kabbinavar F, Vineula F, and Wortel CH: Intra-arterial Cereport (RMP-7) and carboplatin: a dose escalation study for recurrent malignant gliomas. Neurosurgery 1999; 44: 270–8; discussion 278–9.
16. Lee HJ, Zhang Y, and Pardridge WM: Blood-brain barrier disruption following the internal carotid arterial perfusion of alkyl glycerols. J Drug Target 2002; 10: 463–7.

17. Pile-Spellman J and Young WL: Materials and methods for interventional neuroradiology (INR) (Chapter 18.2), Neuroradiology, 3rd ed. Taveras JM (eds). Baltimore: Williams & Wilkins, 1996, pp. 1047–73.
18. Kallmes DF, McGraw JK, Evans AJ, Mathis JM, Hergenrother RW, Jensen ME, Cloft HJ, Lopes M, and Dion JE: Thrombogenicity of hydrophilic and nonhydrophilic microcatheters and guiding catheters. Am J Neuroradiol 1997; 18: 1243–51.
19. Bendszus M and Stoll G: Silent cerebral ischaemia: hidden fingerprints of invasive medical procedures. Lancet Neurol 2006; 5: 364–72.
20. Young WL: Anesthesia for interventional neuroradiologic procedures. 1993: Presented in *Panel on Anesthesia at Remote Locations for Adults*, American Society of Anesthesiologists Annual Meeting, Washington, D.C., October 13
21. Aviv RI, O'Neill R, Patel MC, and Colquhoun IR: Abciximab in patients with ruptured intracranial aneurysms. Am J Neuroradiol 2005; 26: 1744–50.
22. Bendszus M, Koltzenburg M, Bartsch AJ, Goldbrunner R, Gunthner-Lengsfeld T, Weilbach FX, Roosen K, Toyka KV, and Solymosi L: Heparin and air filters reduce embolic events caused by intra-arterial cerebral angiography: a prospective, randomized trial. Circulation 2004; 110: 2210–15.
23. Gobin YP, Cloughesy TF, Chow KL, Duckwiler GR, Sayre JW, Milanese K, and Vinuela F: Intraarterial chemotherapy for brain tumors by using a spatial dose fractionation algorithm and pulsatile delivery. Radiology 2001; 218: 724–32.
24. Pile-Spellman J, Young WL, Joshi S, Duong H, Vang MC, Hartmann A, Kahn RA, Rubin DA, Prestigiacomo CJ, and Ostapkovich ND: Adenosine-induced cardiac pause for endovascular embolization of cerebral arteriovenous malformations: technical case report. Neurosurgery 1999; 44: 881–6; discussion 886–7.
25. Joshi S, Wang M, Etu JJ, Nishanian EV, and Pile-Spellman J: Cerebral blood flow affects dose requirements of intracarotid propofol for electrocerebral silence. Anesthesiology 2006; 104: 290–8; discussion 5A.
26. Joshi S, Wang M, Etu JJ, Nishanian EV, and Pile-Spellman J: Cerebral blood flow affects dose requirements of intracarotid propofol for electrocerebral silence. Anesthesiology 2006; 104: 290–8.
27. Joshi S, Wang M, Etu JJ, and Pile-Spellman J: Bolus configuration affects dose requirements of intracarotid propofol for electroencephalographic silence. Anesth Analg 2006; 102: 1816–22.
28. Kanick SC, Eiseman JL, Joseph E, Guo J, and Parker RS: Noninvasive and nondestructive optical spectroscopic measurement of motexafin gadolinium in mouse tissues: comparison to high-performance liquid chromatography. J Photochem Photobiol B 2007; 88: 90–104.
29. Mourant JR, Johnson TM, Los G, and Bigio IJ: Non-invasive measurement of chemotherapy drug concentrations in tissue: preliminary demonstrations of in vivo measurements. Phy Med Biol 1999; 44: 1397–417.
30. Bigio IJ and Bown SG: Spectroscopic sensing of cancer and cancer therapy: current status of translational research. Cancer Biol Ther 2004; 3: 259–67.
31. Reif R, Wang M, Joshi S, A'Amar O, and Bigio IJ: Optical method for real-time monitoring of drug concentrations facilitates the development of novel methods for drug delivery to brain tissue. J Biomed Opt 2007; 12: 034036.
32. Wada J and Rassmussen T: Intracarotid injection of sodium amytal for the lateralization of cerebral speech dominance: experimental and clinical observations. J Neurosurg 1960; 17: 266–82.
33. Coubes P, Baldy-Moulinier M, Zanca M, Boire JY, Child R, Bourbotte G, and Frerebeau P: Monitoring sodium methohexital distribution with [99mTc]HMPAO with single photon emission computed tomography during Wada test. Epilepsia 1995; 36: 1041–9.
34. Bazin JE, Picard P, Gabrillargues J, and Dordain M: Propofol administered via the carotid artery to achieve a Wada test. Can J Anaesth 1998; 45: 707–8.
35. Bacia T, Bendarzewska-Nawrocka B, Szymanska O, Dowzenko A, and Sadowski Z: A comparison of Wada test for cerebral speech dominance by use of amytal or brietal. Neurologia i Neurochirurgia Polska 1998; 32 Suppl 2: 247–61.
36. Takayama M, Miyamoto S, Ikeda A, Mikuni N, Takahashi JB, Usui K, Satow T, Yamamoto J, Matsuhashi M, Matsumoto R, Nagamine T, Shibasaki H, and Hashimoto N: Intracarotid propofol test for speech and memory dominance in man. Neurology 2004; 63: 510–5.

37. Grote CL, Wierenga C, Smith MC, Kanner AM, Bergen DC, Geremia G, and Greenlee W: Wada difference a day makes: interpretive cautions regarding same-day injections. Neurology 1999; 52: 1577–82.
38. Doolittle ND, Miner ME, Hall WA, Siegal T, Jerome E, Osztie E, McAllister LD, Bubalo JS, Kraemer DF, Fortin D, Nixon R, Muldoon LL, and Neuwelt EA: Safety and efficacy of a multicenter study using intraarterial chemotherapy in conjunction with osmotic opening of the blood-brain barrier for the treatment of patients with malignant brain tumors. Cancer 2000; 88: 637–47.
39. Shibui S: [Treatment for malignant glioma]. No To Shinkei 2005; 57: 1027–33.
40. Weller M, Muller B, Koch R, Bamberg M, and Krauseneck P: Neuro-Oncology Working Group 01 trial of nimustine plus teniposide versus nimustine plus cytarabine chemotherapy in addition to involved-field radiotherapy in the first-line treatment of malignant glioma. J Clin Oncol 2003; 21: 3276–84.
41. Kuratsu J and Ushio Y: [Chemotherapy of brain tumors]. Gan To Kagaku Ryoho 1994; 21 Suppl 3: 377–83.
42. Doolittle ND, Abrey LE, Ferrari N, Hall WA, Laws ER, McLendon RE, Muldoon LL, Peereboom D, Peterson DR, Reynolds CP, Senter P, and Neuwelt EA: Targeted delivery in primary and metastatic brain tumors: summary report of the seventh annual meeting of the Blood-Brain Barrier Disruption Consortium. Clin Cancer Res 2002; 8: 1702–9.
43. Oldfield EH, Dedrick RL, Chatterji DC, Yeager RL, Girton ME, Kornblith PL, and Doppman JL: Arterial drug infusion with extracorporeal removal. II. Internal carotid carmustine in the rhesus monkey. Cancer Treat Rep 1985; 69: 293–303.
44. Dedrick RL, Oldfield EH, and Collins JM: Arterial drug infusion with extracorporeal removal. I. Theoretic basis with particular reference to the brain. Cancer Treat Rep 1984; 68: 373–80.
45. Vance RB, Pittisapu J, and Kapp JP: Experiences with sodium thiosulfate after intracarotid infusion of cisplatin and BCNU for malignant gliomas. J Neurooncol 1986; 4: 151–4.
46. Kovacs AF: Maximized combined modality treatment of an unselected population of oral and oropharyngeal cancer patients. Final results of a pilot study compared with a treatment-dependent prognosis index. J Craniomaxillofac Surg 2006; 34: 74–84.
47. Stewart DJ, Grahovac Z, Hugenholtz H, DaSilva V, Richard MT, Benoit B, Belanger G, and Russell N: Feasibility study of intraarterial vs intravenous cisplatin, BCNU, and teniposide combined with systemic cisplatin, teniposide, cytosine arabinoside, glycerol and mannitol in the treatment of primary and metastatic brain tumors. J Neurooncol 1993; 17: 71–9.
48. Tonn JC, Roosen K, and Schachenmayr W: Brain necroses after intraarterial chemotherapy and irradiation of malignant gliomas – a complication of both ACNU and BCNU? J Neurooncol 1991; 11: 241–2.
49. Origitano TC, Wascher TM, Reichman OH, and Anderson DE: Sustained increased cerebral blood flow with prophylactic hypertensive hypervolemic hemodilution ("Triple-H" Therapy) after subarachnoid hemorrhage. Neurosurgery 1990; 27: 729–40.
50. Joshi S, Young WL, Pile-Spellman J, Duong DH, Vang MC, Hacein-Bey L, Lee HT, and Ostapkovich N: The feasibility of intracarotid adenosine for the manipulation of human cerebrovascular resistance. Anesth Analg 1998; 87: 1291–8.
51. Suzuki S, Ito O, Savama T, Yamaguchi S, Goto K, and Sasaki T: Intraarterial injection of cloforsin daropate hydrochloride for treatment of cerebral vasospasm after aneurysmal subarachnoid hemorrhage: preliminary report of two cases. Neuroradiology 2006; 48: 50–3.
52. Cross DT, III, Moran CJ, Angtuaco EE, Milburn JM, Diringer MN, and Dacey RG, Jr: Intracranial pressure monitoring during intraarterial papaverine infusion for cerebral vasospasm. Am J Neuroradiol 1998; 19: 1319–23.
53. Jahan R: Hyperacute therapy of acute ischemic stroke: intraarterial thrombolysis and mechanical revascularization strategies. Tech Vas Interv Radiol 2005; 8: 87–91.
54. Keris V, Rudnicka S, Vorona V, Enina G, Tilgale B, and Fricbergs J: Combined intraarterial/intravenous thrombolysis for acute ischemic stroke. Am J Neuroradiol 2001; 22: 352–8.
55. Mielke O, Wardlaw J, and Liu M: Thrombolysis (different doses, routes of administration and agents) for acute ischaemic stroke. [update of Cochrane Database Syst Rev. 2000;(2):CD000514; PMID: 10796381]. Cochrane Database Syst Rev 2004: CD000514
56. Bourekas EC, Slivka AP, Shah R, Sunshine J, and Suarez JI: Intraarterial thrombolytic

therapy within 3 hours of the onset of stroke. Neurosurgery 2004; 54: 39–44; discussion 44–6.
57. Mahon BR, Nesbit GM, Barnwell SL, Clark W, Marotta TR, Weill A, Teal PA, and Qureshi AI: North American clinical experience with the EKOS MicroLysUS infusion catheter for the treatment of embolic stroke. Am J Neuroradiol 2003; 24: 534–8.
58. Nesbit GM, Clark WM, O'Neill OR, and Barnwell SL: Intracranial intraarterial thrombolysis facilitated by microcatheter navigation through an occluded cervical internal carotid artery. J Neurosurg 1996; 84: 387–92.
59. Lutsep HL, Clark WM, Nesbit GM, Kuether TA, and Barnwell SL: Intraarterial suction thrombectomy in acute stroke. Am J Neuroradiol 2002; 23: 783–6.
60. Li SM, Miao ZR, Zhu FS, Wang ML, Hua Y, Dong Y, Song QB, and Ling F: Combined intraarterial thrombolysis and intra-cerebral stent for acute ischemic stroke institute of brain vascular diseases. Chung-Hua i Hsueh Tsa Chih [Chinese Medical Journal] 2003; 83: 9–12.
61. Behrens S, Spengos K, Daffertshofer M, Wirth S, and Hennerici M: Potential use of therapeutic ultrasound in ischemic stroke treatment. Echocardiography 2001; 18: 259–63.
62. Kerber CW, Barr JD, Berger RM, and Chopko BW: Snare retrieval of intracranial thrombus in patients with acute stroke. J Vasc Interv Radiol 2002; 13: 1269–74.
63. Domanskaia IA, Kukushkin AA, Mazurok II, and Il'in KL: Regionarnoe primenenie antibiotikov pri lechenii cherepno-mozgovoi travmy i gnoinykh oslozhnenii. Vestnik Khirurgii Imeni i - i - Grekova 1993; 150: 53–6.
64. Perkins BA and Strausbaugh LJ: Effect of mannitol infusions into the internal carotid artery on entry of two antibiotics into the cerebrospinal fluid and brains of normal rabbits. Antimicrob Agents Chemother 1983; 24: 339–42.
65. Kudriavtsev AE and Shikov VG: Primenenie karotidnoi infuzii lekarstvennykh veshchestv dlia lecheniia bol'nykh gnoinym meningitom. Sovetskaia Meditsina 1976: 55–7
66. Draliuk NS and Draliuk MG: Dlitel'naia intrakarotidnaia infuziia v neirokhirurgii. Voprosy Neirokhirurgii 1986: 54–7
67. Mozhaev SV, Zubkov Iu N, Ponomarev AM, and Shimchenko P: Nekotorye voprosy kliniki i kompleksnogo lecheniia septicheskikh trombozov kavernoznykh sinusov. Vestnik Khirurgii Imeni i - i - Grekova 1980; 125: 43–50.
68. Yokota H, Nakabayashi M, Fuse A, Mashiko K, Yamamoto Y, Henmi H, and Otsuka T: [Continuous intracarotid infusion of mannitol in severe head injury Japanese]. No Shinkei Geka [Neurological Surgery] 1993; 21: 205–11.
69. Rainov NG, Ikeda K, Qureshi NH, Grover S, Herrlinger U, Pechan P, Chiocca EA, Breakefield XO, and Barnett FH: Intraarterial delivery of adenovirus vectors and liposome-DNA complexes to experimental brain neoplasms. Hum Gene Ther 1999; 10: 311–8.
70. Chauvet AE, Kesava PP, Goh CS, and Badie B: Selective intraarterial gene delivery into a canine meningioma. J Neurosurg 1998; 88: 870–3.
71. Barnett FH, Rainov NG, Ikeda K, Schuback DE, Elliott P, Kramm CM, Chase M, Qureshi NH, Harsh Gt, Chiocca EA, and Breakefield XO: Selective delivery of herpes virus vectors to experimental brain tumors using RMP-7. Cancer Gene Therapy 1999; 6: 14–20.
72. Lu D, Li Y, Wang L, Chen J, Mahmood A, and Chopp M: Intraarterial administration of marrow stromal cells in a rat model of traumatic brain injury. J Neurotrauma 2001; 18: 813–9.
73. Amar AP, Zlokovic BV, and Apuzzo ML: Endovascular restorative neurosurgery: a novel concept for molecular and cellular therapy of the nervous system. Neurosurgery 2003; 52: 402–12; discussion 412–3.
74. Zlokovic BV and Apuzzo ML: Cellular and molecular neurosurgery: pathways from concept to reality – part II: vector systems and delivery methodologies for gene therapy of the central nervous system. Neurosurgery 1997; 40: 805–12; discussion 812–3.
75. Cengelli F, Maysinger D, Tschuddi-Monnet F, Montet X, Corot C, Petri-Fink A, Hofmann H, and Juillerat-Jeanneret L: Interaction of functionalized superparamagnetic iron oxide nanoparticles with brain structures. J Pharmacol Exp Ther 2006; 318(1): 108–16.
76. Mishra V, Mahor S, Rawat A, Gupta PN, Dubey P, Khatri K, and Vyas SP: Targeted brain delivery of AZT via transferrin anchored pegylated albumin nanoparticles. J Drug Target 2006; 14: 45–53.
77. Gao K and Jiang X: Influence of particle size on transport of methotrexate across blood brain barrier by polysorbate 80-coated polybutylcyanoacrylate nanoparticles. Int J Pharm 2006; 310: 213–9.

78. Noble CO, Krauze MT, Drummond DC, Yamashita Y, Saito R, Berger MS, Kirpotin DB, Bankiewicz KS, and Park JW: Novel nanoliposomal CPT-11 infused by convection-enhanced delivery in intracranial tumors: pharmacology and efficacy. Cancer Res 2006; 66: 2801–6.
79. Michaelis K, Hoffmann MM, Dreis S, Herbert E, Alyautdin RN, Michaelis M, Kreuter J, and Langer K: Covalent linkage of apolipoprotein E to albumin-nanoparticles strongly enhances drug transport into the brain. J Pharmacol Exp Ther 2006;317(3): 1246–53.
80. Chen Y, Dalwadi G, and Benson HA: Drug delivery across the blood-brain barrier. Curr Drug Deliv 2004; 1: 361–76.
81. Tiwari SB and Amiji MM: A review of nanocarrier-based CNS delivery systems. Curr Drug Deliv 2006; 3: 219–32.
82. Huynh GH, Deen DF, and Szoka FC, Jr.: Barriers to carrier mediated drug and gene delivery to brain tumors. J Control Release 2006; 110: 236–59.
83. Ishida T, Takanashi Y, and Kiwada H: Safe and efficient drug delivery system with liposomes for intrathecal application of an antivasospastic drug, fasudil. Biol Pharm Bull 2006; 29: 397–402.
84. de Lima MC, da Cruz MT, Cardoso AL, Simoes S, and de Almeida LP: Liposomal and viral vectors for gene therapy of the central nervous system. Curr Drug Targets CNS Neurol Disord 2005; 4: 453–65.
85. Schnyder A and Huwyler J: Drug transport to brain with targeted liposomes. NeuroRx 2005; 2: 99–107.
86. Vail DM, Amantea MA, Colbern GT, Martin FJ, Hilger RA, and Working PK: Pegylated liposomal doxorubicin: proof of principle using preclinical animal models and pharmacokinetic studies. Semin Oncol 2004; 31: 16–35.
87. Rainov NG and Ren H: Gene therapy for human malignant brain tumors. Cancer J 2003; 9: 180–8.
88. Breedveld P, Beijnen JH, and Schellens JH: Use of P-glycoprotein and BCRP inhibitors to improve oral bioavailability and CNS penetration of anticancer drugs. Trends Pharmacol Sci 2006; 27: 17–24.
89. Kerb R: Implications of genetic polymorphisms in drug transporters for pharmacotherapy. Cancer Lett 2006; 234: 4–33.
90. Jolliet-Riant P and Tillement JP: Drug transfer across the blood-brain barrier and improvement of brain delivery. Fundam Clin Pharmacol 1999; 13: 16–26.
91. Dietz GP and Bahr M: Peptide-enhanced cellular internalization of proteins in neuroscience. Brain Res Bull 2005; 68: 103–14.
92. Pardridge WM: Brain drug targeting and gene technologies. Jpn J Pharmacol 2001; 87: 97–103.
93. Doolittle ND, Petrillo A, Bell S, Cummings P, and Eriksen S: Blood-brain barrier disruption for the treatment of malignant brain tumors: The National Program. J Neurosci Nurs 1998; 30: 81–90.
94. Morris P: Interventional neuroradiology in the treatment of brain tumors. Neuroimaging Clin N Am 1999; 9: 767–78.
95. Madajewicz S, Chowhan N, Tfayli A, Roque C, Meek A, Davis R, Wolf W, Cabahug C, Roche P, Manzione J, Iliya A, Shady M, Hentschel P, Atkins H, and Braun A: Therapy for patients with high grade astrocytoma using intraarterial chemotherapy and radiation therapy. Cancer 2000; 88: 2350–6.
96. Haines SJ: Moving targets and ghosts of the past: outcome measurement in brain tumour therapy. J Clin Neurosci 2002; 9: 109–12.
97. Silvani A, Eoli M, Salmaggi A, Erbetta A, Fariselli L, and Boiardi A: Intra-arterial ACNU and carboplatin versus intravenous chemotherapy with cisplatin and BCNU in newly diagnosed patients with glioblastoma. Neurol Sci 2002; 23: 219–24.
98. Fraticelli AT, Cholley BP, Losser MR, Saint Maurice JP, and Payen D: Milrinone for the treatment of cerebral vasospasm after aneurysmal subarachnoid hemorrhage. Stroke 2008; 39: 893–8.
99. Keuskamp J, Murali R, and Chao KH: High-dose intraarterial verapamil in the treatment of cerebral vasospasm after aneurysmal subarachnoid hemorrhage. J Neurosurg 2008; 108: 458–63.
100. Feng L, Fitzsimmons BF, Young WL, Berman MF, Lin E, Aagaard BD, Duong H, and Pile-Spellman J: Intraarterially administered verapamil as adjunct therapy for cerebral vasospasm: safety and 2-year experience. Am J Neuroradiol 2002; 23: 1284–90.
101. Fortin D, Osztie E, and Neuwelt EA: Iatrogenic arterial spasm relieved by intraarterial mannitol infusion. Am J Neuroradiol 2000; 21: 968–70.
102. Yoshida K, Watanabe H, and Nakamura S: Intraarterial injection of amrinone for

vasospasm induced by subarachnoid hemorrhage. Am J Neuroradiol 1997; 18: 492–6.

103. Kaku Y, Yonekawa Y, Tsukahara T, and Kazekawa K: Superselective intra-arterial infusion of papaverine for the treatment of cerebral vasospasm after subarachnoid hemorrhage. J Neurosurg 1992; 77: 842–7.

104. Kassell NF, Helm G, Simmons N, Phillips CD, and Cail WS: Treatment of cerebral vasospasm with intra-arterial papaverine. J Neurosurg 1992; 77: 848–52.

105. Boker DK, Solymosi L, and Wassmann H: Immediate postangiographic intraarterial treatment of cerebral vasospasm after subarachnoid hemorrhage with nimodipine. Report on 3 cases. Neurochirurgia (Stuttg) 1985; 28 Suppl 1: 118–20.

106. Nagel S, Schellinger PD, Hartmann M, Juettler E, Huttner HB, Ringleb P, Schwab S, and Kohrmann M: Therapy of acute basilar artery occlusion: intraarterial thrombolysis alone vs bridging therapy. Stroke 2009; 40: 140–6.

107. Khatri P, Neff J, Broderick JP, Khoury JC, Carrozzella J, and Tomsick T: Revascularization end points in stroke interventional trials: recanalization versus reperfusion in IMS-I. Stroke 2005; 36: 2400–3.

108. Abou-Chebl A, Bajzer CT, Krieger DW, Furlan AJ, and Yadav JS: Multimodal therapy for the treatment of severe ischemic stroke combining GPIIb/IIIa antagonists and angioplasty after failure of thrombolysis. Stroke 2005; 36: 2286–8.

109. Sorimachi T, Fujii Y, Tsuchiya N, Nashimoto T, Saito M, Morita K, Ito Y, and Tanaka R: Blood pressure in the artery distal to an intraarterial embolus during thrombolytic therapy for occlusion of a major artery: a predictor of cerebral infarction following good recanalization. J Neurosurg 2005; 102: 870–8.

110. Zaidat OO, Suarez JI, Sunshine JL, Tarr RW, Alexander MJ, Smith TP, Enterline DS, Selman WR, and Landis DM: Thrombolytic therapy of acute ischemic stroke: correlation of angiographic recanalization with clinical outcome. Am J Neuroradiol 2005; 26: 880–4.

111. Lee DH, Jo KD, Kim HG, Choi SJ, Jung SM, Ryu DS, and Park MS: Local intraarterial urokinase thrombolysis of acute ischemic stroke with or without intravenous abciximab: a pilot study. J Vasc Interv Radiol 2002; 13: 769–74.

112. Rabinstein AA, Wijdicks EF, and Nichols DA: Complete recovery after early intraarterial recombinant tissue plasminogen activator thrombolysis of carotid T occlusion. Am J Neuroradiol 2002; 23: 1596–9.

113. Fukuda I, Gomi S, Meguro K, and Wada M: Impact of immediate cerebral angiography for in-hospital cerebral thromboembolism after cardiovascular surgery. Jpn J Thorac Cardiovasc Surg 2001; 49: 282–6.

114. Kase CS, Furlan AJ, Wechsler LR, Higashida RT, Rowley HA, Hart RG, Molinari GF, Frederick LS, Roberts HC, Gebel JM, Sila CA, Schulz GA, Roberts RS, and Gent M: Cerebral hemorrhage after intra-arterial thrombolysis for ischemic stroke: the PROACT II trial. Neurology 2001; 57: 1603–10.

115. Moazami N, Smedira NG, McCarthy PM, Katzan I, Sila CA, Lytle BW, and Cosgrove DM, 3rd: Safety and efficacy of intraarterial thrombolysis for perioperative stroke after cardiac operation. Ann Thorac Surg 2001; 72: 1933–7; discussion 1937–9.

116. Sylaja PN, Kuruttukulam G, Joseph S, Gupta AK, and Radhakrishnan K: Selective intra arterial thrombolysis in acute carotid territory stroke. Neurol India 2001; 49: 153–7.

117. Suwanwela NC, Phanthumchinda K, Suwanwela N, Tantivatana J, and Janchai A: Thrombolytic treatment for acute ischemic stroke: a 2 year-experience at King Chulalongkorn Memorial Hospital. J Med Assoc Thai 2001; 84 Suppl 1: S428–36.

118. Endo S, Kuwayama N, Hirashima Y, Akai T, Nishijima M, and Takaku A: Results of urgent thrombolysis in patients with major stroke and atherothrombotic occlusion of the cervical internal carotid artery. Am J Neuroradiol 1998; 19: 1169–75.

119. Jahan R, Duckwiler GR, Kidwell CS, Sayre JW, Gobin YP, Villablanca JP, Saver J, Starkman S, Martin N, and Vinuela F: Intraarterial thrombolysis for treatment of acute stroke: experience in 26 patients with long-term follow-up [see comments]. Am J Neuroradiol 1999; 20: 1291–9.

120. Barr JD, Mathis JM, Wildenhain SL, Wechsler L, Jungreis CA, and Horton JA: Acute stroke intervention with intraarterial urokinase infusion. J Vasc Interv Radiol 1994; 5: 705–13.

121. Casto L, Moschini L, Camerlingo M, Gazzaniga G, Partziguain T, Belloni G, and Mamoli A: Local intraarterial thrombolysis for acute stroke in the carotid artery territories. Acta Neurol Scand 1992; 86: 308–11.

Chapter 8

Vesicular Systems for Intranasal Drug Delivery

Ibrahim A. Alsarra, Amel Y. Hamed, Fars K. Alanazi, and Gamal M. El Maghraby

Abstract

Recently, the nasal route for systemic drug delivery has gained great interest. It provides several advantages over other routes of drug administrations. These include rapid absorption, avoidance of the intestinal and hepatic presystemic disposition, and high potential for drug transfer to the cerebrospinal fluid. Unfortunately, the mucociliary clearance, which reduces the residence time of the nasally applied drugs, and the poor nasal permeability made it difficult for many drugs to be delivered through this route. Alternative approaches have been adopted to overcome these problems. These include the use of mucoadhesive formulations or chemical penetration enhancers. Vesicular drug delivery systems provide promising alternative for enhanced and controlled nasal drug delivery.

Alternative terms have been used to describe the vesicular systems. These include liposomes, niosomes, ethosomes, and transfersomes. These systems are morphologically similar but differ in composition and function. Nasal delivery employs liposomes and niosomes, and their corresponding proconcentrates, proliposomes and proniosomes. Encouraging results have been recorded for these systems after nasal application with the possibility of achieving many objectives such as systemic delivery of small and large molecular weight drugs. This review article discusses such systems for intranasal vaccination and for improvement of nasal drug delivery to the central nervous system. The review critically evaluates the potential of such systems for systemic drug delivery after intranasal applications.

Key words: Nasal mucociliary clearance, olfactory route, nose-to-brain delivery, targeted brain delivery, blood–brain barrier, colloidal carriers, polymeric hydrogels, liposomes, mucoadhesion, vaccine delivery.

1. Introduction

The history of nasal drug delivery dates back to earlier topical applications of drugs intended for local effects. The early 1980s saw the introduction of nasal route as a promising systemic delivery alternative to other conventional drug delivery routes (1). Intranasal drug delivery has many advantages over other routes of drug

K.K. Jain (ed.), *Drug Delivery to the Central Nervous System*, Neuromethods 45,
DOI 10.1007/978-1-60761-529-3_8,

administration. It is easily accessible, convenient, and a reliable method, with a porous endothelial membrane, and a highly vascularized epithelium that provides a rapid absorption of compound into the systemic circulation, avoiding the hepatic first pass elimination.

In addition, intranasal drug delivery enables dose reduction, rapid attainment of therapeutic blood levels, quicker onset of pharmacological activity, and fewer side effects (2, 3). It was reported that lipophilic drugs are generally well absorbed from the nasal cavity with pharmacokinetic profiles, which are often identical to those obtained after an intravenous injection with a bioavailability approaching 100% (1). The unique characteristic of intranasal drug delivery is the high potential for drug transfer to the cerebrospinal fluid through the olfactory region which is situated in nasal cavity (4). Recent developments in nasal drug delivery have suggested intranasal administration as a safe and acceptable route for brain targeting, especially for drugs with biological effects on the central nerves system (CNS) and limited blood–brain permeability (BBB) (5). The brain targeting research attempts of large molecular weight molecules investigated nerve growth factor, insulin, desmopressin, cholecystokinin, and insulin-like growth factor-1 demonstrated the potential of nose-to-brain pathway (6–9). These advantages have provided the intranasal route some superiority over the parenteral as well as oral routes (10).

The major problems with nasal delivery are the mucociliary clearance, which reduces the residence time of nasally applied dosage forms and the poor nasal permeability of many drugs (11). Several alternative strategies have been employed to overcome these limitations. Bioadhesive polymers, for example, can be used to achieve long residence time on nasal mucosa which results in higher concentration gradient and subsequent increased absorption of the drugs (12). Polymers may widen the tight junctions (**Fig. 8.1**) producing absorption enhancing effect (13).

Successful nasal delivery has been obtained with solutions, powders, gels, and microspheres as delivery systems (14–16). However, these systems suffer from number of disadvantages; solutions show rapid clearance from nasal cavity and do not allow extended drug release (17). In addition, chemical instability problems may be encountered in solutions, particularly in case of peptide or protein drugs (18). Powders and microspheres require sophisticated delivery devices for deposition and accurate dosing (19), with insufficient wetting at mucosa resulting in low bioadhesive force and incomplete drug release (20).

Vesicular drug delivery systems provide promising alternatives with many advantages over the conventional systems. Various pharmaceutical approaches can be employed to render their final formulation more effective. Liposomes are preferred over other

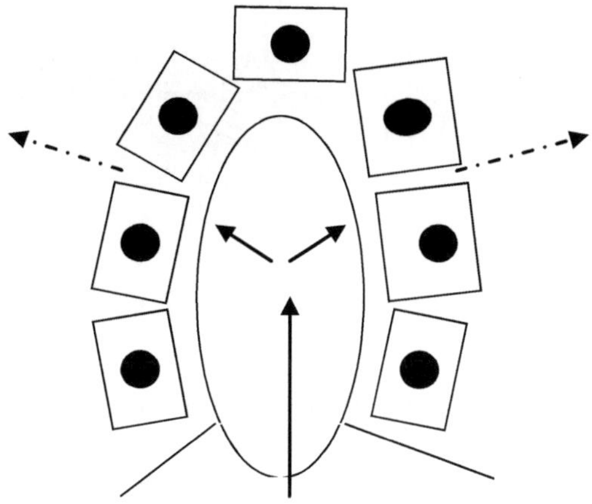

Fig. 8.1. Methods of brain selective drug delivery.

vesicular systems in nasal drug delivery. Liposomes are known to sustain the release of the entrapped drug and in the case of nasal administration are able to decrease mucociliary clearance due to their surface viscosity. The action of liposomes on nasal mucosa is related to the incorporation of phospholipids in the membrane, opening "new pores" in the paracellular tight junctions (21).

2. Nasal Anatomy and Physiology

The nasal cavity (**Fig. 8.2**) is subdivided along the centre into two halves by the nasal septum. The two cavities open to the facial side through the anterior nasal apertures and to the rhinopharynx via the posterior nasal apertures and each of two nasal cavities can be subdivided into different regions: nasal vestibule, inferior turbinate, middle turbinate, superior turbinate, olfactory region, frontal sinus, sphenoidal sinus, and cribriform plate of ethmoid bone. The nasal cavity also contains the nasal associated lymphoid tissue (NALT), which is mainly situated in the nasopharynx (22). The NALT contains specialized M-like cells similar to those present in the Peyer's patches in the gut. However, mucosal lymphoid tissue is located immediately under the nasal mucosa, where B and T lymphocyte follicles, macrophages, and dendritic cells are present (23). The cells are capable of taking up antigen and processing these for immune stimulation. It is generally recognized that

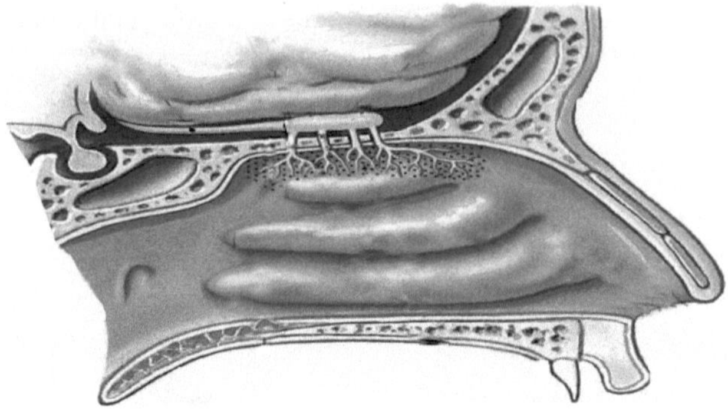

Fig. 8.2. A model for intraarterial drug delivery to the brain. Model of intracarotid drug delivery, C1 C2 and V1 V2 are concentrations and volumes in the brain (1) and remainder of the body (2), respectively. Q is the regional blood flow and CL1 and CL2 are cerebral and remaining body clearances. Modern microcatheters can restrict intraarterial interventions to tumor tissue. See Dedrick 1988 for details.

soluble antigens can penetrate the whole of the nasal mucosa and reach the superficial cervical lymph nodes to produce mainly a systemic immune response (24).

The total surface area of the nasal cavity in human adult is about 150 cm^2 and total volume is about 15 ml. The olfactory region in men covers an area of about 10 cm^2 and is positioned on superior turbinate on opposite septum (25). The respiratory region contains three nasal turbinates: superior, middle, and inferior which project from the lateral wall of each half of the nasal cavity. The presence of these turbinates creates a turbulent airflow through the nasal passages ensuring a better contact between the inhaled air and the mucosal surface (1). The nasal epithelial membrane provides a significant barrier to the free diffusion of substance across them.

2.1. Nasal Epithelium

2.1.1. Respiratory Region

The respiratory region is considered as the major site for drug absorption into systemic circulation. The mucosa consists of an epithelium resting on a basement membrane and a lamina propria. The anterior part of respiratory region is covered with squamous epithelium, while the posterior part covered by a pseudostratified columnar epithelium. The cells of respiratory epithelium are covered by about 300 microvilli per cells (25, 26).

The respiratory epithelium consists of four dominated cell types; ciliated columnar cells, non-ciliated columnar cells, goblet cells, and basal cells. The basal cells are situated on the basal membrane and do not extend to the apical epithelial surface, as do the other three cell types. The main function of the goblet cells is the secretion of mucus. The respiratory cells are covered by layer of long cilia of size 2–4 μm; the cilia move in a coordinated way to propel mucus across the epithelial surface toward the pharynx with a clearance half time of approximately 15–20 min. The mucus layer consists of low viscosity sol layer that surrounds the cilia and a more viscous gel layer forming a layer on the top of the sol layer and covering the tips of the cilia. The epithelial cells are closely packed on apical surface, surrounded by intercellular junction whose specialized sit and structural components are commonly known as junction complex (22, 25). The presence of tight junction between neighboring epithelial cells prevents the free diffusion of hydrophilic molecules across the epithelial by the paracellular route (27). The normal diameter of the tight junctions is considered to be of the order of 3.9–8.4 Å (28). Tight junctions are located at the boundary between apical and boslateral domains in epithelial cell periphery. On a molecular level, tight junctional complex consists of a number of complexes of transmembran proteins (occluding, claudine, and junction adhesion molecule) and cytoplasmic proteins (27).

2.1.2. The Olfactory Region

The olfactory region is situated between the nasal septum and the lateral walls of each of the two nasal cavities and just below the cribriform plate of the ethmoid bone separating the cranial cavity from nasal cavity (25). The olfactory epithelium is a pseudostratified epithelium, comprising olfactory sensory neurons and two types of cells; basal cells that are able to differentiate into neuronal receptor cells and sustentacular cells (supporting cell) that provide mechanical support by ensheathing neuronal receptor cells and maintain the normal extracellular potassium level for neuronal activity (22). The olfactory epithelium is covered by a dense and viscous layer of mucus, which is secreted from the tubuloalveolar Bowman's glands and the supporting cells. The olfactory epithelium constitutes only about 5% of the total area of the nasal cavity in man (26), but is of considerable interest in drug delivery because it bypasses the BBB, delivering therapeutic drugs to CNS (29).

It should be emphasized that the blood supply to the nasal mucosa is pertinent with regards to systemic drug delivery. The arterial blood supply to the nasal cavity is derived from both the external and internal carotid arteries. The blood that is supplied to olfactory region by anterior and posterior ethmoidal branches come from the ophthalmic artery supply, which is a branch of carotid artery. These vessels supply the anterior portion of the nose. The venous drainage is as for respiratory system via sphenopalatine foramen into the pterygoid plexus or via superior ophthalmic vein (30). When the drug is administered intranasally, it can enter into the brain via three different paths (31). The first one is the systemic pathway by which the drug is absorbed into the systemic circulation and subsequently reaches the brain by crossing BBB (especially lipophilic drug). The others are the olfactory region and the trigeminal neural pathway by which the drug is transported directly from the nasal cavity to CNS (cerebrospinal fluid and brain tissue) (32). The trigeminal nerve receptors which are present in the nasal cavity are responsible for most chemoperception and are suggested to transport the drug directly to CNS (33).

The deep and superficial cervical lymph nodes were of special interest in intranasal drug delivery because they are known to receive lymphatic afferents from portions of the nasal passages and nasolabial areas, respectively (34). The exact pathway from nasal cavity to lymph node is uncertain, but similar delivery from nasal cavity to lymphatics has been observed. This pathway is thought to mediate the efflux of large molecules and/or immune cells from sites within the CNS to the lymphatic system (35). The connection between the brain and nasal lymphatics may offer a direct pathway from the brain interstitial fluid to the nasal submucosa that excludes direct contact with the cerebrospinal fluid (36).

There are different mechanisms by which the drugs cross the olfactory membrane to reach CNS. The first mechanism involves direct transfer of the drug to primary neurons of the olfactory epithelium and transport to the olfactory bulb by intracellular axonal transport with subsequent possible distribution into more distant brain tissues. The second mechanism depends on drug permeation across the olfactory sustentacular epithelial cells, either by transcellular or paracellular mechanisms followed by uptake into CNS. The last one employs pinocytosis by olfactory neurons (26). The drug can cross olfactory lobe by one or combination of pathways.

3. Factors Influencing the Absorption of Drugs Across the Nasal Epithelium

The factors influencing nasal absorption are related to nasal physiology, the physicochemical characteristics of the drug, and the properties of specific drug formulation.

3.1. Physiological Barrier

3.1.1. Mucociliary Clearance

The function of mucociliary clearance (MCC) system is to remove foreign substances and particles from the nasal cavity, consequently preventing them from reaching the lower airways (37). Nasal clearance proceeds at an average rate of about 5–6 mm/min (4). Nasally administered formulation can be cleared from the nasal cavity with a half-life of clearance of about 15 min with the result of limiting the time available for absorption (38). The normal mucociliary transit time in humans has been reported to be 12–15 min (11).

Rapid mucociliary clearance of drug formulations that are deposited in the nasal cavity is thought to be an important factor underlying the low bioavailability of intranasally administered drugs (11). Some drugs, hormonal changes in the body, pathological conditions, environmental conditions, and formulation factors especially rheology are reported to affect the mucociliary clearances and in turn exert significant influence on drug permeability (39). In isotonic solution, the ciliary beat frequency is best preserved. Optimal ciliary beat frequency was observed between pH values of 7 and 11. Values outside this range can result in a larger decrease in the beat frequency (40). Ciliary beat frequency measurements have shown to be a good indicator of the effects of substances on nasal tissue morphology (11).

3.1.2. Enzymes

Despite avoiding the hepatic first pass metabolism, nasally administered drugs can be subjected to a broad range of metabolic enzymes in nasal mucosa with possible reduction in the bioavailability of some drugs, especially those containing peptides or proteins (41). In spite of this possibility, the nasal route is still considered to be superior to the oral route. Various approaches have been used to overcome these degradations, which include the use enzymatic inhibitors of protease and peptidase such as bacitracin, amastatin, boroleucin, and puromycin which have been reported to improve the absorption of many drugs (42) or design prodrug (43).

3.2. Physicochemical Characteristics of the Drug

The physicochemical characteristics of the administered drug include molecular weight, solubility, dissolution rate, charge, partition coefficient, pKa, particle size, and the presence of polymorphism and these can influence drug absorption.

The permeation of drugs having molecular weight of less than 300 Da is not significantly influenced by the physicochemical properties of the drug as they will mostly permeate through aqueous channels of the membrane. In contrast, the rate of permeation is highly sensitive to molecular size for compounds with molecular weight >300 Da (44). The bioavailability of intranasally administered peptides and proteins including insulin may be low because of high molecular weight and hydrophilicity (45).

As for other routes of administration, the nasal absorption can take place only after the drug's dissolution. The dissolution rate is important in determining nasal absorption of powder and suspensions dosage forms. Rapid dissolution is critical for the drug particles after nasal administration otherwise the particles will be subjected to rapid clearance from the airway with subsequent reduction of the bioavailability (30).

The rate and extent of absorption of a drug across a biological membrane is influenced by its lipophilicity. Normally, the permeation of the compound through nasal mucosa increases with increasing the lipophilicity (46). Low molecular weight lipophilic drugs are absorbed quite efficiently across the nasal epithelium, whereas larger hydrophilic drugs, such as peptides and proteins, have substantially lower bioavailability (45).

Highly lipophilic corticosteroids drugs are absorbed more quickly by the nasal mucosa and may preferentially partition into the systemic tissue providing a large volume of distribution at steady state (47). Prodrug technique has been employed to increase the lipophilicity. The aliphatic prodrug of acyclovir provides a classical example of this process, which resulted in an increased drug bioavailability. However, it should be noted that the 140-fold increase in partition coefficient of the drug was only associated with 30% increase in bioavailability. It should be also emphasized that the ester form of the prodrug can show greater increase in transnasal drug transport but premature hydrolysis of such ester in the nasal cavity provides the main limitation of this technique (48).

Water-soluble prodrugs of 17β-estradiol have been evaluated after intranasal administration. These prodrugs were capable of producing high levels of estradiol in the cerebrospinal fluid (CSF), compared to an equivalent intravenous dose. These data suggest that the drug can reach the CSF via a direct pathway through the nasal cavity and as a result may have a significant value in the treatment of Alzheimer's disease (49).

4. Types of Vesicular Drug Delivery Systems

Vesicular systems have been employed as drug delivery carriers for many decades. They were adopted to achieve many objectives which included targeted drug delivery, enhanced drug transport through various biological membranes or prolonging and controlling drug release. Alternative terminologies have been used to describe such vesicular systems throughout these investigations (50). These included liposomes, niosomes, transfersomes, ethosomes, vesosomes, colloidosomes, and pharmacosomes. These vesicles are similar to the standard liposomes in morphology but may differ in function and composition. Only vesicles employed in the nasal drug delivery will be discussed in the subsequent paragraphs, and for detailed methods of preparation and characterizations, the readers are strongly encouraged to refer to the constructive practical approaches published in 1990 by New (51).

4.1. Liposomes

Liposomes are spherical microscopic vesicles composed of one (unilamellar) or more (multilamellar) concentric lipid bilayers, arranged around a central aqueous core (**Fig. 8.3**). They are made of natural, biodegradable, nontoxic, and natural constituents such as phospholipids and may mimic naturally occurring cell membranes. They may contain cholesterol as a membrane stabilizer and may include trace amounts of charging agents (51, 52). Having these desirable structure features, liposomes can encapsulate drugs with widely varying lipophilicities, with the lipophilic ones being located in the lipid bilayer and the hydrophilic ones

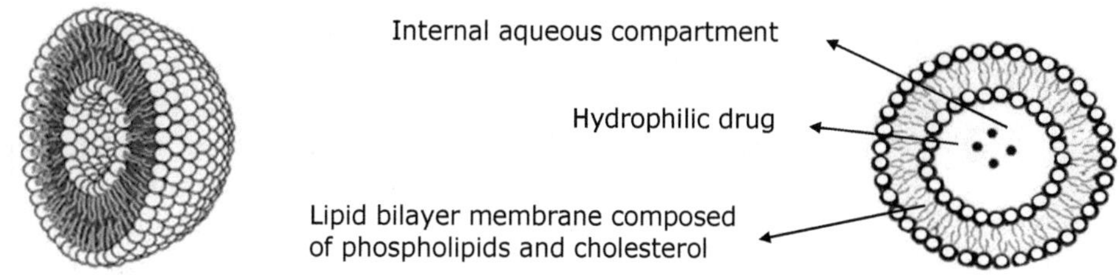

Fig. 8.3. Liposomes structure.

being retained in the aqueous core. Amphiphilic drugs can be adsorbed at the head group region of the bilayers (53). Liposomes have been investigated as carriers of various pharmacologically active agents such as antineoplastic, antimicrobial drugs, chelating agents, steroids, vaccines, and genetic materials (54). Liposomes provide an efficient drug delivery system because they can alter the pharmacokinetics and pharmacodynamics of the entrapped drugs. Table **8.1** provides an overview of the reported liposomal administered drugs.

Table 8.1.
Liposome-encapsulated drugs studied for nasal administration

Drug	Results	Reference
Diphenhydramine	Increased drug retention in the nasal	(55)
HIVgp160-encapsulated hemagglutinating virus	HIV-specific humoral and cellular immunity in mucosal and systemic sites	(56)
Meningococcal OpaB and OpaJ proteins	Induced highly significant anti-Opa responses	(57)
Influenza virus hemagglutinin from 3 viral strains	Provides an almost total prevention of virus shedding combined with a high level of immunological protection against homologous virus challenge	(58)
Trivalent influenza A/H1N1-proteosome	Produced high antibody titers in serum as well as in nasal secretions	(59)
Salmon calcitonin	Ultra-flexible liposomes significantly enhanced the hypocalcemia effect than conventional liposomes	(60)
Ovalbumin in an archaeal lipid mucosal vaccine adjuvant and delivery (AMVAD)	Eliciting robust antigen-specific mucosal and systemic immune responses	(61)
Tetanus toxoid antigen	Effective mucosal immune responses and high mucosal secretory IgA titers	(62)
M. tuberculosis vaccines (DNA-hsp65)	Effective protection against TB with a single dose vaccination	(63)

According to their size, liposomes can be classified as either small unilamellar vesicles (SUV) 10–100 nm or large unilamellar vesicles (LUV) 100–3000 nm. If more than one bilayers are present, then they are referred to MLV (64). These characteristics can be controlled by proper selection of the method of preparations (51). Liposomal formulations should have high entrapment efficiencies, narrow size of distributions, long-term stabilities, and ideal release properties (based on the intended application). These require the preparation method to have the potential to produce liposomes using a wide range of ingredient molecules, e.g., lipids/phospholipids that promote liposome stability (65).

Liposomes can be formulated as dry powder or a suspension, as an aerosol or in a semisolid form such as a gel or cream. In vivo, they can be administered topically or parenterally. In the systemic circulation, liposomes can be recognized as foreign particles and consequently endocytosed by reticuloendothelial system reaching the liver and spleen (66). MLV are considered the largest type of liposomes which are capable of entrapping large percent of drugs. Once they are infused, they are rapidly recognized by the immune system and taken up by macrophages which subsequently remove them from the circulation. LUVs (intermediate size liposomes) have a better opportunity of escaping the reticuloendothelial system (RES) and so have the ability to stay in the circulation for a longer period. The small liposomes SUV show the shortest circulation time in blood due to capillary extravasations (52).

The activity of liposomes as carriers for drugs depends upon various factors such as encapsulation efficiency, stability, release rates, body distribution after administration, size surface charge, and rigidity. The properties of liposomes can be varied and controlled by incorporating different types of lipids and by varying the preparation methods (67). Poor liposomal stability is the major problem in liposome research. The instability problem arises from chemical degradation of the liposome components in addition to physical stability problems which are manifested as loss of entrapped drug and size change upon storage. Loss of entrapped material can be minimized by increasing the rigidity of the bilayer membrane or reducing the water content of liposome formulations producing the so-called proliposomes (52). Furthermore, addition of appropriate cryoprotectants allows storage of liposomes in frozen or lyophilized state.

4.2. Niosomes

These are similar to liposomes in morphology but with different compositions; they are formed from the self-assembly of non-ionic amphiphilic in combination with other lipidic surfactants in aqueous medium (68–71). Niosomes or non-ionic surfactant vesicles are microscopic lamellar structures formed from admixture of non-ionic surfactant of the alkyl or dialkyl polyglycerol ether class and cholesterol with subsequent hydration in aqueous media resulting in closed bilayer structures (72). Like liposomes, aqueous dispersions of niosomes may exhibit fusion, aggregation, leaking, or hydrolysis of encapsulated drugs, therefore limiting the shelf life of the dispersion (73).

Niosomes are widely studied as an inexpensive alternative of non-biological origin to liposomes. Niosomal surfactants are biodegradable, biocompatible, and non-immunogenic. It has greater accessibility, superior chemical stability, and relatively low cost of niosomes compared with liposomes, resulting in easier storage leading to the exploitation niosomes as alternatives to phospholipids. Theoretically, niosomal formulation requires presence of a particular class of amphiphile and an aqueous system (74).

Niosomes improve the therapeutic performance of drug molecules by delaying clearance from the circulation, protecting the drug from biological environment and restricting effects to target cells (75). Like liposomes, the surface of niosomes can be attached to hydrophilic moieties by incorporation of hydrophilic groups such as PEG, concanavalin A, and polysaccharide producing stealth or long circulating niosomes (76). Similar to liposomes, niosomes can be classified into MLV, SUV, and LUV. Non-ionic surfactant-based vesicles were also investigated as potential non-viral carriers for non-invasive topical delivery of plasmid DNA encoding HBsAg (77).

4.3. Proliposomes and Proniosomes

These are dry, free-flowing particles which immediately form vesicular dispersions upon addition of water. Their free-flowing particulate properties permit the fabrication of these nano-aggregates into solid dosage forms, which then is converted to liposomes/niosomes on contact with water or biological fluids (69, 71, 78). In general, they are prepared by penetrating a solution of drugs and phospholipids in volatile organic solvents into the microporous matrix of water-soluble carrier particles, followed by evaporation of the organic solvents. Drugs and phospholipids are thus deposited in the microporous structure of the carrier materials, thus maintaining the free-flowing surface characteristics of the carrier materials. Because of the characteristics of these preparations, the sterilization of proliposomes can be achieved without influencing their intrinsic characteristics (79). They have several advantages over their corresponding liquid formulations. These include the minimization of physical instability problems, such as aggregation, fusion, and leakage. In addition, they provide ease of transportation, distribution, storage, and dosage. Proniosomes have shown equal or greater efficacy in drug release performance when compared with conventional niosomes (78).

5. Pharmaceutical Applications

5.1. Drug Therapy

Vesicular systems play an important role in nasal drug delivery into the systemic circulation by overcoming limitations of the nasal route such as ciliary clearance and breakdown by nasal peptidase enzyme. They showed promising results not only with small molecular but also with large molecules. Liposomes are one of the vesicular systems that offer better absorption and drug retention in nasal mucosa, e.g., desmopressin and insulin (80, 81). The superiority of liposomes was indicated after comparing the permeability of liposomes entrapping insulin through nasal mucosa of rabbit with the permeability of insulin from solution with or without pretreatment with sodium glycocholate (82). Intranasal administration of insulin in liposomes composed of dipalmitoylphosphatidylcholine and sterylglucoside showed a greater reduction in blood glucose level with the effect lasting for 8 h (80). Jain et al. studied the usefulness of multivesicular liposomes as a mucoadhesive to prolong the release of insulin via nasal and ocular route. In their study, the multivesicular liposomes were shown to be marginally effective after nasal administration compared to ocular route although better therapeutic profile as the hypoglycemic effects were prolonged until 72 h (83).

The loading and leakage characteristics of the desmopressin-containing liposomes and the effect of liposomes on the nasal mucosal permeation were investigated. The increase of permeation of the antidiuretic, desmopressin, through the nasal mucosa was in the order of positively charged liposomes > negatively charged liposomes > solution (81).

Calcitonin liposomes with different charges were administered to rabbits to evaluate the effect of liposomes charges on nasal absorption. The bioavailability of intranasally administered calcitonin liposomes was in the following order: positively charged liposomes > negatively charged liposomes > calcitonin solution. The significant bioavailability enhancement of the positively charged calcitonin liposomes may be due to interaction of positively charged liposomes with the negatively charged mucosal surface. The retention of positively charged liposomes on the negatively charged nasal mucosa resulted in an increase in the residence time of calcitonin and thus increased bioavailability (84).

Nasal administration of acyclovir mucoadhesive liposomes has been demonstrated to have good permeability characteristics with enhanced nasal penetration of acyclovir in comparison to free drug suspended in gel. Fifteen rabbits were used in this study and were divided into three groups. The first group received acyclovir liposomes as nasal gel. The second group received the free drug as nasal gel (control). The third group received an intravenous injection of acyclovir solution. Intranasal preparations were administered by nasal droppers with a wide orifice inserted about 5 mm into the nostril of the rabbit while in a supine position. The AUC values of acyclovir mucoadhesive liposomal gel and acyclovir suspended in gel were 1.91175 and 21.90264 (μg h ml), respectively. The differences in the AUC values are due to the variations in the drug profile of acyclovir suspended in gel (**Fig. 8.4**). The improvement of bioavailability of the prepared liposomal formulations in comparison to free drug suspended in gel could be attributed to both encapsulation and incorporation of acyclovir in nasal mucoadhesive gel. Nasal bioavailability of acyclovir was 60.72% calculated relative to the serum acyclovir levels over a period of 8 h after intravenous injection of acyclovir. Liposomes have been demonstrated to have good permeability characteristics to enhance nasal penetration of many drugs (85).

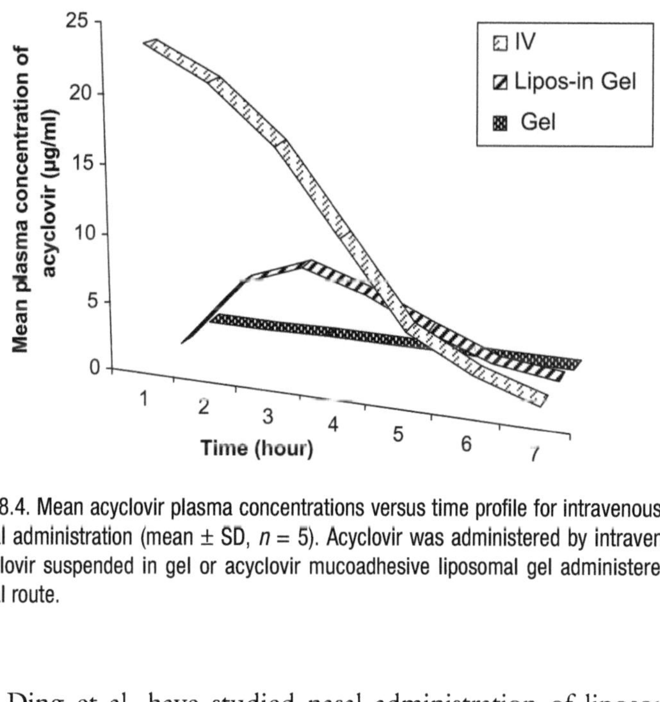

Fig. 8.4. Mean acyclovir plasma concentrations versus time profile for intravenous and nasal administration (mean ± SD, $n = 5$). Acyclovir was administered by intravenous, acyclovir suspended in gel or acyclovir mucoadhesive liposomal gel administered by nasal route.

Ding et al. have studied nasal administration of liposomes formulation of levonorgestrel and evaluated their pharmacokinetic properties; they found that administration of levonorgestrel liposomes via nasal route increases its bioavailability when compared with levonorgestrel suspension by oral route. It was concluded that the liposomes greatly facilitated levonorgestrel nasal absorption and may provide a rapid onset of action of levonorgestrel for emergency contraception (86).

Nasal administration of a liposomal leuprorelin acetate formulation with chitosan produced contraception in rats when compared to subcutaneous (87). Liposomes also provided protection of entrapped drugs from enzymatic degradation (80) and disrupted the mucosal membrane to increase absorption (88). Taking into considerations that leuprorelin acetate was administered by nasal route at low dose, it would be expected that the nasal route can increase the therapeutic index and reduce the adverse effects of the drug (87).

Zhang and coworkers concluded that the encapsulation of recombinant hirudin-2 in liposomes provided high entrapment capacity, greater stability, and enhancement of nasal absorption of recombinant hirudin-2 (89).

In a comparative study, nicotine base-proliposomes and nicotine hydrogen tartarate-proliposomes and a mixture of powdered nicotine hydrogen tartarate salt and sorbitol were administered intranasally to rats at dose of 1 mg/kg. Nasal absorption of nicotine from these formulations was very rapid (less than 10 min) and showed substantially sustained plasma nicotine levels compared to saline solutions of nicotine base and salt. Nicotine base-proliposomes demonstrated the best characteristics in terms of the area under the plasma concentration, mean residence. It was found that nasal application of proliposomes containing nicotine base could provide a very rapid absorption with prolonged delivery to the systemic circulation (90).

Proliposomes containing propranolol hydrochloride were also evaluated for their potential as a nasal drug delivery system to sustain the plasma concentration of the drug. The proliposomes were administered intranasally to the rats and plasma concentrations of the propranolol hydrochloride obtained after nasal administration of proliposomes were compared with those after nasal, oral, and intravenous administrations of aqueous solution. Nasal administration of the proliposomes resulted in low propranolol concentration at the initial phase and sustained at the terminal phase. Plasma concentrations of oral propranolol solution were much lower than those after the intravenous and nasal administrations. The absolute bioavailability of the orally administered propranolol was only 14.2%, the bioavailability obtained from nasal proliposomes reaching 97.5% (91).

5.2. Gene Therapy

Gene delivery is a challenging task in the treatment of genetic disorders. In this case, the plasmid DNA has to be introduced into the target cells, which should get transcribed and the genetic information should ultimately be translated into the corresponding protein. To accomplish this objective, a number of hurdles are to be overcome by the gene delivery system. Transfection is affected by (1) targeting the delivery system to the target cell, (2) transport through the cell membrane, (3) uptake and degradation in the endolysosomes, and (4) intracellular trafficking of plasmid DNA to the nucleus (92).

Macrophages play an important role in host immune functions; therefore, several strategies have been developed to transfer genes directly into macrophages but most of them use viral vectors (93). Despite the high transfection efficiency of viral vectors, questions remain regarding to their potential toxicity (94).

The use of nonviral vectors is attractive for in vivo gene delivery because it is safer than using simple viral systems. Cationic liposomes were considered to be one of the most promising nonviral gene delivery systems. Various kinds of cationic lipids have been synthesized and shown to be able to deliver genes into cells both in vitro and in vivo (95). Introduction of ligands for cell-surface receptors into liposomes has been attempted for achieving optimum transfection efficiency in vivo (96).

Dioleoylphosphatidyl-ethanolamine (DOPE) is one of liposomes constituent; this type phospholipids is known to accelerate the endosomal escape of plasmid DNA due to its pH-sensitivity and high transfection efficiency (97). Plasmid DNA complex with mannosylated liposomes is recognized and taken up by mannose receptors, mannosylated system exhibited a higher transfection in macrophages based on a receptor-mediated mechanism. A mannosylated cholesterol derivative itself has a positive charge, and a high density of mannose residues that can be deposited on the liposome surface without affecting the binding ability of cationic liposomes to DNA. These properties reflected in their superior in vivo gene transfection and it can be effectively introduced in cell-specific ligand structures to liposomes.

Goncharova et al. demonstrated the importance of nasal mucosa for the immunization against Tick-Borne encephalitis. To study intranasal immunization against TBE virus, they chose biodegradable micelles, cationic liposomes, and live attenuated bacterial/viral vectors. Their results showed the expression of the gene in transfected cells, thereby indicating that the liposomal formulations are suitable for mucosal immunization (98). Complexes of cationic lipid and plasmid DNA (lipoplexes) are the most widely used nonviral vectors for gene delivery. To improve the efficiency of nonviral gene delivery, pharmacological agents such as sex hormones and glucocorticoids have been shown to enhance liposome-mediated gene uptake (99). Estradiol is a steroidal female sex hormone that can enhance liposome-mediated gene delivery by increasing gene uptake in vitro and promote nuclear accumulation of the transgene (100). Incorporating β-estradiol into lipoplexes consistently increased gene expression in the lungs and nasal epithelia (sub-confluent and confluent) to human airway epithelial cells, in particular the bronchial epithelial cell line. The greatest enhancement was found using polarized cells (101). These data indicated that β-estradiol and methyl-prednisolone are promising adjuvents for improving gene delivery. Methyl-prednisolone has also been incorporated into lipoplexes and assessed on confluent and polarized cells where it increased delivery 70-fold and 48-fold, respectively, compared with lipoplex alone. This represents a considerably larger enhance calcium phosphate based delivery to human myoblasts (102).

Tanaka et al. reported that nasal administration of insulin gene packed in cationic liposome produced a sufficiently high expression, processing, and secretion of insulin in mice with streptozocin-induced type 1 diabetes. Within 8 days of treatment, plasma insulin levels in streptozocin-injected mice given the luciferase gene decreased to <100 pg/ml. In comparison, streptozocin-injected mice given the insulin gene exhibited plasma insulin levels at least ten times higher and is actually greater than the mean level in normal nondiabetic mice. These high levels of total insulin corrected hyperglycemia without producing hypoglycemia even after a 16-h fast and devoid of adverse effect of insulin gene expression in treated mice. Using fluorescence in situ hybridization (FISH) analysis, intracellular plasmids observed in alveolar epithelial cells of the lung, where extracellular plasmids were observed in muscle, kidney, gastrointestinal system, and immune cells, as well as the respiratory system, suggesting plasmid dissemination, possibly due to hematogeneous spread (103).

Inclusion of free liposomes into DNA/lipid complexes may be important to accomplish optimal transfection activity in vivo (104). Liposome/DNA complexes can be delivered into organs with high cationic lipid to DNA ratios when administered intravenously. However, through nasal inoculation of liposome/DNA complexes, liposomes are trapped within the lung epithelium. Thus, the cationic lipid to DNA ratio is changed during transit through the lung epithelial cells. In addition, there are inhibitors which inhibit the transfection activity of DNA/lipid complexes in serum. Therefore, in other organs, transfection efficiency was very low, and plasmid DNA was detected only in the extracellular spaces. Liposome-mediated in vivo gene transfer via nasal administration may provide an efficacious route for delivery of hormonal and other gene products into the blood stream.

The major cause of mortality in patients with cystic fibrosis is a lung malfunction. Therefore, gene transfer to correct the underlying genetic defect is a potential treatment for cystic fibrosis. A DNA-liposome formulation was delivered to the patients with no immune tolerance, in repeated doses (three doses). It was concluded that the DNA containing liposomes can be successfully re-administered without apparent loss of efficacy for cystic fibrosis treatment (105).

5.3. Vaccine Delivery

Many of the available vaccines including protein antigens and DNA vaccines are very unstable and need to be protected from degradation in the biologic environment. In addition, their efficacy is limited by their poor capacity to cross biologic barriers and reach the target sites. Usually parenteral vaccination involves intramuscular administration of antigens. This route of immunization stimulates the immune system to produce IgG antibody in the serum but fails to generate a mucosal antibody response. In contrast, intranasally administered vaccines stimulate IgA antibody response along the mucosal surfaces. IgG facilitates the phagocytosis of bacteria and activates the complement, whereas IgA principally acts by preventing attachment and colonization of bacteria on mucosal surfaces (106).

Mucosal delivery is a highly effective route for the stimulation of local and systemic immunity. However, soluble drugs usually provide poor immunization when administered by mucosal routes and require the adjunct of a mucosal adjuvant or a drug delivery system (107). Mucosal immunization required five times less DNA than epidermal inoculation to induce the same level of protection against rotavirus challenge. This result indicated that mucosal immunization was superior to epidermal inoculation using the same vaccine dose (108). Many diseases such as measles, pertussis, tuberculosis, meningitis, and influenza are associated with the entry of pathogenic microorganisms across the respiratory mucosal surfaces. Therefore nasal vaccines delivery system is a good candidate for induction of both mucosal and systemic immune responses.

A range of different vaccine systems have been described in literature using either whole cells, spilt cells surface antigens, or DNA vaccine (genetic immunization) with or without adjuvant (109). Genetic immunization works by using host cells as protein factories to produce the plasmid encoded antigen. The translated protein is then processed and presented by the immune system in a mode similar to that, which occurs following a natural infection (110). The DNA was administered either directly in saline solution or in combination with carriers or adjuvants such as saponin, liposomes, and cochleates (108). The efficacy of DNA vaccines was monitored by humoral or cellular immune responses or resistance to virus challenges (111). Additionally, DNA vaccine was combined with liposomes or bioadhesive polymers and delivered by the mucosal route. These polymers can form highly viscous aqueous solutions that are thought to attach to mucosal surfaces (112).

Nasal vaccination has received a lot of attention since the nasal cavity is rich in NALT through which viral infections can be acquired. Intranasal vaccination has proven to be safe, easy, with less antigens being required via this route as compared to that needed for oral immunization. Nasal vaccination is thus a cost-effective means for controlling viral and bacterial diseases. The mucosal surfaces are rich in B-cells, T-cells, and plasma cells and such antigen-reactive cells are essential for the induction and maintenance of specific immune responses. Concerning both nasal vaccination and delivery of therapeutics, it has been shown that nanoscale drug carriers exhibiting mucoadhesive and permeation-enhancing properties have a great potential for improving the delivery through the nasal route (113).

Liposomes might be taken in by M-cell, and transported across the mucosal boundary, thereby transfect immune cells within NALT (114). Liposomes constitute attractive immunoadjuvants that can provide a vehicle or a carrier system into which antigens and coadjuvants can be incorporated (115). The adjuvant potential of liposomes in enhancing the systemic and mucosal immune responses was compared to immunopotentiating adjuvant which could directly activate the cells of the immune system. Intranasal administration of liposomes encapsulated with tetanus toxoid produced high mucosal IgA responses compared to liquid formulation. While liposomes loaded with tetanus toxoid and CpG-ODN produced higher serum IgG, antitoxin titers, and lower nasal IgA titers, compared with tetanus liposomes without adjuvant (116).

It is well established that liposomes possess immunoadjuvant activity and have a potential for the intranasal and oral delivery of protein antigen. Immunogenicity of tetanus toxoid has been improved when delivered mucosally (orally and nasally) associated with liposomes producing anti-toxoid IgG antibody titer similar to those obtained via the intramuscular delivery. Nasal delivery of tetanus toxoid entrapped in liposomes improved the immune response compared to delivery of free antigen. Furthermore, if the liposomes were taken up intact, the superficial layer of nose is highly vascularized and therefore a quantity of liposomes will pass directly into systemic circulation resulting in a systemic immune response and some liposomes would probably be taken up and delivered to underlying lymphoid cells of nasal associated lymphoid tissue (117). Antigen uptake by NALT is very important for stimulation immune responses.

Intranasal immunization with liposome-encapsulated influenza hemagglutinin (HA) DNA vaccine-induced T cell proliferation, indicative of $CD4^+$ activity and humoral immune responses, in addition to increasing serum IgG and IgA titres (118). Intranasal immunization with the liposome-supplemented vaccine conferred a better protection against an influenza infection than did intradermal immunization with the antigen alone (119).

Mucosal immunization required five times less DNA than epidermal inoculation to induce the same level of protection against rotavirus challenge. This result indicated that mucosal immunization was superior to epidermal inoculation using the same vaccine dose (108).

Liposome mediated DNA immunization by promoting the disruption of the endosomal membrane after endocytosis/phagocytosis of liposomal-DNA systems and ensuring escape of the plasmid DNA into the cytoplasm was recorded (120). Entrapment of plasmid DNA into liposomes will protect the nucleic acids against desoxyribonuclease attack which led to enhancing the efficacy the DNA vaccine (121).

Aramaki et al. studied the activation and mucosal response following nasal administration of liposomes in mice and found that IgG level was significantly elevated when bovine serum albumin (BSA)-associated liposomes were administrated intranasally twice with 4-week intervals. They also observed that the contribution of antigen-presenting cells in activation of systemic and mucosal immunity following intranasal administration was different (122).

Intranasal administration of glycol chitosan coated liposomes encapsulating plasmid DNA encoding surface protein of Hepatitis B virus induced significantly higher systemic, humoral, mucosal, and cellular immune responses when compared to naked DNA. This could have resulted from coating liposomes with glycol chitosan which may be able to remain homogeneously dispersed in the mucus, allowing good contact with respiratory mucosa at physiological pH due to electrostatic interaction by chitosan and mucosal surface. In addition, DNA encapsulated into both plain and chitosan-coated liposomes was protected against degradation by DNase, presumably because of the inability of the enzyme to reach its substrate, whereas naked DNA was completely degraded after 15 min (123). It has been reported that intranasal administration of liposomes can provide a promising adjuvant system for stimulation of antibody responses in general and mucosal secretory immunoglobulin (sIgA) responses in particular (124).

The effectiveness of immunizing humans by the intranasal route with *Streptococcus mutans* antigens, either incorporated into liposomes or in free form was investigated, in order to design a more effective approach to prevent oral diseases. Intranasal immunization resulted in primarily a nasal response and the liposomal antigen vaccine induced higher nasal but similar salivary IgA responses, when compared to responses induced with the free antigen vaccine. There may be a mucosal IgA inductive site which preferentially promoted a salivary response resulting in immune responses in nasal secretions, parotid saliva and serum (125).

Effective enhancement of mucosal immune responses was also observed with single intranasal immunization with liposome-formulated *Yersinia pestis* vaccine (formaldehyde-killed whole cell vaccine; KWC). Liposomes significantly enhanced the mucosal and systemic immune responses which were assessed 14 days following a single immunization. Immune responses were characterized by increased levels of specific IgA and IgG in mucosal secretions compared to *Y. pestis* KWC vaccine alone which induced low antibody titers (126). The ability of liposomes to enhance immune responses to vaccine antigens has been attributed primarily to an increased antigen uptake by antigen presenting cells and consequently increased antigen presentation to T-cells (127). In another study, it was illustrated that single nasal vaccinations with heat-labile toxin (HLT) adjuvant virosomal influenza vaccine can elicit humoral immune response that was comparable to that obtained after a single parenteral vaccination with the same total influenza virus hemagglutinin (HA) content (128).

Liposomes prepared from conventional ester lipids are usually ineffective as mucosal adjuvants, leading to the use of additional known adjuvants or targeting molecules such cholera toxin B subunit, lipid A, or interleukin-2 to improve the liposomal vaccine effectiveness (126, 129). Therefore, polar archaeal lipids have potential advantages for developing a non-replicating mucosal adjuvant and vaccine delivery system. Intranasal immunization of unilamellar archaeosomes (liposomes made from archaeal polar lipids) with encapsulated ovalbumin (OVA/archaeosomes), induced anti-OVA IgG, IgG1, and IgG2a antibody responses in sera and OVA-specific mucocal IgA in several mucosal sites. Calcium was added in the formulation to interact with the negatively charged archaeal polar lipids and to convert OVA/archaeosomes into an archaeal lipid mucosal vaccine adjuvant and delivery (AMVAD) vaccine (OVA/AMVAD). The ability to induce mucosal immune responses was demonstrated with OVA/AMVAD formulations prepared from a range of different polar lipid compositions (archaeol *Halobacterium salinarum*) or caldarchaeol (*Thermoplasma acidophilum*) or a mixture of these core lipids (*Methanobrevibacter smithii*), suggesting a broad applicability of archaeal polar lipids for intranasal immunization (130). AMVAD vaccines consisting of archaeal polar lipids could have potential advantages over the use of other vesicular delivery systems such as those based on ester lipids. Fatty acyl chains of many ester lipids used for preparing liposomes and cochleate formulations were shown to have some degree of unsaturation. Consequently, manufacturing and storage of theses formulations should be conducted under nitrogen to prevent lipid oxidation (131).

5.4. CNS Delivery

Intranasal administration offers a non-invasive alternative route to deliver drugs to the central nervous system, effectively by passing the BBB (**Fig. 8.5**) (132). It was proposed to be an excellent route of administration to target drugs directly to the brain via the olfactory neurons, which provide extracellular and intracellular pathways into CNS (133). It was suggested that substances could be absorbed via the olfactory route by two different mechanisms: the olfactory nerve pathway (axonal transport) and the olfactory epithelial pathway (26). The neural connections between the nasal mucosa and the brain provide a unique pathway for the non-invasive delivery of therapeutic agents to the CNS (134). The

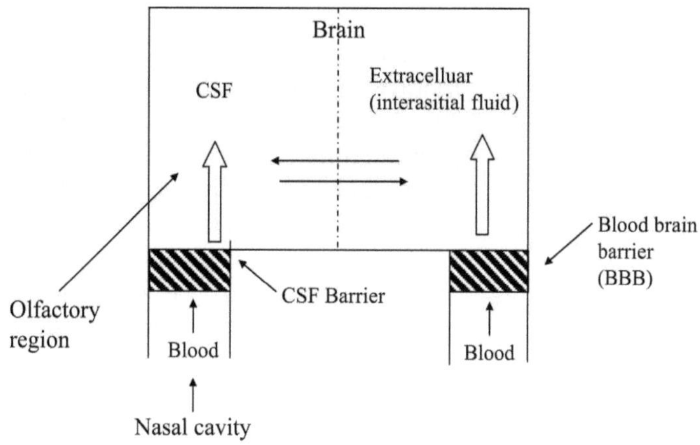

Fig. 8.5. Suggested transport pathways from nose to central nervous system.

olfactory neural pathway provides both an intraneuronal and extraneuronal pathway into the brain (134). The intraneuronal pathway involves axonal transport (olfactory nerve pathway) and it is considered a slow route where substance enters the olfactory neuron via endocytotic or pinocytotic mechanisms and diffuses to the olfactory bulb by utilizing the same mechanisms the cell uses to transport endogenous substances to the rest of the brain (135). The extraneuronal pathway (epithelial pathway) is a faster route for direct nose-to-brain transfer as compounds pass paracellularly across the olfactory epithelium into the perineural space, which is continuous with the subarachnoid space before transport to basolatral side of the olfactory epithelium which delivers drugs directly to the brain parenchymal tissue and/or CSF. Then the molecules can diffuse into the brain tissue or will be cleared by the CSF flow into the lymphatic vessels and subsequently into the systemic circulation (136).

Depending on the substance administered, axonal transport rates range from 20–400 mm/day to a slower 0.1–4 mm/day (137). Lipophilicity, molecular size, degree of dissociation, and route of administration are very important physicochemical factors must be considered when designing intranasal delivery for brain targeting (31). Formulation factors are also to be considered while designing brain targeted nasal drug delivery systems. The liquid formulations, liquid spray, and drops are the most widely used preparations for intranasal drug delivery. The nasal spray deposits anteriorly in the nasal atrium provide greater residence time, while the drops are dispersed throughout the length of the nasal cavity. Nasal sprays deposit more anteriorly, having more potential for brain delivery. The permeability of the posterior nasal passage is generally higher than the anterior passage (30).

Intranasal delivery is a promising route for delivery of drugs into the CNS and there are 35–40 drugs, which have been reported to reach CNS after nasal administration. Example of theses drugs that rapidly pass into CSF include estradiol and progesterone (138), carbamazepine (139), dopamine (140), risperidone (141), tacrine (142), and neuropeptide (143).

Nasal mucosa in the olfactory region is likely to be a promising target for mucosal immunization to protect the CNS from neurotropic viral infections. Intranasal immunization inducing mucosal and systemic immune responses blocks the propagation of neurotropic virus into the brain via the olfactory pathway and neutralizes the multiplication of virus in visceral organs, allowing more effective protection against neurotropic infections (98).

Rivastigmine is an acetyl cholinesterase which can be rapidly absorbed after oral administration but extensively metabolized by cholinesterase-mediated hydrolysis. Liposomes my provide carrier system for this drug through nasal route to CNS. In a comparative study intranasal liposome was compared with the oral free drug and it was recorded that liposomal formulation can provide ten times higher C_{max}, higher systemic AUC, and higher concentration in the brain compared to oral administration. The liposomal formulation provided better absorption into the brain following intranasal administration compared to the free drug. This might also be due to direct transfer of the drug from nasal mucosa to the brain via the olfactory route (144).

The intranasal administration of quercetin liposome to rats provided opportunity for the drug to enter the central nervous system and act on the central nervous system to promote anxiolytic activity and cognitive enhancing effect with high efficiency (145). The anxiolytic activity of oral quercetin liposomes was compared with intranasal quercetin liposomes, both routes showed anxiolytic and cognitive-enhancing effects. A lower dose and a faster rate were observed with intranasal quercetin liposomes when compared with oral quercetin liposomes. The intranasal quercetin liposome was thus considered as effective in the delivery of quercetin to the central nervous system (146).

References

1. Illum, L. (2003) Nasal drug delivery-possibilities, problems and solutions. *J Control Release* **87**, 187–198.
2. Ugwoke, M. I., Verbek, N., and Kinget, R. (2001) The biopharmaceutical aspects of nasal mucoadhesion drug delivery. *J Pharm Pharmacol* **53**, 3–22.
3. Arora, P., Sharma, S., and Gary, S. S. (2002) Permeability issues in nasal drug delivery. *Drug Discov Today* 7, 967–975.
4. Chien, Y. W., Su, K. S. E., and Chang, S. F. (1989) Anatomy and physiology of the nose. In: Swarbrick, J., (eds.), Nasal systemic drug delivery: drugs and pharmaceutical science, vol. 39. Marcel Dekker, New York, pp. 1–19.
5. Illum, L., Watts, P., Fisher, A. N., Hinchcliffe, M., Norbury, H., Jabbal-Gill, I., Nankervis, R., and Davis, S. S. (2002) Intranasal delivery of morphine. *J Pharmacol Exp Ther* **301**, 391–400.

6. Chen, X. Q., Fawcett, J. R., Rahman, Y. E., Ala, T. A., and Frey 2nd, W. H. (1998) Delivery of nerve growth factor to the brain via the olfactory pathway. *J Alzheimer Dis* **1**, 35–44.
7. Liu, J. X. F., Fawcett, R., Thorne, R. G., DeFor, T. A., and Frey 2nd, W. H. (2001) Intranasal administration of insulin-like growth factor-I bypasses the blood-brain barrier and protects against focal cerebral ischemicdamage. *J Neurol Sci* **187**, 91–97.
8. Fehm, H. L., Perras, B., Smolink, R., Kern, W., and Born, J. (2000) Manipulating neuropeptidegric pathways in humans: a novel approach in neuropharmacology? *Eur J Pharmacol* **405**, 43–54.
9. Giacobini, P., Kopin, A. S., Beart, P. M., Mercer, L. D., Fasolo, A., and Wray, S. (2004) Cholecystokinin modulates migration of gonadotropin-releasing hormone-1 neurons. *J Neurosci* **24**, 4737–4748.
10. Turker, S., Onur, E., and Ozer, Y. (2004) Nasal route and drug delivery systems. *Pharm World Sci* **26**, 137–142.
11. Merkus, F. W., Verhoef, J. C., Schipper, N. G., and Marttin, E. (1998) Nasal mucociliary clearance as a factor in nasal drug delivery. *Adv Drug Deliv Rev* **29**, 13–38.
12. Alsarra, I. A., Hamed, A. Y., Mahrous, G. M., El Maghraby, G. M., Al-Robayan, A. A., and Alanazi, F. K. (2008) Mucoadhesive polymeric hydrogels for nasal delivery of acyclovir. *Drug Dev Ind Pharm* **3**, 1–11.
13. Guo, J., Ping, Q., Jiang, G., Dong, J. Q. S., Feng, L., Li, Z., and Li, C. (2004) Transport of leuprolide across rat intestine, rabbit intestine and caco-2 cell monolayer. *Int J Pharm* **278**, 415–422.
14. Varshosaz, J., Sadrai, H., and Heidar, A. (2006) Nasal delivery of insulin using bioadhesive chitosan gels. *Drug Deliv* **13**, 31–38.
15. Callens, G., and Remon, J. P. (2000) Evaluation of starch-maltodextrin-carpobol 974 for nasal delivery of insulin in rabbits. *J Control Release* **66**, 215–220.
16. Lindhardt, K., Ravn, G., Gizurarson, S., and Bechgaard, E. (2000) Intranasal absorption of buprenorphine in-vivo bioavailability study in sheep. *Int J Pharm* **205**, 159–163.
17. Soan, R. J., Hinchcliff, M., Davis, S. S., and Illum, L. (2001) Clearance characteristics of chitosan based formulations in sheep nasal cavity. *Int J Pharm* **217**, 183–191.
18. Wang, W. (1999) Instability, stabilization and formulation of liquid protein pharmaceuticals. *Int J Pharm* **85**, 129–188.
19. Kublik, H., and Vidgren, M. T. (1998) Nasal delivery systems and their effect on deposition and absorption. *Adv Drug Deliv Rev* **29**, 157–177.
20. Jiang, W., and Schwendeman, S. P. (2001) Stabilization and controlled release of bovin serum albumin encapsulated in poly(D-L-lactide) and poly(ethylene glycol) microspheres blends. *Pharm Res* **18**, 878–885.
21. Mainardes, R. M., Urban, M. C., Cinto, P. O., Chaud, M. V., Evangelista, R. C., and Daflon Gremião, M. P. (2006) Liposomes and micro/nanoparticles as colloidal carriers for nasal drug delivery. *Curr Drug Deliv* **3**, 275–285.
22. Illum, I. (2007) Nanoparticulate systems for nasal delivery of drugs: a real improvement over simple systems? *J Pharm Sci* **96**, 473–483.
23. Stanleya, A. C., Huntleya, J. F., Jeffreyb, M., and Buxtona, D. (2001) Characterization of ovine nasal-associated lymphoid tissue and identification of M cells in the overlying follicle-associated epithelium. *J Comp Pathol* **125**, 262–270.
24. Behrens, I., Pena, A. V. I., Alonso, M. J., and Kissel, T. (2002) Comparative uptake studies of bioadhesive nanoparticles in human intestinal cell lines and rats: the effect of mucus on particle adsorption and transport. *Pharm Res* **19**, 1185–1193.
25. Illum, L. (2000) Transport of drug from the nasal cavity to central nervous system. *Eur J Pharm Sci* **11**, 1–18.
26. Illum, L. (2004) Is nose-to-brain transport of drugs in man a reality? *J Pharm Pharmacol* **56**, 3–17.
27. Cerijido, M., Shoskani, L., and Contreres, R. (2000) Molecular physiology and pathophysiology of tight junction I. Biogenesis of tight junction and epithelial polarity. *Am J Physiol* **279**, 477–482.
28. Hayashi, M., Hirasawa, T., Muraoka, T., Shiga, M., and Awaza, S. (1985) Comparison of water influx and sieving coefficient in rat jejunal, rectal and nasal absorption of antipyrine. *Chem Pharm Bull* **33**, 2149–2152.
29. Frey 2nd, W. H. (2002) Intranasal delivery: bypassing the blood-brain barrier to deliver therapeutic agents to the brain and spinal cord. *Drug Deliv Technol* **2**, 46–49.
30. Behl, C. R., Pimplaskar, H. K., Sileno, A. P., Demeireles, J., and Remo, V. D. (1998) Effect of physiochemical properties and other factors on systemic nasal drug delivery. *Adv Drug Deliv Rev* **29**, 89–116.

31. Vyas, T. K., Shahiwala, A., Marathe, S., and Misra, A. (2005) Intranasal drug delivery for central nervous system. *Curr Drug Deliv* **2**, 165–175.
32. Thorne, R. G., Pronk, G. J., and Padmanabhan, V. (2004) Delivery of insulin like growth factor-I to the rat brain and spinal cord along olfactory and trigeminal pathways following intranasal administration. *Neuroscience* **127**, 481–496.
33. Thorne, R. G., Pronk, G., and Frey, W. H. (2000) Delivery of insulin like growth factor-I to the brain and spinal cord along olfactory and trigeminal pathways following intranasal administration: a non-invasive method for bypassing the blood-brain barrier. *Soc Neurosci Abstr* **26**, 1365.
34. Yang, J. P., Liub, H. J., Chenga, S. M., Wanga, Z. L., Chenga, X., Yuc, H. X., and Liua, X. F. (2009) Direct transport of VEGF from the nasal cavity to brain. *Neurosci Lett* **449**, 108–111.
35. Ross, T. M., Martinez, P. M., Renner, J. C., Thorne, R. G., Hanson, L. R., and Frey 2nd, W. H. (2004) Intranasal administration of interferon beta bypasses the blood-brain barrier to target the central nervous system and cervical lymph nodes: a non-invasive treatment strategy for multiple sclerosis. *J Neuroimmunol* **151**, 66–77.
36. Földi, M. (1996) The brain and the lymphatic system (I). *Lymphology* **29**, 1–9.
37. Ugwoke, M. I., Agu, R. U., Verbeke, N., and Kinget, R. (2005) Nasal mucoadhesive drug delivery: background, applications, trends and future perspectives. *Adv Drug Deliv Rev* **57**, 1640–1665.
38. Soane, R. J., Frier, M., Perkins, A. C., Jones, N. S., Davis, S. S., and Illum, L. (1999) Evaluation of the clearance characteristics of bioadhesive systems in humans. *Int J Pharm* **178**, 55–65.
39. Cornaz, A. L., and Buri, P. (1994) Nasal mucosa as an absorption barrier. *Eur J Pharm Biopharm* **40**, 261–270.
40. Green, A., Smallman, L. A., Logan, A. C. M., and Darke-lee, A. B. (1995) The effect of temperature on nasal ciliary beat frequency. *Clin Otolaryngol* **20**, 178–180.
41. Chung, F. Y., and Donovan, M. D. (1996) Nasal presystemic metabolism of peptide drugs: substance P metabolism in the sheep nasal cavity. *Int J Pharm* **128**, 229–237.
42. Bernkop-schnurch, A. (1998) Use of inhibitory agents to overcome the enzymatic barrier to perorally administered therapeutic peptides and proteins. *J Control Release* **52**, 1–16.
43. Krishnamoorthy, R., and Mitra, AK. (1998) Prodrugs for nasal drug delivery. *Adv Drug Deliv Rev* **29**, 135–146.
44. Yamamoto, A., Iseki, T., Ochi-Sugiyama, M., Okada, N., Fujita, T., and Muranishi, S. (2001) Absorption of water-soluble compounds with different molecular weight. *J Control Release* **76**, 363–374.
45. Hinchcliffe, M., and Illum, L. (1999) Intranasal insulin delivery and therapy. *Adv Drug Deliv Rev* **35**, 199–234.
46. Corbo, D. C., Liu, J. C., and Chien, Y. W. (1990) Characterization of the barrier properties of mucosal membrane. *J Pharm Sci* **79**, 202–206.
47. Derendorf, H., and Meltzer, E. O. (2008) Molecular and clinical pharmacology of intranasal corticosteroids: clinical and therapeutic implications. *Allergy* **63**, 1292–1300.
48. Yang, C., Gao, H., and Mitra, A. (2000) Chemical stability, enzymatic hydrolysis, and nasal uptake of amino acid ester prodrugs of acyclovir. *J Pharm Sci* **90**, 617–624.
49. Al-Ghananeem, A. M., Traboulsi, A. A., Dittert, L. W., and Hussain, A. A. (2002) Targeted brain delivery of 17β-estradiol via nasally administered water soluble prodrugs. *AAPS Pharm Sci Tech* **3**, 1–8.
50. El Maghraby, G. M., Barry, B. W., and Williams, A. C. (2008) Liposomes and skin. From drug delivery to model membranes. *Eur J Pharm Sci* **34**, 203–222.
51. New, R. R. C. (1990) Introduction. In: New, R. (ed.), Liposomes a practical approach, 1st Ed//. Oxford University Press, Oxford, pp. 1–32.
52. Brandl, M. (2001) Liposomes as drug carriers: a technological approach. *Biotechnol Annu Rev* **7**, 59–85.
53. El Maghraby, G. M., Williams, A. C., and Barry, B. W. (2005) Drug interaction and location in liposomes: correlation with polar surface area. *Int J Pharm* **292**, 179–185.
54. Gregoriadis, G., and Florence, A. T. (1993) Liposomes in drug delivery, clinical, diagnostic and ophthalmic potential. *Drugs* **45**, 15–28.
55. Iwanaga, K., Matsumoto, S., Morimoto, K., Kakemi, M., Yamashita, S., and Kimura, T. (2000) Usefulness of liposomes as an intranasal dosage formulation for topical drug application. *Biol Pharm Bull* **23**, 323–326.
56. Sakaue, G., Hiroi, T., Nakagawa, Y., Someya, K., Iwatani, K., Sawa, Y.,

Takahashi, H., Honda, M., Kunisawa, J., and Kiyono, H. (2003) HIV mucosal vaccine: Nasal immunization with gp160-encapsulated hemagglutinating virus of Japan-liposome induces antigen-specific CTLs and neutralizing antibody responses. *J Immunol* **170**, 495–502.

57. De Jonge, M. I., Hamstra, H. J., Jiskoot, W., Roholl, P., Williams, N. A., Dankert, J., Van Alphen, L., and Van der Ley, P. (2004) Intranasal immunisation of mice with liposomes containing recombinant meningococcal OpaB and OpaJ proteins. *Vaccine* **22**, 4021–4028.
58. Lambkin, R., Oxford, J. S., Bossuyt, S., Mann, A., Metcalfe, I. C., Herzog, C., Viret, J. F., and Gluck, R. (2004) Strong local and systemic protective immunity induced in the ferret model by an intranasal virosome formulated influenza subunit vaccine. *Vaccine* **22**, 4390–4396.
59. Langley, J. M., Halperin, S. A., McNeil, S., Smith, B., Jones, T., Burt, D., Mallett, C. P., Lowell, G. H., and Fries, L. (2006) Safety and immunogenicity of a proteosome(TM)-trivalent inactivated influenza vaccine, given nasally to healthy adults. *Vaccine* **24**, 1601–1608.
60. Chen, M., Deng, Q., Li, X. R., and Liu, Y. (2007) The hypoglycaemia effect of salmon calcitonin ultra-fexible liposomes after nasal administration in rats. *Yao Xue Xue Bao* **42**, 681–686.
61. Patel, G. B., Ponce, A., Zhou, H., and Chen, W. (2008) Safety of intranasal administration archaeal lipid mucosal vaccine aduvant and delivery (AMWAD) vaccine in mice. *Int J Toxicol* **27**, 329–339.
62. Tafaghodi, M., Jaafari, M. R., and Sajadi-Tabassi, S. A. (2008) Nasal immunization studies by cationic, fusogenic and cationic-fusogenic liposomes encapsulated with tetanus toxoid. *Curr Drug Deliv* **5**, 108–113.
63. Rosada, R. S., de la Torre, L. G., Frantz, F. G., Trombone, A. P., Zárate-Bladés, C. R., Fonseca, D. M., Souza, P. R., Brandão, I. T., Masson, A. P., Soares, E. G., Ramos, S. G., Faccioli, L. H., Silva, C. L., Santana, M. H., and Coelho-Castelo, A. A. (2008) Protection against tuberculosis by a single intranasal administration of DNA-hsp65 vaccine complexed with cationic liposomes. *BMC Immunol* **9**, 1–13.
64. Kaur, I. P., Garg, A., Singla, A. K., and Aggarwal, D. (2004) Vesicular systems in ocular drug delivery: an overview. *Int J Pharm* **269**, 1–14.
65. Mozafari, M. R. (2005) Liposomes: an overview of manufacture techniques. *Cell Mol Biol Lett* **10**, 711–719.
66. Gregoriadis, G. (1995) Engineering liposomes for drug delivery: progress and problems. *Trends Biotechnol* **13**, 527–537.
67. Nii, T., and Ishii, F. (2005) Encapsulation efficiency of water-soluble and insoluble drugs in liposomes prepared by the microencapsulation vesicle method. *Int J Pharm* **298**, 198–205.
68. Uchegbu, I. F., and Vyas, S. P. (1998) Non-ionic surfactant based vesicles (niosomes) in drug delivery. *Int J Pharm* **172**, 33–70.
69. Alsarra, I. A. (2008) Evaluation of proniosomes as an alternative strategy to optimize piroxicam transdermal delivery. *J Microencapsul* **11**, 1–7.
70. Alsarra, I. A., Bosela, A. A., Al-Mohizea, A. M., Mahrous, J. M., and Neau, S. H. (2005) Modulating intestinal uptake of atenolol using niosomes as drug permeation enhancers. *Sci Pharm* **73**, 81–93.
71. Alsarra, I. A., Bosela, A. A., Ahmed, S. M., and Mahrous, J. M. (2005) Proniosomes as a drug carrier for transdermal delivery of ketorolac. *Eur J Pharm Biopharm* **59**, 485–490.
72. Manosroi, A., Chutoprapat, R., Abe, M., and Manosroi, J. (2008) Characteristics of niosomes prepared by supercritical carbon dioxide (scCO2) fluid. *Int J Pharm* **352**, 248–255.
73. Hu, C., and Rhodes, D. G. (1999) Proniosomes: a novel drug carrier preparation. *Int J Pharm* **185**, 23–35.
74. Carafa, M., Santucci, E., and Lucania, G. (2002) Lidocaine-loaded non-ionic surfactant vesicles: characterization and in vitro permeation studies. *Int J Pharm* **231**, 21–32.
75. Biju, S. S., Talegaonkar, S., Mishra, P. R., and Khar, R. K. (2006) Vesicular systems: an overview. *Indian J Pharm Sci* **68**, 141–153.
76. Dufes, C., Schatzlein, A. G., Tetley, L., Gray, A. I., Watson, D. G., Olivier, J. C., Couet, W., and Uchegbu, I. F. (2000) Niosomes and polymeric chitosan based vesicles bearing transferring and glucose ligands for drug targeting. *Pharm Res* **17**, 1250–1258.
77. Vyas, S. P., Singh, R. P., Jain, S., Mishra, V., Mahor, S., Singh, P., Gupta, P. N., Rawat, A., and Dubey, P. (2005) Non-ionic surfactant based vesicles (niosomes) for non-invasive topical genetic immunization against hepatitis B. *Int J Pharm* **296**, 80–86.

78. Hu, C., and Rhodes, D. G. (2000) Proniosomes: a novel drug carrier preparation. *Int J Pharm* **206**, 110–122.
79. Katare, O. P., Vyas, S. P., and Dixit, V. K. (1991) Preparation and performance evaluation of plain proliposome systems for cytoprotection. *J Microencapsul* **8**, 295–300.
80. Muramatsu, K., Maitani, Y., Takayama, K., and Nagai, T. (1999) The relationship between the liposomal membrane of insulin after nasal administration of liposomes modified with an enhancer containing insulin in rabbits. *Drug Dev Ind Pharm* **25**, 1099–1105.
81. Law, S. L., Huang, K. J., and Chou, H. Y. (2001) Preparation of desmopressin-containing liposomes for intranasal delivery. *J Control Release* **70**, 375–382.
82. Maitani, Y., Asano, S., Takahashi, S., Nasayuki, M., and Nagai, T. (1992) Permeability of insulin in liposomes through the nasal mucosa of rabbits. *Chem Pharm Bull* **40**, 1569–1572.
83. Jain, A. K., Chalasani, K. B., Khar, R. K., Ahamed, F. J., and Diwan, P. V. (2007) Mucoadhesive multivesicular liposomes as an effective carrier for transmucosal insulin delivery. *J Drug Target* **15**, 417–427.
84. Law, S. L., Huang, K. J., Chou, H. Y., and Cherng, J. Y. (2001) Enhancement of nasal absorption of calcitoninloded in liposomes. *J Liposome Res* **11**, 165–174.
85. Alsarra, I. A., Hamed, A. Y., and Alanazi, F. K. (2008) Acyclovir liposomes for intranasal systemic delivery: development and pharmacokinetics evaluation. *Drug Deliv* **15**, 313–321.
86. Ding, W. X., Qi, X. R., Fu, Q., and Piao, H. S. (2007) Pharmacokinetics and pharmacodynamics of sterylglucoside-modified liposomes for levonorgestrel delivery via nasal route. *Drug Deliv* **14**, 101–104.
87. Shahiwala, A., and Misra, A. (2006) Preliminary investigation of the nasal delivery of liposomal leuprorelin acetate fro contraception in rats. *J Pharm Pharmacol* **58**, 19–26.
88. Lee, V. H. I., Yamamoto, A., and Kompella, U. B. (1991) Mucosal penetration enhancers for facilitation of peptide and protein drug absorption. *Crit Rev Ther Drug Carrier Syst* **8**, 191–192.
89. Zhang, Y. J., Wang, X. L., Wu, J. M., and Cheng, M. X. (2007) Studies on preparation of recombinant hirudin-2 liposome and its pharmacokinetics by nasal delivery in rats. *Zhongguo Zhong Yao Za Zhi* **32**, 801–804.
90. Jung, B. H., Chung, B. C., Chung, S. J., Lee, M. H., and Shim, C. K. (2000) Prolonged delivery of nicotine in rats via nasal administration of proliposomes. *J Control Release* **66**, 73–79.
91. Ahn, B. N., Kim, S. K., and Shim, C. K. (1995) Proliposomes as an intranasal dosage form for the sustained delivery of propranolol. *J Control Release* **34**, 203–210.
92. Kaparissides, C., Alexandridou, S., Kotti, K., and Chaitidou, S. (2006) Recent advances in novel drug delivery systems. *J Nanosci Nanotechnol* **2**, 1–11.
93. Schneider, S. D., Rusconi, S., Seger, R. A., and Hossle, J. P. (1997) Adenovirusmediated gene transfer into monocyte-derived macrophages of patients with X-linked chronic granulomatous disease: ex vivo correction of feficient respiratory burst. *Gene Ther* **4**, 524–532.
94. Alanazi, F., Fu, Z. F., and Lu, D. R. (2004) Effective transfection of rabies DNA vaccine in cell culture using an artificial lipoprotein carrier system. *Pharm Res* **21**, 675–682.
95. Oudrhiri, N., Vigneron, J. P., Peuchmaur, M., Leclerc, T., Lehn, J. M., and Lehn, P. (1997) Gene transfer by guanidinium-cholesterol cationic lipids into airway epithelial cells in vitro and in vivo. *Proc Natl Acad Sci USA* **94**, 1651–1656.
96. Simoes, S., Slepushkin, V., Pretzer, E., Dazin, P., Gaspar, R., de Lima, M. C., and Duzgunes, N. (1999) Transfection of human macrophages by lipoplexes via the combined use of transferrin and pH-sensitive peptides. *J Leuk Biol* **65**, 270–279.
97. Kawakami, S., Sato, A., Nishikawa, M., Yamashita, F., and Hashida, M. (2000) Mannose receptor-mediated gene transfer into macrophages using novel mannosylated cationic liposomes. *Gene Ther* **7**, 292–299.
98. Goncharova, E. P., Ryzhikov, A. B., Bulychev, L. E., Sizov, A. A., Lebedev, L. R., Poryvaev, V. D., Karpenko, L. I., and Il'ichev, A. A. (2002) A study of systems for delivering antigens and plasmid DNA for intranasal immunization against tick-borne encephalitis virus. *Wien Klin Wochenschr* **31**, 630–635.
99. Jain, P. T., Seth, P., and Gewirtz, D. A. (1999) Estradiol enhances liposomemediated uptake, preferential nuclear accumulation and functional expression of exogenous genes in MDA-MB231 breast tumor cells. *Biochim Biophys Acta* **1451**, 224–232.
100. Jain, P. T., and Gewirtz, D. A. (1998) Estradiol enhances gene delivery to human breast tumor cells. *J Mol Med* **76**, 709–714.

101. Wiseman, J. W., Goddard, C. A., and Colledge, W. H. (2001) Steroid hormone enhancement of gene delivery to a human airway epithelial cell line in vitro and mouse airways in vivo. *Gene Ther* **8**, 1562–1571.

102. Braun, S, Jenny, C., Thioudellet, C., Perraud, F., Claudepierre, M. C., Langle-Rouault, F., Ali-Hadji, D., Schughart, K., and Pavirani, A. (1999) In vitro and in vivo effects of glucocorticoids on gene transfer to skeletal muscle. *FEBS Lett* **454**, 277–282.

103. Tanaka, S. I., Yamakawa, T., Kimura, M., Aok, I., Kameie, J., Okudac, K., and Mobbsg, C. (2004) Daily nasal inoculation with the insulin gene ameliorates diabetes in mice. *Diabetes Res Clin Pract* **63**, 1–9.

104. Song, Y. K., and Liu, D. (1998) Free liposomes enhance the transfection activity of DNA/lipid complexes in vivo by intravenous administration. *Biochim Biophys Acta* **1372**, 141–150.

105. Hyde, S. C., Southern, K. W., Gileadi, U., Fitzjohn, E. M., Mofford, K. A., Waddell, B. E., Gooi, H. C., Goddard, C. A., Hannavy, K., Smyth, S. E., Egan, J. J., Sorgi, F. L., Huang, L., Cuthbert, A. W., Evans, M. J., Colledge, W. H., Higgins, C. F., Webb, A. K., and Gill, D. R. (2000) Repeat administration of DNA/liposomes to the nasal epithelium of patients with cystic fibrosis. *Gene Ther* **7**, 1156–1165.

106. Suckow, M. A., Jarvinen, L. Z., HogenEsch, H., Park, K., and Bowersock, T. L. (2002) Immunization of rabbits against a bacterial pathogen with and alginate microparticle vaccine. *J Control Release* **85**, 227–235.

107. Czerkinsky, C., Anjuere, F., McGhee, J. R., George-Chandy, A., Holmgren, J., and Kieny, M. P. (1999) Mucosal immunity and tolerance: relevance to vaccine development. *Immunol Rev* **170**, 197–222.

108. Chen, S. C., Fynan, E. F., Greenberg, H. B., and Herrmann, J. E. (1999) Immunity obtained by gene-gun inoculation of a rotavirus DNA vaccine to the abdominal epidermis or anorectal epithelium. *Vaccine* **17**, 3171–3176.

109. FitzGerald, D., and Mrsny, R. J. (2000) New approaches to antigen delivery. *Crit Rev Ther Drug Carrier Syst* **17**, 1405–1412.

110. Donnelly, J. J., Ulmer, J. B., Shiver, J. W., and Liu, M. A. (1997) DNA vaccines. *Annu Rev Immunol* **15**, 617–648.

111. Schultz, J., Dollenmaier, G., and Mölling, K. (2000) Update on antiviral DNA vaccine research (1998–2000). *Intervirology* **43**, 197–217.

112. Sha, Z., Vincent, M. J., and Compans, R. W. (1999) Enhancement of mucosal immune responses to the influenza virus HA protein by alternative approaches to DNA immunization. *Immunobiology* **200**, 21–30.

113. Manochaa, M., Pala, P. C., Chitralekhaa, K. T., Thomasa, B. E., Tripathia, V., Guptab, S. D., Paranjapec, R., Kulkarnic, S., and Rao, D. N. (2005) Enhanced mucosal and systemic immune response with intranasal immunization of mice with HIV peptides entrapped in PLG microparticles in combination with Ulex Europaeus-I lectin as M cell target. *Vaccine* **23**, 5599–5617.

114. Janes, K. A., Calvo, P., and Alonso, M. J. (2001) Polysaccharide colloidal particles as delivery systems for macromolecules. *Adv Drug Deliv Rev* **47**, 83–97.

115. Isaka, M., Yasuda, Y., Mizokami, M., Kozuka, S., Taniguchi, T., Matano, K., Maeyama, J., Mizuno, K., Morokuma, K., Ohkuma, K., Goto, N., and Tochikubo, K. (2001) Mucosal immunization against hepatitis B virus by intranasal co-administration of recombinant hepatitis B surface antigen and recombinant cholera toxin B subunit as an adjuvant. *Vaccine* **19**, 1460–1466.

116. Tafaghodi, M., Jaafari, M. R., and Tabassi, S. A. S. (2006) Nasal immunization studies using liposomes loaded with tetanus toxoid and CpG-ODN. *Eur J Pharm Biopharm* **64**, 138–145.

117. Alpar, H. O., Bowen, J. C., and Brown, M. R. W. (1992) Effectiveness of liposomes as adjuvant of orally and nasally administered tetanus toxoid. *Int J Pharm* **88**, 335–344.

118. Wang, D., Christopher, M. E., Nagata, L. P., Zabielski, M. A., Li, H., Wong, J. P., and Samuel, J. (2004) Intranasal immunization with liposome encapsulated plasmid DNA encoding influenza virus hemagglutinin elicits mucosal, cellular and humoral immune responses. *Clin Virol* **31**, 99–106.

119. Ninomiya, A., Ogasawara, K., Kajino, K., Takada, A., and Kida, H. (2002) Intranasal administration of a synthetic peptide vaccine encapsulated in liposome together with an anti-CD40 antibody induces protective immunity against influenza A virus in mice. *Vaccine* **20**, 3123–3129.

120. Perrie, Y., Frederik, P. M., and Gregoriadis, G. (2001) Liposome-mediated DNA vaccination: the effect of vesicle composition. *Vaccine* **19**, 3301–3310.

121. Mannino, R. J., Canki, M., Feketeova, E., Scolpino, A. J., Wang, Z., Zhang, F., Kheiri, M. T., and Gould-Fogerite, S. (1998) Targeting immune response induction with cochleate and liposome-based vaccines. *Adv Drug Deliv Rev* **32**, 273–287.
122. Aramaki, Y., Fuiji, Y., Yachi, K., Kikuchi, H., and Tsuchiya, S. (1994) Activation of systemic and mucosal immune response following nasal administration of liposomes. *Vaccine* **12**, 1241–1245.
123. Khatri, K., Goyal, A. K., Gupta, P. N., Mishra, N., Mehta, A., and Vyas, S. P. (2008) Surface modified liposomes for nasal delivery of DNA vaccine. *Vaccine* **26**, 2225–2233.
124. de Haan, A., Geerligs, H. J., Huchshorn, J. P., van Scharrenburg, G. J. M., Palache, A. M., and Wilschut, J. (1995) Mucosal immunoadjuvant activity of liposomes: induction of systemic IgG and secretory IgA responses in mice by intranasal immunization with an influenza subunit vaccine and coadministered liposomes. *Vaccine* **13**, 155–162.
125. Childers, N. K., Tong, G., Mitchell, S., Kirk, K., Russell, M. W., and Michalek, S. M. A. (1999) Controlled clinical study of the effect of nasal immunization with a *Streptococcus mutans* antigen alone or incorporated into liposomes on induction of immune responses. *Infect Immun* **67**, 618–623.
126. Baca-Estrada, M. E., Foldvari, M., Snider, M., Harding, K., Kournikakisc, B., Babiuka, L. A., and Griebela, P. (2000) Intranasal immunization with liposome-formulated *Yersinia pestis* vaccine enhances mucosal immune responses. *Vaccine* **18**, 2203–2211.
127. Gregoriadis, G. (1990) Immunological adjuvants: a role for liposomes. *Immunol Today* **11**, 89–97.
128. Glück, U., Gebbers, J. O., and Glück, R. (1999) Phase 1 evaluation of intranasal virosomal influenza vaccine with and without escherichia coli heat-labile toxin in adult volunteers. *J Virol* **73**, 7780–7786.
129. Harokopakis, E., Hajishengallis, G., and Michalek, S. M. (1998) Effectiveness of liposomes possessing surface-linked recombinant B subunit of cholera toxin as an oral antigen delivery system. *Infect Immun* **66**, 4299–4304.
130. Patel, G. B., Zhou, H., Ponce, A., and Chen, W. (2007) Mucosal and systemic immune responses by intranasal immunization using archaeal lipid-adjuvanted vaccines. *Vaccine* **25**, 8622–8636.
131. Gould-Fogerite, S., and Mannino, R. J. (2000) Cochleates for induction of mucosal and systemic immune responses. In: O'Hagan D. T., (ed.), Vaccine adjuvants: preparation methods and research protocols. Humana Press, Inc., Totowa, NJ, 179–196.
132. Graff, C. L., and Pollack, G. M. (2005) Nasal drug administration: potential for targeted central nervous system delivery. *J Pharm Sci* **94**, 1187–1195.
133. Thorne, R. G., Emory, C. R., Ala, T. A., and Frey, W. H. (1995) Quantitative analysis of the olfactory pathway for drug delivery to the brain. *Brain Res* **692**, 278–282.
134. Mathison, S., Nagilla, R., and Kompella, U. B. (1998) Nasal route for direct delivery of solutes to the central nervous system: factor fiction? *J Drug Target* **5**, 415–441.
135. Yamada, T. (2004) The potential of the nasal mucosa route for emergency drug administration via a high-pressure needleless injection system. *Anesth Prog* **51**, 56–61.
136. Costantino, H. R., Lisbeth, I., Brandt, G., Johnson, P. H., and Quay, S. C. (2007) Intranasal delivery: physicochemical and therapeutic aspects. *Int J Pharm* **337**, 1–24.
137. Bleske, B. E., Warren, E. W., Rice, T. L., Shea, M. J., Amidon, G., and Knight, P. (1992) Comparison of intravenous and intranasal administration of epinephrine during CPR in a canine model. *Ann Emerg Med* **21**, 1125–1130.
138. van den Berg, M. P., Verhoef, J. C., Romeijn, S. G., and Merkus, F. W. H. M. (2004) Uptake of estradiol or progesterone into the CSF following intranasal and intravenous delivery in rats. *Eur J Pharm Sci* **58**, 131–135.
139. Gavini, E., Hegge, A. B., Rassu, G., Sanna, V., Testa, C., Pirisino, G., Karlsen, J., and Giunchedi, P. (2006) Nasal administration of carbamazepine using chitosan microspheres: in vitro/in vivo studies. *Int J Pharm* **3**, 9–15.
140. Chemuturi, N. V., and Donovan, M. D. (2007) Role of organic cation transporters in dopamine uptake across olfactory and nasal respiratory tissues. *Mol Pharm* **4**, 936–942.
141. Kumar, M., Misra, A., Babbar, A. K., Mishra, A. K., Mishra, P., and Pathak, K. (2008) Intranasal nanoemulsion based brain targeting drug delivery system of risperidone. *Int J Pharm* **24**, 285–291.
142. Jogani, V. V., Shah, P. J., Mishra, P., Mishra, A. K., and Misra, A. R. (2008) Intranasal mucoadhesive microemulsion of tacrine to improve brain targeting. *Alzheimer Dis Assoc Disord* **22**, 116–124.

143. Dhuria, S. V., Hanson, L. R., and Frey, W. H. (2009) Novel vasoconstrictor formulation to enhance intranasal targeting of neuropeptide therapeutics to the central nervous system. *J Pharmacol Exp Ther* **328**, 312–320.
144. Arumugam, K., Subramanian, G. S., Mallayasamy, S. R., Averineni, R. K., Reddy, M. S., and Udupa, N. (2008) A study of rivastigmine liposomes for delivery into the brain through intranasal route. *Acta Pharm* **58**, 287–297.
145. Wattanathorn, J., Phachonpai, W., Priprem, A., and Suthiparinyanont, S. (2007) Intranasal administration of quercetin liposome decreases anxiety-like behaviour and increases spatial memory. *Am J Agric Biol Sci* **2**, 31–35.
146. Priprem, A., Watanatorn, J., Sutthiparinyanont, S., Phachonpai, W., and, Muchimapura, S. (2008) Anxiety and cognitive effects of quercetin liposomes in rats. *Nanomedicine* **4**, 70–78.

Chapter 9

Convection-Enhanced Drug Delivery and Monitoring in a Rat Model

Yael Mardor and Zvi Ram

Abstract

Convection-enhanced drug delivery is a novel technology used to deliver drugs into brain tissue and which is currently evaluated in clinical trials. Drugs are delivered continuously via intracranial catheters and enable to achieve large volume of distributions at high drug concentrations with minimum systemic toxicity.

Accumulated clinical experience has shown that convection does not always form and the efficiency of distribution varies among patients and different drugs. The efficiency of drug distribution depends on multiple treatment and physiological parameters, such as infusate characteristics, catheter type, flow rate, volume of distribution, and tissue characterization along the catheter path and at the catheter tip. Therefore, extensive research is ongoing focusing on optimizing convection to enhance the therapeutic effect of drugs in various brain disorders.

This chapter describes various aspects of performing convection-enhanced drug delivery experiments in a rat brain model.

Key words: Convection-enhanced drug delivery (CED), rat brain, imaging, MRI, brain tumors, infusion, nanoparticles.

1. Introduction

Local delivery of drugs into the brain and/or brain tumors enables maximizing exposure of the target tissue to the administered drug while minimizing systemic exposure. Convection-enhanced drug delivery (CED), established by creating an infusion-mediated pressure gradient through intracranial catheters, evenly distributes therapeutic agents in brain or tumor tissue, enabling reaching orders of magnitude higher drug concentrations within large tissue volumes than can be achieved by systemic delivery (1, 2).

K.K. Jain (ed.), *Drug Delivery to the Central Nervous System*, Neuromethods 45,
DOI 10.1007/978-1-60761-529-3_9, © Humana Press, a part of Springer Science+Business Media, LLC 2010

Efficient distribution of the convected drug depends on many parameters, such as infusate characteristics, catheter type, flow rate, volume of distribution, and tissue characterization along the catheter path and at the catheter tip. (3–7). Extensive research is ongoing focusing on optimizing these parameters and developing new methodologies and devices in order to increase the probability of forming efficient convection in a controlled manner, thus obtaining significant treatment effects in various brain disorders.

Recent innovations in this field include real-time imaging methodologies, improved catheter designs, software packages for treatment planning, improved therapeutic agents, and improved solvents. The implementation of these improvements enables increased treatment effects as well as improved understanding of the technique.

This chapter will describe various aspects of performing convection-enhanced drug delivery (CED) experiments in a rat brain model.

2. Materials

2.1. Animals

Since a key issue in convection-based experiments is the ability to distribute a substance over significant tissue volumes, it is ineffective to work in mice brains, which are too small. The smallest animal in which meaningful CED data may be obtained is the rat, preferably rats weighing 275–300 g in which the brain has reached its maximum volume of about 1.5 ml. Note that in smaller brains the distance between the ventricles and the catheter tip may be short, resulting in leakage of the infusate into the CSF, thus decreased distribution efficiency into the tissue. For this reason, when conducting an experiment on several groups of rats, it is important to pay attention that the weight-range of the rats in the different groups is the same at the time of the CED treatment.

When working with a normal brain model, the most common rat species is the Sprague–Dawley rat. Since changes in the rat physiology may affect the efficiency of distribution and clearance of the infused substances, it is preferable to work with male rats to avoid menstrual cycle effects.

When working with a brain tumor model, other species may be required. For example, when working with the 9L gliosarcoma brain tumor model, it is best to work with Fisher rats.

2.2. Anesthesia

The surgical procedure in which the catheter is implanted in the brain, the infusion and the MRI monitoring are all performed under general anesthesia. The most common methodology is an intra peritoneal (IP), or intra muscular (IM) injection of drugs such as Ketamin, Ketalar, Pental, and Rompun, although inhalational anesthesia such as isofloran may be used as well. In cases of long infusions and/or MRI monitoring, it is necessary to keep the animals anesthetized for long periods of time (hours), which may require more than one dose of anesthesia and control of temperature using heating pads. In these cases it is recommended to wait at least 15 min between anesthetic doses to avoid toxicity.

2.3. Infusates

The CED infusion is performed by pressuring the infusate through a catheter into the extracellular region of the target tissue. Therefore, one has to take into account that the infused material may be affected by this pressure. For example, certain molecules may aggregate under the pressure. Large aggregations will block the catheter or accumulate around the catheter tip, thus blocking the entrance into the tissue and eventually impeding convection. Large/fragile particles may be compressed or break, such as certain types of liposomes or viral vectors, and lose their contents. In general, convection of large particles is more challenging than that of small molecules, and optimization of the infusate characteristics and/or the infusion parameters may be required. For the purpose of drug distribution studies and/or toxicity studies in preparation for clinical trials, it is recommended to use the same infusate formulation which will be used in the clinical study. Various biophysical parameters of the drugs and the solvents may affect the distribution efficacy and the toxicity of the convected infusate, therefore changing these parameters (such as increasing or decreasing the concentration of the drug/solvent) may have significant effects (4, 6, 8–10).

Chemical and physical characteristics of the therapeutic agent and the solvent, such as molecular (particle) size, polarity, density, viscosity, and concentration, may affect the distribution efficiency in the tissue and the toxicity of the treatment. Therefore, when preparing a new infusate, it is important to optimize these parameters in order to obtain efficient distributions on one hand and minimize toxicity to normal tissue on the other.

2.4. Contrast Agents

Real-time non-invasive imaging of the drug distribution in the brain is crucial, to assure that the drug is delivered efficiently to the target tissue and to avoid delivery to sensitive brain regions which may be damaged by exposure to the drug. Various methods have been suggested for real-time imaging of CED. In cases where the delivered molecules are carried by large particles such as liposomes, the particles can be loaded with a contrast agent, such as Gd-DTPA for MR (11). When infusing smaller molecules (up to a few hundred KDaltons in size), it is possible to co-infuse small contrast agent molecule, such as Gd-DPTA for MRI and iopamidol for CT (6, 12, 13). It is also possible to use particles which may be directly detected by the imaging modality, such as iron oxide particles which may be directly depicted by MRI and CT (14, 15). These particles may be loaded with the therapeutic agent and simultaneously serve for therapeutic and imaging purposes.

2.5. Pumps and Catheters

The infusion rate used to form convection is in the range of a few microliter per minute (3). For this purpose it is convenient to use syringe pumps, enabling flexible control over the flow rate while secured to an arm of a stereotactic frame. This enables a precise stereotactic placement of the syringe needle in the brain target. An example of a convection station consisting of a stereotactic frame and a BASI Bee Hive syringe pump is shown in **Fig. 9.1**.

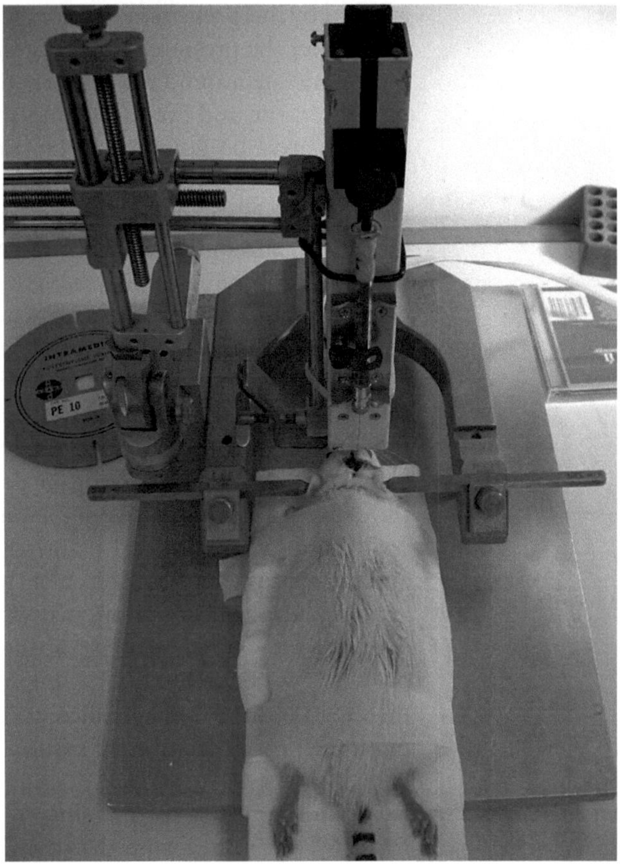

Fig. 9.1. An example of a convection station consisting of a stereotactic frame and a BASI Bee Hive system with a 100-μl Gastight Hamilton syringe. The syringe drive is mounted on the stereotactic arm, enabling accurate navigation of the catheter (*syringe needle*) into the brain with controlled infusion flow rates.

In the rat model, the catheter implanted intracranially may be the needle of the syringe itself. To obtain efficient convection it is possible to use thin needles (32–36 gauge) and flow rates of 1–4 μl/min. If the infusion is performed in the MR system, the pump should be either MR compatible or kept out of the magnetic field. It is also recommended to use MR compatible catheters in order to avoid significant image artifacts.

3. Methods

3.1. Overall Experimental Design

A typical convection-based rat study is performed on several groups of rats, each infused with a different infusate and/or infusion parameters.

The surgical procedure, the infusion itself, and the imaging modality used are all performed under general anesthesia.

When using one infusion pump, it is recommended to infuse the rats from the different groups in an alternating manner. When using two or more infusion pumps it is recommended to infuse rats from different groups in parallel. This is done to avoid differences between the groups originating from the dependence of the experimental parameters on time. Various parameters which may change throughout the experiment may affect the distribution efficiency. These include the time of day (rats are night animals; therefore parameters related to their awake–sleep cycle may affect the experimental results. For example, the anesthesia is less effective toward the end of the day), room temperature (which may affect the infusate characteristics as well as the rat physiology), and the time that passes between infusate preparation and infusion (which may result in degradation or aggregation of substances in the infusate).

Imaging for assessment of infusate distribution in the brain should be performed at the end, or soon after the end of the infusion, especially in the case of small molecules which are cleared from the tissue relatively fast.

When using MRI to assess the drug distribution, it is possible to apply additional imaging sequences in order to depict procedure-related toxicity such as hydrocephalus or tissue damage.

Some experiments, designed to assess toxicity, may require additional follow-up of days to weeks after the infusion. Variables such as periodic weighing, periodic imaging, and subjection of brain samples to histological examination should be evaluated. The information acquired in the MRI scan performed immediately post-infusion may add useful information in differentiating procedure-related toxicity from cytotoxic response to the delivered substance.

3.2. CED Procedure

The surgical procedure, performed under full anesthesia, consists of a midline scalp incision made to identify the bregma. In order to target the striatum a 1-mm burr hole can be made 2 mm posterior and 2 mm lateral to the bregma. In order to infuse in frontal white matter, the burr hole can be made 2 mm anterior and 2 mm lateral to the bregma. When using syringe pumps such as the BASI Bee Hive system, a thin needle attached to a 10- to 1000-μl syringe will be placed stereotactically, typically 3–5.5 mm deep into the brain. Note that the dura is thick and most catheters will not be able to penetrate it, therefore it is recommended to use a large-gauge needle to puncture a hole in the dura before inserting the catheter. The infusion will be performed typically at a rate of 0.1–10 μl/min for duration of 15–90 min.

If possible, it is recommended to insert the catheter while infusing in order to avoid the catheter from being blocked by tissue fragments.

3.2.1. CED into Gray/White Matter

When CED is performed in white matter, the infusate tends to distribute along the white matter tracks. When infusing gray matter, the distribution tends to be more homogenous. White matter infusions are typically performed in the frontal white matter region of the rat brain. In this case the burr hole can be made 2 mm anterior and 2 mm lateral to the bregma and the catheter would typically be placed 3–3.5 mm deep. Gray matter infusions are typically performed in the striatum region. In this case the burr hole can be made 2 mm posterior and 2 mm lateral to the bregma and the catheter would typically be placed 5–5.5 mm deep.

3.2.2. Catheter Position Considerations

In order to obtain efficient convection, it is best to place the catheter tip as deep as possible in homogenous tissue, as far as possible from the brain surface on one hand and the ventricular system on the other. If the tip is too close to the ventricular system or to other pial/ependymal surfaces, the probability for leakage into the CSF is high, impeding the formation of efficient convection. Leakage into the CSF will also occur if the catheter tip is not deep enough and the backflow along the path of the catheter will reach the brain surface and subarachnoid space. Therefore, if the experiment is designed to obtain maximal infusate distributions it would be preferred to target the catheter tip in the striatum region, 5–5.5 mm deep from the brain surface, use a thin catheter and low flow rates. Challenging the formation of convection may be easily obtained by decreasing the depth of the catheter tip, increasing the catheter diameter or increasing the flow rate.

3.2.3. Flow Considerations

Convection is usually obtained by using flow rates in the range of 0.1–10 μl/min. Low flow rates tend to leak less into the CSF thus forming more efficient convection. However, low flow rates result in longer infusion times. Therefore, optimizing other parameters which affect the efficient formation of convection, such as catheter position, catheter size, and chemical–physical parameters of the infusate may enable using higher flow rates with reasonable distribution efficiencies, thus minimizing the infusion time. In some cases, such as in the case of large particles, higher flow rates may be necessary to enable larger distributions. Still, high flow rates, in the order of 10 μl/min, may cause irreversible tissue damage.

3.2.4. Infused Volume Considerations

The volume of the rat brain is in the range of 1–1.5 ml. Infusions of up to 100 μl in healthy rats' brains are well tolerated. Minimal hydrocephalus and local edema may be noted in the imaging performed immediately after infusion with no damage to the tissue (as depicted by later MRI and histological examination) and no clinical abnormalities. Infusion of larger volumes may cause increased intracranial pressure (ICP), significant hydrocephalus, and subsequent tissue toxicity.

In rats bearing cerebral tumors, it is recommended to infuse smaller volumes, since these rats may be less efficient in clearing CSF, thus more sensitive to increased ICP.

3.2.5. Clearance from the Brain

Small molecules tend to be cleared from the tissue within minutes to hours post-infusion, while large particles may persist in the tissue for days, weeks, and even months (15, 16). CED distributes the drug molecules in the extracellular space of the target tissue. When planning the infusion duration it is important to make sure that the exposure of the cells to the distributed drug is long enough for the drug to penetrate the cells. For example, an infustion of 15 min may be inefficient for a drug which takes 20 min to penetrate the cells.

3.2.6. CED into Brain Tumor Models

Intracranial tumors are usually implanted in the region of the striatum or in the frontal white matter region of the rat brain. Frontal tumors are relatively superficial (typically 3–3.5 mm deep in contrast to 5–5.5 mm in the striatum region), due to the shape of the rat brain, thus forming efficient convection in these tumors is more challenging. Survival tends to be longer at these locations as the tumors are further away from essential brain organs.

Tumor inoculation is normally performed in a similar manner to the infusion procedure. Under full anesthesia the bregma is identified through a midline scalp incision, and a 1-mm burr hole can be made 2 mm posterior and 2 mm lateral to the bregma in order to target the striatum or 2 mm anterior and 2 mm lateral to the bregma in order to target the frontal white matter. When using syringe pumps such as the BASI Bee Hive system, a thin needle attached to a 10-1000-μl syringe is placed stereotactically, typically 5–5.5 mm deep in the case of the striatum and 3–3.5 mm deep in the case of frontal white matter. The cancer cells are infused at a rate of 1–2 μl/min for 1–5 min, depending on the infusion volume. It is recommended to seal the burr hole with bone wax to prevent the tumor from growing out of the skull.

For the CED treatment the burr hole used to implant the tumors can be re-opened and the infusion can be performed in the same location of tumor implantation.

It is important to note that most brain tumor models in rats tend to grow into dense high-pressure tumors, hampering formation of convection within the tumor tissue, resulting in inefficient drug distributions in the tumor and high drug concentrations in normal brain tissue surrounding the tumor. An example of CED performed in a 9L gliosarcoma tumor model is shown in **Fig. 9.4**.

3.3. Imaging CED

3.3.1. Real-Time Imaging of the Drug Distribution

Convection is known to depend on various treatment parameters, such as catheter size, flow rate, and infusate characterization (3, 4, 6–10, 14, 17). The variability in these parameters may explain the variability in tumor response among different drugs and different treatment protocols seen in clinical studies. Physiological parameters such as tissue consistency, blood pressure, and blood–brain barrier function may have a significant effect as well (5, 18, 19). The later may explain the significant variability in the extent of convection seen among patients treated under the same protocol. Therefore, individual real-time, non-invasive monitoring of the extent of convection and its early effects on the tissue are essential.

Various approaches have demonstrated the feasibility of imaging the drug distribution by co-infusion of liposomes or macromolecules (albumin) containing contrast agents that are visible on images obtained using non-invasive techniques, such as iopanoic acid (IPA) for CT and Gd-diethylenetriamine pentaacetic acid (Gd-DTPA) for MRI (11, 20–22). These markers are co-infused with the therapeutic agent to enable real-time monitoring of the distributed drugs.

It is also possible to co-infuse low molecular weight contrast agents, such as Gd-DTPA for MRI and iopamidol for CT as surrogate markers for real-time monitoring of the distributed drugs. The advantage of using low molecular weight contrast agents is that there is no need to bind them to the therapeutic agent or to load them on other particles, thus they may be simply mixed with the infusate just before starting the infusion. The limitation is that the low molecular weight tracers can only be used for depicting distribution of drugs up to ~100 kDa in size (6, 12). Examples of efficient, moderate, and poor distributions obtained by CED and depicted by MRI are shown in **Fig. 9.2**.

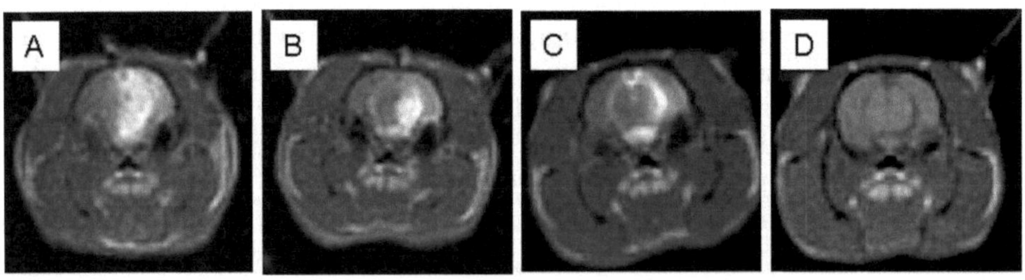

Fig. 9.2. Examples of T1-weighted MR images acquired immediately post-infusion. All rats were infused with saline solutions containing 1:70 (in volume) of Gd-DTPA. The rats in images (**a**) and (**b**) were infused with a high-viscosity solution obtained by adding 12% sucrose. (**a**) Efficient distribution; (**b**) moderate distribution; (**c**) full backflow into the ventricular system; (**d**) no penetration of the infusate into the brain (the catheter did not penetrate the dura).

3.3.2. Imaging Tissue Response

Diffusion-weighted MR images (DWMRIs) enable non-invasive characterization of biological tissues based on their water diffusion characteristics. It has been shown that there is an order of magnitude difference between the diffusion of slow/intracellular and fast/extracellular water molecules in vitro (23, 24). Therefore, DWMRIs are sensitive to early changes in morphology and physiology of tissues associated with changes in water content such as changes in the permeability of cell membranes, cell swelling, and cell lysis. Note that the contrast in DWMRIs consists of diffusion components as well as magnetic (T2-weighted) components. In order to differentiate between these two components it is possible to calculated from the DWMRIs apparent diffusion coefficient (ADC) maps, which reflect the water diffusion properties of the tissue separated from the magnetic components.

In previous CED-based clinical studies with various drugs, it has been shown that DWMRI enables depiction of cytotoxic tissue response between 12 h and 2 weeks after initiation of treatment. Different drugs presented different tissue toxicity patterns (temporal and spatial), depicted in DWMR images several weeks before any changes could be observed in conventional MRI (6, 25–27). In clinical studies, early depiction of the cytotoxic tissue response may enable later differentiation between treatment induced necrosis to progressive tumor necrosis. In animal studies, tissue toxicity is usually depicted in DWMRI and T2-weighted MRI up to 24 h post-treatment, enabling rapid/efficient screening of candidate drugs and solvents based on their toxicity to normal tissue. Examples are shown in **Fig. 9.3**.

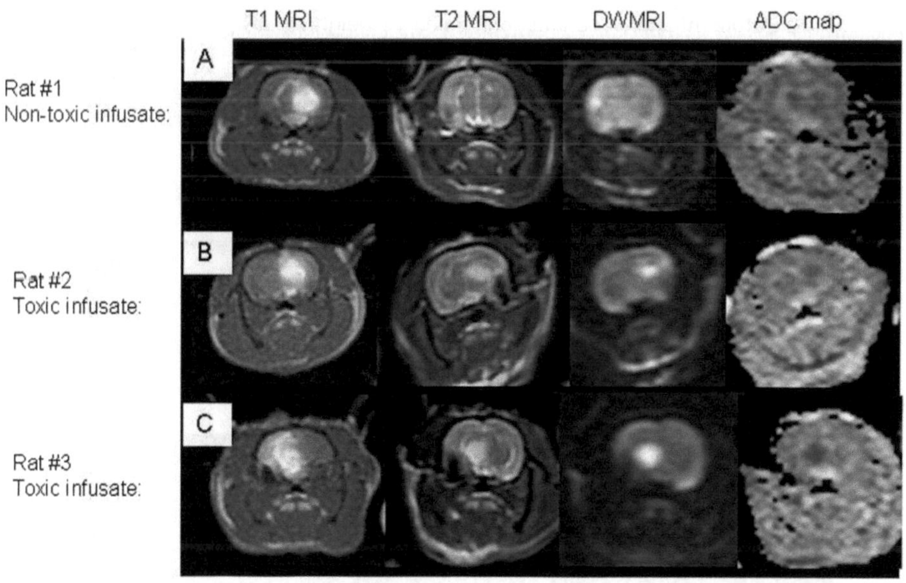

Fig. 9.3. Examples of T1-weighted MR images acquired immediately after CED and T2-weighted and diffusion-weighted MRI acquired 24 h after CED of toxic and non-toxic infusates in normal rat brains. Gd-DTPA was co-infused in all rats. (**a**) T1 MRI acquired after infusion of a sucrose solution, depicting efficient infusate distribution; Later T2 MRI and DWMRI depict no tissue toxicity; (**b**) T1 MRI acquired after infusion of Cremaphore depicting moderate distribution; later T2 MRI and DWMRI depict a moderate region of cytotoxic tissue response (*bright*). Note that the toxic region in the T2 image seems larger than in the DWMRI due to edema; (**c**) T1 MRI acquired after infusion of Cremaphore depicting efficient distribution; later T2 MRI and DWMRI depict a large region of cytotoxic tissue response.

3.4. Calculation of CED Extent

When co-infusing with a surrogate marker, the infusate distribution volume (in mm^3) may be calculated from images acquired immediately after the CED treatment. Regions of interest (ROIs) are defined over the entire enhancing region in each slice (excluding the ventricles). The number of pixels in the ROIs are then counted and multiplied by the volume of a single pixel to obtain the final volume of distribution. An example of an ROI plotted on an MR image acquired immediately after CED is shown in **Fig. 9.5**.

4. Examples

4.1. Depiction of Drug Distribution and Cytotoic Tissue Response

Examples for MRI monitoring of the infusate distribution and cytotoxic tissue response in CED treatments are shown in **Figs. 9.2, 9.3, 9.4 and 9.5**. In these experiments the solution was infused into the rat striatum at a rate of 1 μl/min over 40 min. Immediate assessment of CED formation and extent was performed using T1-weighted MRI acquired with a General Electric 0.5 T interventional MRI machine (Signa SP/i (special proceeding/interventional)) with the lx operating system and gradients intensity of up to 1 gauss/cm using a specially designed animal

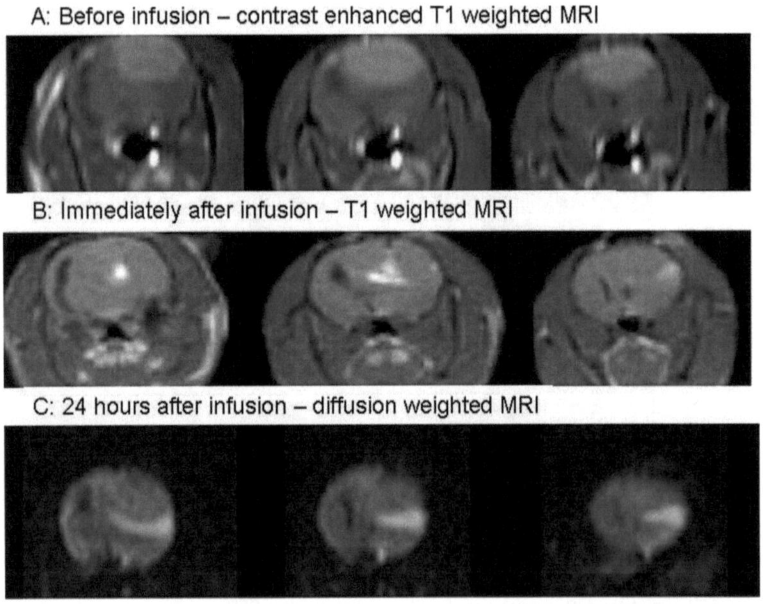

Fig. 9.4. Examples of MR images acquired before, immediately after and 24 h after convection-enhanced Taxol delivery in a rat bearing an intracranial 9L gliosarcoma tumor. (**a**) Contrast-enhanced T1 MRI of a rat with a large 9L gliosarcoma tumor acquired 24 h prior to infusion; (**b**) T1 MRI of the same rat, immediately after infusion of Taxol with Gd-DTPA. The catheter was placed in the center of the tumor, but the infusate leaked out, forming convection in the normal brain surrounding the tumor; (**c**) DWMRI of the same rat 24 h post-treatment, showing significant tissue toxicity (*bright*) in normal tissue surrounding the tumor with no toxicity in the tumor itself.

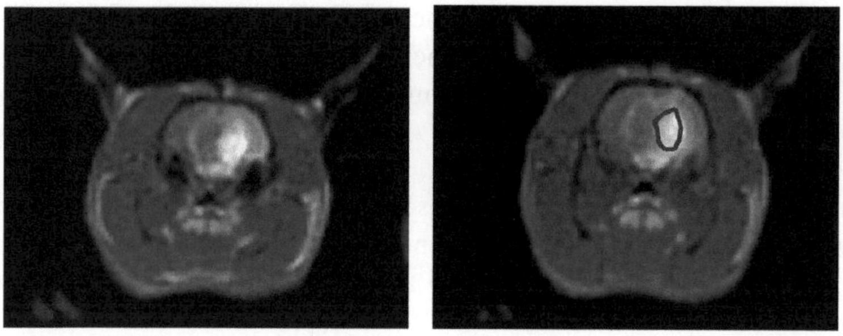

Fig. 9.5. Example of an MR image acquired immediately after CED of a solution containing Gd-DTPA. An ROI was plotted over the entire enhancing region excluding the ventricles. The number of pixels in the ROI are then counted and multiplied by the volume of a single pixel and summed over the enhancing slices to obtain the final volume of distribution.

volume coil of 5 cm diameter. Early assessment of cytotoxic tissue response was performed 24 h later. T2-weighted and line scan diffusion-weighted MR images were acquired on a clinical General Electric (Waukesha, WI) 3.0 T MR system, equipped with the HD 12 × operating system and at a field-gradient intensity of up to 4 gauss/cm, using a clinical phased-array Knee coil.

4.2. Depiction of Procedure-Related Toxicity

When performing the MRI immediately post-infusion for assessment of infusate distribution, it is possible to add an acquisition of T2-weighted MRI for assessment of procedure-related toxicity, which may be reflected by significant hydrocephalus or tissue damage. Examples are shown in **Fig. 9.6**. This information may

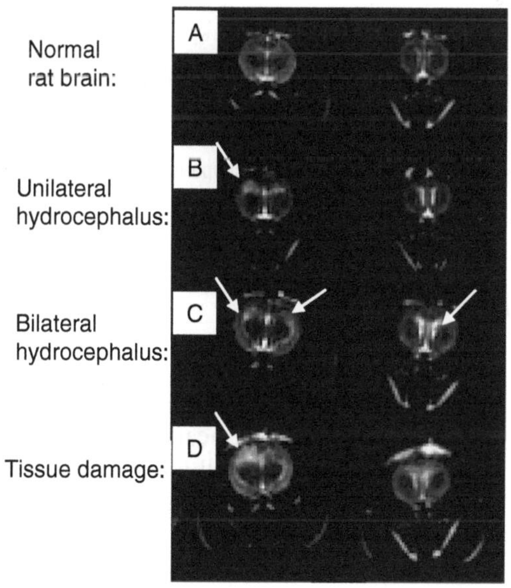

Fig. 9.6. T2-weighted MRI acquired immediately after infusion with non-toxic infusates. (**a**) Two adjacent slices of a rat depicting no procedure-related toxicity; (**b**) unilateral hydrocephalus; (**c**) bilateral hydrocephalus; (**d**) tissue damage at the catheter entrance point.

become valuable when the experiment is designed to assess the toxicity of the treatment. The immediate T2-weighted images may help in differentiating drug-related toxicity from procedure-related toxicity.

4.3. CED of Iron Oxide Nano-particles

CED of nanoparticles was performed with a solution of 0.2 mg/ml dextrane-coated monodispersed maghemite nanoparticles in 3% sucrose infused into the rat striatum at a rate of 4 μl/min over 15 min.

Due to the superparamagnetic properties of the monodispersed maghemite nanoparticles, it is possible to image the particles distribution directly using MRI. Superparamagnetic monodispersed maghemite nanoparticles produce a dramatic long-range disturbance in magnetic field homogeneity, causing marked shortening of T2* relaxation time. This results in a rapid loss of phase coherence and the loss of MRI signal in the area of the particles. Examples of MR images acquired on the 0.5 T MR system after CED with monodispersed maghemite nanoparticles are shown in **Fig. 9.7**.

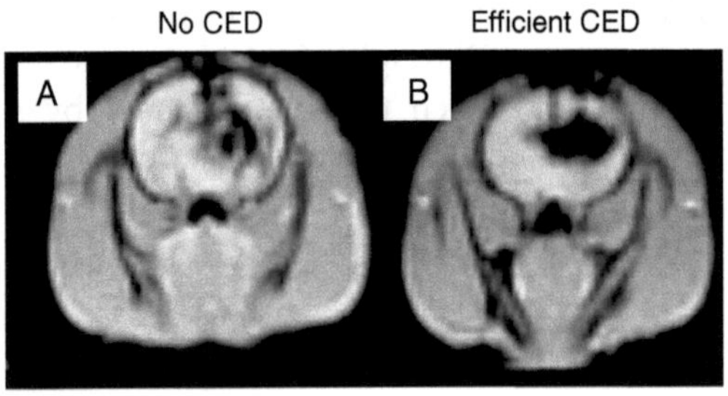

Fig. 9.7. Gradient-echo MRI acquired immediately after infusion with monodispersed maghemite nanoparticles. (**a**) Full backflow into the CSF; (**b**) efficient distribution.

Acknowledgments

The work in our lab was supported by grants from The Israeli ministry of health, Israel Science Foundation, Israel Cancer Association, Goldhirsh Foundation, and Adams Brain Research Center at Tel-Aviv University.The authors would like to thank the MRI Research Group at The Advanced Technology Center, Sheba Medical Center – Dianne Daniels, David Last, Ran Shneor, Gregory Tamar and Sharona Salomon – for their help in obtaining the figures and the unpublished data that have been incorporated in this chapter.

References

1. Bobo, R.H., Laske, D.W., Akbasak, A., Morrison, P.F., Dedrick, R.L., and Oldfield, E.H. (1994) Convection-enhanced delivery of macromolecules in the brain. *Proc Natl Acad Sci USA* 91(6), 2076–2080.
2. Lieberman, D.M., Laske, D.W., Morrison, P.F., Bankiewicz, K.S., and Oldfield, E.H. (1995) Convection-enhanced distribution of large molecules in gray matter during interstitial drug infusion. *J Neurosurg* 82(6), 1021–1029.
3. Chen, M.Y., Lonser, R.R., Morrison, P.F., Governale, L.S., and Oldfield, E.H. (1999) Variables affecting CED to the striatum: a systematic examination of rate of infusion, cannula size, infusate concentration, and tissue-cannula sealing time. *J Neurosurg* 90, 315–320.
4. Chen, M.Y., Hoffer, A., Morrison, P.F., Hamilton, J.F., Hughes, J., Schlageter, K.S., Lee, J., Kelly, B.R., and Oldfield, E.H. (2005) Surface properties, more than size, limiting convective distribution of virus-sized particles and viruses in the central nervous system. *J Neurosurg* 103(2), 311–319.
5. Sampson, J.H., Brady, M.L., Petry, N.A., Croteau, D., Friedman, A.H., Friedman, H.S., Wong, T., Bigner, D.D., Pastan, I., Puri, R.K., and Pedain, C. (2007) Intracerebral infusate distribution by convection-enhanced delivery in humans with malignant gliomas: descriptive effects of target anatomy and catheter positioning. *Neurosurgery* 60(2 Suppl 1), ONS89–98; discussion ONS98–9.
6. Mardor, Y., Rahav, O., Zauberman, Y., Lidar, Z., Ocherashvilli, A., Daniels, D., Roth, Y., Maier, S.E., Orenstein, A., and Ram, Z. (2005) Convection-enhanced drug delivery: increased efficacy and magnetic resonance image monitoring. *Cancer Res* 65(15), 6858–6863.
7. Jagannathan, J., Walbridge, S., Butman, J.A., Oldfield, E.H., and Lonser, R.R. (2008) Effect of ependymal and pial surfaces on convection-enhanced delivery. *J Neurosurg* 109(3), 547–552.
8. MacKay, J.A., Deen, D.F., and Szoka, F.C. Jr. (2005) Distribution in brain of liposomes after convection enhanced delivery; modulation by particle charge, particle diameter, and presence of steric coating. *Brain Res* 1035(2), 139–153.
9. Saito, R., Krauze, M.T., Noble, C.O., Tamas, M., Drummond, D.C., Kirpotin, D.B., Berger, M.S., Park, J.W., and Bankiewicz, K.S. (2006) Tissue affinity of the infusate affects the distribution volume during convection-enhanced delivery into rodent brains: implications for local drug delivery. *J Neurosci Methods* 154(1–2), 225–232.
10. Neeves, K.B., Sawyer, A.J., Foley, C.P., Saltzman, W.M., and Olbricht, W.L. (2007) Dilation and degradation of the brain extracellular matrix enhances penetration of infused polymer nanoparticles. *Brain Res* 1180, 121–132.
11. Saito, R., Bringas, J.R., McKnight, T.R., Wendland, M.F., Mamot, C., Drummond, D.C., Kirpotin, D.B., Park, J.W., Berger, M.S., and Bankiewicz, K.S. (2004) Distribution of liposomes into brain and rat brain tumor models by convection-enhanced delivery monitored with magnetic resonance imaging. *Cancer Res* 64(7), 2572–2579.
12. Croteau, D., Walbridge, S., Morrison, P.F., Butman, J.A., Vortmeyer, A.O., Johnson, D., Oldfield, E.H., and Lonser, R.R. (2005) Real-time in vivo imaging of the convective distribution of a low-molecular-weight tracer. *J Neurosurg* 102(1), 90–97.
13. Lonser, R.R., Warren, K.E., Butman, J.A., Quezado, Z., Robison, R.A., Walbridge, S., Schiffman, R., Merrill, M., Walker, M.L., Park, D.M., Croteau, D., Brady, R.O., and Oldfield, E.H. (2007) Real-time image-guided direct convective perfusion of intrinsic brainstem lesions. Technical note. *J Neurosurg* 107(1), 190–197.
14. Kroll, R.A., Michael, A., Muldoon, L.L., Roman-Goldstein, S., and Neuwelt, E.A. (1996) Increasing volume of distribution to the brain with interstitial infusion: dose, rather than convection, might be the most important factor. *Neurosurg Online* 38(4), 746–754.
15. Perlstein, B., Ram, Z., Daniels, D., Ocherashvilli, A., Roth, Y., Margel, S., and Mardor, Y. (2008) Convection-enhanced delivery of maghemite nanoparticles: Increased efficacy and MRI monitoring. *Neuro Oncol* 10(2), 153–161.
16. Noble, C.O., Krauze, M.T., Drummond, D.C., Yamashita, Y., Saito, R., Berger, M.S., Kirpotin, D.B., Bankiewicz, K.S., and Park, J.W. (2006) Novel nanoliposomal CPT-11 infused by convection-enhanced delivery in intracranial tumors: pharmacology and efficacy. *Cancer Res* 66(5), 2801–2806.

17. Chen, Z.J., Broaddus, W.C., Viswanathan, R.R., Raghavan, R., and Gillies, G.T. (2002) Intraparenchymal drug delivery via positive-pressure infusion: experimental and modeling studies of poroelasticity in brain phantom gels. *IEEE Trans Biomed Eng* 49(2), 85–96.
18. Raghavan, R., Brady, M.L., Rodríguez-Ponce, M.I., Hartlep, A., Pedain, C., and Sampson, J.H. (2006) Convection-enhanced delivery of therapeutics for brain disease, and its optimization. *Neurosurg Focus* 20(4), E12.
19. Hadaczek, P., Yamashita, Y., Mirek, H., Tamas, L., Bohn, M.C., Noble, C., Park, J.W., and Bankiewicz, K. (2006) The "perivascular pump" driven by arterial pulsation is a powerful mechanism for the distribution of therapeutic molecules within the brain. *Mol Ther* 14(1), 69–78.
20. Mamot, C., Nguyen, J.B., Pourdehnad, M., Hadaczek, P., Saito, R., Bringas, J.R., Drummond, D.C., Hong, K., Kirpotin, D.B., McKnight, T., Berger, M.S., Park, J.W., and Bankiewicz, K.S. (2004) Extensive distribution of liposomes in rodent brains and brain tumors following convection-enhanced delivery. *J Neurooncol* 68(1), 1–9.
21. Nguyen, T.T., Pannu, Y.S., Sung, C., Dedrick, R.L., Walbridge, S., Brechbiel, M.W., Garmestani, K., Beitzel, M., Yordanov, A.T., and Oldfield, E.H. (2003) Convective distribution of macromolecules in the primate brain demonstrated using computerized tomography and magnetic resonance imaging. *J Neurosurg* 98(3), 584–590.
22. Lonser, R.R., Walbridge, S., Garmestani, K., Butman, J.A., Walters, H.A., Vortmeyer, A.O., Morrison, P.F., Brechbiel, M.W., and Oldfield, E.H. (2002) Successful and sage perfusion of the primate brainstem: in vivo MRI of macromolecular distribution during infusion. *J Neurosurg* 97, 905–913.
23. Van Zijl, P.C., Moonen, C.T., Faustino, P., Pekar, J., Kaplan, O., and Cohen, J.S. (1991) Complete separation of intracellular, and extracellular information in NMR spectra by diffusion weighting. *Proc Natl Acad Sci USA* 88, 3228–3232.
24. Pilatus, U., Shim, H., Artemov, D., Davis, D., van Zijl, P.C., and Glickson, J.D. (1997) Intracellular volume and apparent diffusion constants of perfused cancer cell cultures, as measured by NMR. *Magn Reson Med* 37, 825–832.
25. Mardor, Y., Roth, Y., Lidar, Z., Jonas, T., Pfeffer, R., Maier, S.E., Faibel, M., Nass, D., Hadani, M., Orenstein, A., Cohen, J.S., and Ram, Z. (2001) Monitoring response to convection-enhanced taxol delivery in brain tumor patients using diffusion-weighted magnetic resonance imaging. *Cancer Res* 61(13), 4971–4973.
26. Lidar, Z., Mardor, Y., Jonas, T., Pfeffer, R., Faibel, M., Nass, D., Hadani, M., and Ram, Z. (2004) Convection-enhanced delivery of paclitaxel for the treatment of recurrent malignant glioma: a phase I/II clinical study. *J Neurosurg* 100(3), 472–479.
27. Tanner, P.G., Holtmannspötter, M., Tonn, J.C., and Goldbrunner, R. (2007) Effects of drug efflux on convection-enhanced paclitaxel delivery to malignant gliomas: technical note. *Neurosurgery* 61(4), E880–2; discussion E882.

Chapter 10

Intrathecal Delivery of Stem Cells to the Spinal Cord

Birgit Neuhuber and Itzhak Fischer

Abstract

Cellular transplantation is a promising therapeutic strategy for spinal cord injury (SCI); the clinical application of transplantation, however, will require safe and efficient protocols of cell delivery. Lumbar puncture (LP) is a minimally invasive delivery method that allows multiple cell deliveries into the intrathecal space, resulting in cell accumulation at the site of injury. Here we review the effectiveness of the LP technique in rodent models of SCI, the advantages of LP compared to traditional parenchymal injection and other minimally invasive methods, and provide detailed instructions for LP delivery of stem cells to the injured rat spinal cord.

Key words: Lumbar puncture, sub-arachnoid space, minimally invasive, spinal cord injury, cell transplant.

1. Introduction

1.1. Spinal Cord Injury and Cell Therapy

Spinal cord injury (SCI) is a complex disorder that progresses in distinct stages beginning as spinal trauma, which in human patients is usually a contusion. The initial injury is exacerbated by a cascade of events including excitotoxicity, oxidative stress, and inflammation that leads to additional tissue damage/loss in a process called secondary injury (1). The direct loss of neurons locally, the indirect loss of neurons by retrograde degeneration in supraspinal control centers, and the demyelination of long-distance axonal connections ultimately lead to persistent deficits in motor and sensory function. Secondary injury leaves behind a large fluid-filled cyst surrounded by a glial scar at the site of the initial injury, which presents a chemical and physical barrier to axonal growth (2). In addition, recovery is limited by the poor regenerative potential of adult CNS neurons and by the inhibitory

K.K. Jain (ed.), *Drug Delivery to the Central Nervous System*, Neuromethods 45,
DOI 10.1007/978-1-60761-529-3_10, © Humana Press, a part of Springer Science+Business Media, LLC 2010

environment of the adult CNS, where a lack of trophic factors, guidance cues, and the presence of scar-derived proteoglycans and myelin-associated growth-inhibitory molecules, e.g., Nogo, MAG, OMgp, hinder axon growth and regeneration (3).

Because of this complexity, effective strategies for SCI repair must consist of a combination of different strategies directed at multiple components of the injury. These approaches include neuroprotection, modulation of inflammation, initiation of a regenerative program in endogenous neurons, reduction of physical barriers (cyst and scar), provision of a permissive environment, and cell replacement. Cell transplantation is a promising therapeutic strategy for SCI because it tackles a number of these issues resulting in anatomical and functional recovery in animal models of SCI (4, 5). Although cell transplants are often used to replace neural tissue (e.g., neurons and glial cells), they can also be utilized as biological mini pumps that provide therapeutic factors over an extended amount of time. Most cell transplants also provide a bridge that may fill the cyst-generated gap; however, the specific beneficial effects depend on the cell type (**Table 10.1**).

Table 10.1
Candidate cells for treatment of spinal cord injury and their potential mode of action

Cell type	Major mode of action	Reviewed in
Neural stem and precursor cells	Cell replacement, remyelination	(5)
OEC or Schwann cells	Permissive growth environment, remyelination	(55)
Mesenchymal stem cells	Permissive growth environment, secretion of trophic factors and immune modulators	(56)
Genetically modified fibroblasts	Permissive growth environment, secretion of therapeutic factors	(57)
Activated macrophages	Immune modulation	(58)
ES or iPS cell-derived neural precursor cells	Cell replacement, remyelination	(5)

1.2. Cell Delivery to the Injured Spinal Cord

While cell transplantation strategies hold significant promise for the treatment of SCI, their clinical translation will require a safe and efficient method of delivery. The most common method for cell transplantation into the clinically relevant animal contusion model of SCI is direct injection into the injury epicenter or into the parenchyma adjacent to the injury (6–13). However, direct injection may further damage an already compromised spinal cord, risking additional functional deterioration, as well as complications due to anesthesia or the neurosurgical procedure. In addition, it may be advantageous to deliver cells at several time points to achieve maximal efficacy of cell transplants. While direct injection strategies, especially those using multiple deliveries, are acceptable in animal models, they are far from optimal for clinical translation to human patients. For this reason, minimally invasive methods have been proposed as preferable alternative for delivering cells to the injured spinal cord. This chapter will address the status of intrathecal cell delivery via lumbar puncture (LP) as a minimally invasive method used in both rodent models of SCI and clinical trials. We will compare LP to other minimally invasive methods, discuss the effectiveness and advantages of the LP technique, and provide detailed instructions for LP delivery of cell transplants to SCI models in the rat.

1.3. Comparison of LP to Other Minimally Invasive Delivery Methods

The least invasive method for cell delivery is intravascular delivery (intra-arterial or intra-venous). Because of the multi-segmented arterial supply of the spinal cord, intra-arterial delivery requires risky cannulation procedures, making it less convenient than intravenous delivery. Intravenous delivery is easy and safe and some labs have reported promising results (14–16). Intravenous delivery is also used as a mode of delivery in clinical trials in SCI (17, 18). However, our studies (19, 20) and those of others (21, 22) have shown a significantly decreased efficiency of engraftment compared to direct or LP delivery. Systemic delivery also carries a high risk of cell implantation in undesired locations in various organs. Cells can also lodge in the capillaries of the lung or be filtered out by the liver. Finally, systemic delivery exposes cells to the immune system for an extended period of time. Taken together, these factors may account for the lower engraftment efficiency observed after intra-venous cell delivery. In contrast, intra-ventricular cell delivery into the cerebrospinal fluid (CSF) is more efficient than intravascular delivery (23–25); however, considerably more invasive than intrathecal delivery by LP.

1.4. Intrathecal Cell Delivery by Lumbar Puncture

Lumbar puncture (LP), also commonly known as spinal tap, is a procedure that in human patients is generally used to collect samples of CSF or to deliver drugs into the lumbar sub-arachnoid space. An optimized LP procedure for CSF sampling and intrathecal drug and cell delivery in the rat was first introduced by De la Calle and Paino (26). De la Calle and others (27–30) successfully used LP to deliver cells secreting anti-nociceptive molecules to rat models (excitotoxic and chronic sciatic constriction) of neuropathic pain. Our laboratory adapted the LP procedure to deliver cell transplants to sites of surgical and contusive SCI (19, 31–33).

In our studies, we used two different stem cell types, adult mesenchymal stem cells from bone marrow (MSC) and fetal neuronal progenitor cells (NPC), to establish the feasibility of cell delivery via LP to a surgical hemisection model of SCI at the cervical level (19, 32). We found that both cell types accumulated in large numbers in the injury cavity with the greatest cell concentration at the interface of injury and meninges. A few cells could be observed in the linings of the ventricular system in the brain, but no cells were detected in portions of the intact spinal cord away from the injury site or in brain parenchyma. In contrast, fluorescent-labeled polystyrene beads with diameters comparable to those of cells, which were used as controls, only rarely accumulated in the injury cavity but were loosely attached to the meninges surrounding the entire spinal cord. These results suggested that cell accumulation at the injury site was at least in part due to selective homing mechanisms rather than simple non-specific processes, e..g., diffusion into the physically disrupted spinal cord parenchyma. Several factors have been implicated in the selective attraction or homing of cells. Cells could be attracted by chemotactic signals expressed at the injury site. A major candidate factor is stromal cell-derived factor 1α (SDF-1α or CXCL12), which is expressed in injured tissue. Both MSC (34) and NPC (35, 36) express CXCR4, the receptor for SDF-1α, and this signaling pathway has been shown to be involved in the migration of MSC (37–39) and NPC (40, 41). Platelet-derived growth factor (PDGF), transforming growth factor alpha (TGFα), insulin growth factor (IGF), and hepatocyte growth factor are among other candidate factors produced in injured tissue that consistently enhance the migration of MSC and NPC (42–46). Adhesion molecules present on cells and in connective tissue at the site of injury could also provide an anchor for injected cells that enter the disrupted spinal cord parenchyma. After accumulating at the injury site, some cells can then migrate further into the spinal cord parenchyma; however, the extent of migration appears to be dependent on the cell type, with NPC migrating extensively (32) and MSC showing limited migration (31).

The demonstration that cells delivered at the lumbar spinal level can accumulate at a cervical level injury site was an important step in evaluating the value of LP-based cell delivery as an alternative to direct injection protocols. In a next step, we delivered MSC and NPC to a clinically relevant thoracic contusion injury (31, 33). Contusion injuries are considerably more complex than surgical SCI or spinal cord penetration. The inflammatory response in the first 1–2 weeks after contusion produces a toxic environment that limits survival of cell transplants. On the other hand, it has been shown that cell transplants delivered in that early sub-acute phase can have significant neuroprotective effects (7, 9, 10, 33, 47, 48). In the clinic, delivering cells during this early sub-acute time period by direct injection is difficult and involves high risks that could be circumvented by using LP delivery instead. As in our previous experiment using hemisections, we found that MSC and NPC accumulated at the site of contusion. Most cells were found in the dorsal aspect of the contusion, which is the region that is most severely damaged and most easily accessible. To evaluate, if this homing is affected by the time of delivery, we injected MSC at different time points within the first 2 weeks and after approximately 3 and 4 weeks and found that engraftment efficiency is highest within the first 2 weeks, although cells can still enter the injury site at later time points (31). This is in accordance with the inflammatory profile, where secretion of chemokines potentially involved in attracting cells to the injury is increased in the initial phase of the injury and levels off in later stages (49). As chemotactic signals decrease in later stages, physical access to the injury may also become more limited because of the development of a glial scar and tissue healing. For these reasons, LP delivery appears to have a window of opportunity at the acute and sub-acute stage and may not be the optimal choice for cell delivery in chronic SCI, unless treatment is initiated by debriding the scar and thereby creating a controlled new injury that will allow homing and access of LP-delivered cells to the injury site.

The most critical question to address is whether LP-delivered cells can result in anatomical and functional recovery that would be comparable to directly injected cell transplants. With both MSC and NPC, we found significant improvements in anatomical measures for neuroprotection (spared white matter, reduced cyst and injury size) when cells were delivered by LP in the sub-acute phase after injury (31, 33). We also found significant improvements in bladder function in LP-treated animals compared to untreated controls (33). These improvements were comparable to those of animals with directly injected cell grafts. While motor function appeared somewhat improved in LP-treated animals, motor scores did not reach significant levels seen in animals that received intra-spinal delivery. Ide's group (23) reported similar results regarding reduced cyst size but described significantly improved motor function after injection of MSC into the 4th ventricle. Neuroprotection and improved motor function were also observed in an ischemia model of SCI (50).

In an attempt to improve engraftment efficiency relative to the number of cells, we delivered single doses of 1×10^6, 2×10^6, or 4×10^6 cells by LP. We found that increased cell numbers in single deliveries do not translate to increased graft volumes (31). However, this finding may reflect the limitations in flow and capacity of CSF in the rat system. Injecting a greater number of cells may simply clog the rat CSF and optimization of cell dosage for clinically relevant data will need to be obtained in larger animal models of SCI (e.g., piglet or cat).

An important potential advantage of LP delivery is the ability to make multiple cell deliveries. We have shown that multiple cell deliveries by LP result to some extent in additive graft volumes (31, 32), but it appears again that the rat CSF system is pushed to capacity by delivering multiple doses of 1×10^6 cells (unpublished results). Future studies using multiple LP deliveries may require fewer cells per delivery or testing in larger animal models.

Together, these results suggest that cell delivery to SCI using LP is a compelling alternative to direct injection. Further optimization of cell dosage, timing of delivery, and number of deliveries are needed to further improve the grafting efficiency and thereby functional recovery. Even if LP delivery is not as efficient as direct injection, LP has a number of advantages. It can be easily translated to the clinic since it is already well established. It limits patient risk, side effects, and cost and can be used to deliver multiple doses of cells. In fact, LP delivery has already been used in a number of clinical trials for the treatment of SCI (51–53). So far, these groups report minimal to no adverse effects and some degree of recovery of function after cell delivery via LP. One study (54) used magnetically labeled cells to track the cell location in patients using MRI and reported cells at the site of injury 1 month after delivery to chronically injured patients.

In summary, LP has been used to deliver cells secreting antinociceptive factors to the spinal cord and successfully alleviate neuropathic pain. It also is a compelling alternative to direct injection for cell delivery in SCI. Moreover, the fact that cells delivered by LP accumulate at the site of injury and that LP deliveries can safely be repeated multiple times makes this technique potentially useful for a variety of CNS disorders with diffuse injuries like multiple sclerosis and amyotrophic lateral sclerosis.

2. Materials

1. Gas anesthesia set-up (including O_2 and Isoflurane inhalation anesthetic)
2. Surgery board
3. Small plastic box or plastic test tube rack 2.5–5 cm in height to “flex” the animal’s back
4. Scalpel, with extra blades
5. Forceps, scissors
6. 100 μl Hamilton glass syringe, with ground glass Luer tip
7. 30 1/2" needle (for loading saline and cells)

8. Gauze pads
9. Cotton tip applicators
10. Sterile Saline
11. Cells, in individual aliquots in 40 μl basal medium (or other cell suspension medium) held on melting ice until use
12. Needle holder and suture material
13. Wound clips
14. Wound clip applicator

Note: All surgical equipment has to be sterilized before use.

3. Method

1. Place the rat in anesthesia chamber with O_2 flow set at 2.0 l/min and isoflurane at 4%.
2. When anesthetized, remove the animal from the anesthesia chamber, and shave and clean the back with iodine solution.
3. Return the animal to the anesthesia chamber and continue anesthesia with 4% ISOFLURANE. The animal is completely anesthetized when it does not respond to a foot pinch or eye touch. At this point, reduce Isoflurane to 2%, apply the anesthesia face mask, and transfer the animal to the surgery board.
4. The animal should be maintained at 2% ISOFLURANE throughout the surgery. Anesthesia can be increased if the animal becomes responsive.
5. On the surgery board, flex the animal's back at the lumbar region by placing the plastic box underneath the animal. Use tape to secure the animal and the mask if needed (**Fig. 10.1**). Cover the animal with sterile drapes leaving only the lumbar area exposed.
6. Make a midline incision from the last rib (T13) to the beginning of the iliac crest (L6).
7. Retract the skin to expose the lumbar vertebral column.
8. Identify L3–L5 vertebrae by counting back from T13, the vertebra to which the last rib is attached.
9. Identify the space between L3 and L4 or L4 and L5. Either location will work – use the space that is most easily accessible.
10. Make one longitudinal incision on both sides of the vertebral bodies from approximately L2–L5 to isolate the vertebral bodies.

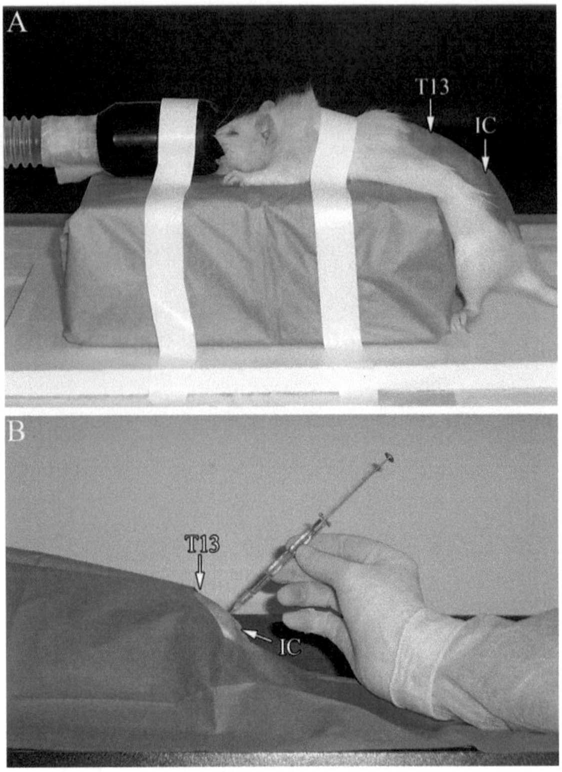

Fig. 10.1. **Panel a** shows the optimal position of the rat with its back flexed at the lumbar region (Point 5 in **Section 3**). **Panel b** shows placement of the syringe after advancing the needle through the inter-vertebral space (L4–L5) and into the spinal canal (Point 13 in **Section 3**). T13, thoracic vertebrae 13; IC, iliac crest.

11. Remove the supra-spinal musculature and ligaments that cover the inter-vertebral space (L3–L4 or L4–L5) that will be used for LP.
12. Load a 100-µl Hamilton syringe using a 30-gauge needle:
    a. First, load 20–40 µl of sterile saline (or other washing fluid such as suspension medium).
    b. Then, load the cell suspension (40 µl).
    c. Finally, load an additional 5–10 µl of sterile saline or other wash. 5–10 µl is the approximate volume of the needle hub. This volume is added to ensure complete delivery of the cell volume. The final volume will be 65–90 µl.
    d. Be sure to draw all the liquid into the Hamilton syringe and do not leave any in the needle hub, then detach the needle.
13. Advance the 30 gauge needle through the inter-vertebral space (L3–L4 or L4–L5) and into the spinal canal.
    a. Ensure that the needle is centered on the midline.
    b. To further increase the inter-vertebral space (if needed) flex the spine by carefully bending the animals' lumbar region.

14. Proper needle placement is defined by three signs (not all of the signs will be seen every time):
    a. A feeling of "give" at the time of entry (soft sign)
    b. A tail flick (more definitive sign)
    c. Observation of CSF in the needle hub (definite sign)
15. If collecting CSF, gently remove the CSF that collects in the needle hub with a pipette being sure not to introduce direct suction on the intrathecal space:
    a. The rostral end of the animal can be elevated to induce more CSF flow if necessary.
    b. If this is a terminal procedure to collect the maximum amount of CSF, massage at the neck will increase CSF flow.
16. Once proper needle placement has been confirmed, ensure that the needle hub is full of either CSF or sterile saline (to avoid air bubbles). Then attach the loaded 100-μl Hamilton syringe to the needle, being careful not to move the needle (this is a very delicate task and a balance must be met between tight needle seating and minimal movement of the needle).
17. Inject the fluid from the Hamilton syringe slowly over a 1- to 2-min period without using too much pressure.
18. When all fluid has been injected, hold the needle in place for approximately 1 min to allow some CSF circulation around the injection site.
19. Carefully remove the needle and apply moderate pressure to the area of injection with a cotton tip applicator to prevent bleeding.
20. Suture the muscle and staple the skin closed.
21. Place the animal in a cage on a heating pad and monitor until the animal awakens from anesthesia.
22. Thoroughly flush the syringe after the animal has been closed to prevent clogging of the syringe and sterilize to prevent cross-contamination.

4. Notes

1. Timing of LP delivery:
    a. Optimal time for engraftment likely depends on cell type and injury.
    b. For optimal engraftment in a rat SCI, cells should be delivered within the first 2 weeks of injury.

   c. Early delivery (within 3–4 days) after contusive injury may result in poor cell survival and should be followed by subsequent deliveries to restore graft volume.

2. Cell dosage for LP delivery:
   a. The optimal cell dosage depends on the model used. For the rat, we have used 1×10^6 cells for single LP delivery. This dose appears to represent the upper limit for the capacity of the rat CSF. For multiple deliveries, lower doses may be advantageous.

3. When and where to expect LP-delivered cells in the spinal cord:
   a. We found that cells can be detected at the site of injury after 3 days; however, graft volumes increase considerably after 10–14 days.
   b. LP-delivered cell grafts survive up to 8 weeks. Longer time points have not been tested in our laboratory.
   c. Cells can also be detected attached to the surface of the spinal cord starting as early as 4 days after LP delivery. A similar distribution was reported by Ohta and colleagues after MSC infusion into the 4th ventricle (23).

4. Difficulty finding an open inter-vertebral space for LP:
   a. Use a larger plastic box to place under the lumbar region of the animal to induce more flexion into the lumbar spine and open the inter-vertebral spaces.
   b. Remove more supraspinal musculature and ligaments.
   c. Perform a partial laminectomy to remove a small portion of the dorsal laminae to open the space.

5. Difficulty with needle placement:
   a. Sure signs of needle placement within the sub-arachnoid space are any of the following:
      i. Feeling of a "give" or "pop" when advancing the needle.
      ii. Visible tail, hindlimb, or abdominal muscle twitches upon entry of the needle.
      iii. Visible CSF flow observed in the needle up (which can be persuaded by elevating the head and thorax of the animal).
      iv. Note: Not all of these signs will be visible each time a LP is performed and it is possible that none of these signs is present during a successful LP.
   b. A small amount of cell suspension or fluid can be injected to ensure that leakage does not occur, indicating proper needle placement in the fully enclosed sub-arachnoid space.

6. Seating the syringe into the needle hub after proper needle placement:

   a. Dislodging the needle when attaching the loaded syringe for injection can happen. Minimal movement of the needle is an absolute requirement, when attaching the syringe.
   b. It is possible to perform the procedure with a loaded syringe attached to the needle. This procedure, however, precludes observation of CSF flow into the needle hub, which is the most definite sign of proper needle placement. It also precludes removal of CSF for analysis.

7. Leaking of fluid during injection:
   a. This signals improper needle placement.
   b. Stop the injection and reposition the needle.
8. Cross contamination between animals:
   a. Clean the syringe thoroughly with sterile saline and sterilize between animals.
   b. Change or sterilize the needle between animals.
9. Difficulty with syringe function:
   a. Syringes and needles can clog easily.
   b. Clean the syringe thoroughly with sterile saline and change or sterilize the needle between animals.
10. Cell clumping:
    a. If cells clump at the bottom of the tube, lightly flick or tap the tube to gently resuspend the cells.

Acknowledgments

The authors would like to express their gratitude to Drs. Marion Murray and Timothy B. Himes for critical discussion and editorial comments. We thank Courtney Paul and Joseph F. Bonner for help with technical aspects of the review. Dr. Amer Samdani, Maryla Obrocka, and Carla Tyler-Polsz contributed to individual studies summarized in this review. This work was supported by NIH grant NS049429 as well as grants from the Craig H. Neilsen Foundation and Shriners Hospital for Children.

References

1. Hall, E.D. and Springer, J.E. (2004) Neuroprotection and acute spinal cord injury: a reappraisal. *NeuroRx* 1, 80–100.
2. Fawcett, J.W. and Asher, R.A. (1999) The glial scar and central nervous system repair. *Brain Res Bull* 49, 377–391.
3. McGee, A.W. and Strittmatter, S.M. (2003) The Nogo-66 receptor: focusing myelin inhibition of axon regeneration. *Trends Neurosci* 26, 193–198.
4. Murray, M. (2004) Cellular transplants: steps toward restoration of function in spinal injured animals. *Prog Brain Res* 143, 133–146.
5. Fischer, I., Lepore, A.C., Han, S.S.W., and Tessler, A. (2006) Cellular replacement in

spinal cord injury. in *Textbook of neural repair and rehablilitation* (Selzer, M., Clarke, S., Cohen, L., Duncan, P., and Gage, F.H., Eds.), pp. 445–467, Cambridge University Press, Cambridge.
6. Parr, A.M., Kulbatski, I., and Tator, C.H. (2007) Transplantation of adult rat spinal cord stem/progenitor cells for spinal cord injury. *J Neurotrauma* 24, 835–845.
7. Himes, B.T., Neuhuber, B., Coleman, C., Kushner, R., Swanger, S.A., Kopen, G.C., Wagner, J., Shumsky, J.S., and Fischer, I. (2006) Recovery of function following grafting of human bone marrow-derived stromal cells into the injured spinal cord. *Neurorehabil Neural Repair* 20, 278–296.
8. Carvalho, K.A., Cunha, R.C., Vialle, E.N., Osiecki, R., Moreira, G.H., Simeoni, R.B., Francisco, J.C., Guarita-Souza, L.C., Oliveira, L., Zocche, L., and Olandoski, M. (2008) Functional outcome of bone marrow stem cells (CD45(+)/CD34(−)) after cell therapy in acute spinal cord injury: in exercise training and in sedentary rats. *Transplant Proc* 40, 847–849.
9. Mitsui, T., Shumsky, J.S., Lepore, A.C., Murray, M., and Fischer, I. (2005) Transplantation of neuronal and glial restricted precursors into contused spinal cord improves bladder and motor functions, decreases thermal hypersensitivity, and modifies intraspinal circuitry. *J Neurosci* 25, 9624–9636.
10. Mitsui, T., Fischer, I., Shumsky, J.S., and Murray, M. (2005) Transplants of fibroblasts expressing BDNF and NT-3 promote recovery of bladder and hindlimb function following spinal contusion injury in rats. *Exp Neurol* 194, 410–431.
11. Ohta, Y., Takenaga, M., Tokura, Y., Hamaguchi, A., Matsumoto, T., Kano, K., Mugishima, H., Okano, H., and Igarashi, R. (2008) Mature adipocyte-derived cells, dedifferentiated fat cells (DFAT), promoted functional recovery from spinal cord injury-induced motor dysfunction in rats. *Cell Transplant* 17, 877–886.
12. Eaton, M.J., Pearse, D.D., McBroom, J.S., and Berrocal, Y.A. (2008) The combination of human neuronal serotonergic cell implants and environmental enrichment after contusive SCI improves motor recovery over each individual strategy. *Behav Brain Res* 194, 236–241.
13. Hofstetter, C.P., Schwarz, E.J., Hess, D., Widenfalk, J., El Manira, A., Prockop, D.J., and Olson, L. (2002) Marrow stromal cells form guiding strands in the injured spinal cord and promote recovery. *Proc Natl Acad Sci USA* 99, 2199–2204.
14. Akiyama, Y., Radtke, C., Honmou, O., and Kocsis, J.D. (2002) Remyelination of the spinal cord following intravenous delivery of bone marrow cells. *Glia* 39, 229–236.
15. Takeuchi, H., Natsume, A., Wakabayashi, T., Aoshima, C., Shimato, S., Ito, M., Ishii, J., Maeda, Y., Hara, M., Kim, S.U., and Yoshida, J. (2007) Intravenously transplanted human neural stem cells migrate to the injured spinal cord in adult mice in an SDF-1- and HGF-dependent manner. *Neurosci Lett* 426, 69–74.
16. Urdzikova, L., Jendelova, P., Glogarova, K., Burian, M., Hajek, M., and Sykova, E. (2006) Transplantation of bone marrow stem cells as well as mobilization by granulocyte-colony stimulating factor promotes recovery after spinal cord injury in rats. *J Neurotrauma* 23, 1379–1391.
17. Chernykh, E.R., Stupak, V.V., Muradov, G.M., Sizikov, M.Y., Shevela, E.Y., Leplina, O.Y., Tikhonova, M.A., Kulagin, A.D., Lisukov, I.A., Ostanin, A.A., and Kozlov, V.A. (2007) Application of autologous bone marrow stem cells in the therapy of spinal cord injury patients. *Bull Exp Biol Med* 143, 543–547.
18. Sykova, E., Homola, A., Mazanec, R., Lachmann, H., Konradova, S.L., Kobylka, P., Padr, R., Neuwirth, J., Komrska, V., Vavra, V., Stulik, J., and Bojar, M. (2006) Autologous bone marrow transplantation in patients with subacute and chronic spinal cord injury. *Cell Transplant* 15, 675–687.
19. Bakshi, A., Hunter, C., Swanger, S., Lepore, A., and Fischer, I. (2004) Minimally invasive delivery of stem cells for spinal cord injury: advantages of the lumbar puncture technique. *J Neurosurg Spine* 1, 330–337.
20. Paul, C., Samdani, A.F., Betz, R.R., Fischer, I., and Neuhuber, B. (2009) Grafting of human bone marrow stromal cells into spinal cord injury: a comparison of delivery methods. *Spine* 34, 328–334.
21. Vaquero, J., Zurita, M., Oya, S., and Santos, M. (2006) Cell therapy using bone marrow stromal cells in chronic paraplegic rats: systemic or local administration? *Neurosci Lett* 398, 129–134.
22. Ban, D.X., Kong, X.H., Feng, S.Q., Ning, G.Z., Chen, J.T., and Guo, S.F. (2009) Intraspinal cord graft of autologous activated Schwann cells efficiently promotes axonal regeneration and functional recovery after rat's spinal cord injury. *Brain Res* 1256, 149–161.

23. Ohta, M., Suzuki, Y., Noda, T., Ejiri, Y., Dezawa, M., Kataoka, K., Chou, H., Ishikawa, N., Matsumoto, N., Iwashita, Y., Mizuta, E., Kuno, S., and Ide, C. (2004) Bone marrow stromal cells infused into the cerebrospinal fluid promote functional recovery of the injured rat spinal cord with reduced cavity formation. *Exp Neurol* 187, 266–278.
24. Wu, S., Suzuki, Y., Kitada, M., Kataoka, K., Kitaura, M., Chou, H., Nishimura, Y., and Ide, C. (2002) New method for transplantation of neurosphere cells into injured spinal cord through cerebrospinal fluid in rat. *Neurosci Lett* 318, 81–84.
25. Wu, S., Suzuki, Y., Noda, T., Bai, H., Kitada, M., Kataoka, K., Nishimura, Y., and Ide, C. (2002) Immunohistochemical and electron microscopic study of invasion and differentiation in spinal cord lesion of neural stem cells grafted through cerebrospinal fluid in rat. *J Neurosci Res* 69, 940–945.
26. De la Calle, J.L., and Paino, C.L. (2002) A procedure for direct lumbar puncture in rats. *Brain Res Bull* 59, 245–250.
27. De la Calle, J.L., Mena, M.A., Gonzalez-Escalada, J.R., and Paino, C.L. (2002) Intrathecal transplantation of neuroblastoma cells decreases heat hyperalgesia and cold allodynia in a rat model of neuropathic pain. *Brain Res Bull* 59, 205–211.
28. Eaton, M.J., Wolfe, S.Q., Martinez, M., Hernandez, M., Furst, C., Huang, J., Frydel, B.R., and Gomez-Marin, O. (2007) Subarachnoid transplant of a human neuronal cell line attenuates chronic allodynia and hyperalgesia after excitotoxic spinal cord injury in the rat. *J Pain* 8, 33–50.
29. Wolfe, S.Q., Garg, M., Cumberbatch, N.M., Furst, C., Martinez, M., Hernandez, M., Reimers, R., Berrocal, Y., Gomez-Marin, O., and Eaton, M.J. (2007) Optimizing the transplant dose of a human neuronal cell line graft to treat SCI pain in the rat. *Neurosci Lett* 414, 121–125.
30. Lin, C.R., Wu, P.C., Shih, H.C., Cheng, J.T., Lu, C.Y., Chou, A.K., and Yang, L.C. (2002) Intrathecal spinal progenitor cell transplantation for the treatment of neuropathic pain. *Cell Transplant* 11, 17–24.
31. Bakshi, A., Barshinger, A.L., Swanger, S.A., Madhavani, V., Shumsky, J.S., Neuhuber, B., and Fischer, I. (2006) Lumbar puncture delivery of bone marrow stromal cells in spinal cord contusion: a novel method for minimally invasive cell transplantation. *J Neurotrauma* 23, 55–65.
32. Lepore, A.C., Bakshi, A., Swanger, S.A., Rao, M.S., and Fischer, I. (2005) Neural precursor cells can be delivered into the injured cervical spinal cord by intrathecal injection at the lumbar cord. *Brain Res* 1045, 206–216.
33. Neuhuber, B., Barshinger, A.L., Paul, C., Shumsky, J.S., Mitsui, T., and Fischer, I. (2008) Stem cell delivery by lumbar puncture as a therapeutic alternative to direct injection into injured spinal cord. *J Neurosurg Spine* 9, 390–399.
34. Shi, M., Li, J., Liao, L., Chen, B., Li, B., Chen, L., Jia, H., and Zhao, R.C. (2007) Regulation of CXCR4 expression in human mesenchymal stem cells by cytokine treatment: role in homing efficiency in NOD/SCID mice. *Haematologica* 92, 897–904.
35. Ji, J.F., He, B.P., Dheen, S.T., and Tay, S.S. (2004) Expression of chemokine receptors CXCR4, CCR2, CCR5 and CX3CR1 in neural progenitor cells isolated from the subventricular zone of the adult rat brain. *Neurosci Lett* 355, 236–240.
36. Tran, P.B., Ren, D., Veldhouse, T.J., and Miller, R.J. (2004) Chemokine receptors are expressed widely by embryonic and adult neural progenitor cells. *J Neurosci Res* 76, 20–34.
37. Shichinohe, H., Kuroda, S., Yano, S., Hida, K., and Iwasaki, Y. (2007) Role of SDF-1/CXCR4 system in survival and migration of bone marrow stromal cells after transplantation into mice cerebral infarct. *Brain Res* 1183, 138–147.
38. Ji, J.F., He, B.P., Dheen, S.T., and Tay, S.S. (2004) Interactions of chemokines and chemokine receptors mediate the migration of mesenchymal stem cells to the impaired site in the brain after hypoglossal nerve injury. *Stem Cells* 22, 415–427.
39. Dar, A., Kollet, O., and Lapidot, T. (2006) Mutual, reciprocal SDF-1/CXCR4 interactions between hematopoietic and bone marrow stromal cells regulate human stem cell migration and development in NOD/SCID chimeric mice. *Exp Hematol* 34, 967–975.
40. Tiveron, M.C. and Cremer, H. (2008) CXCL12/CXCR4 signalling in neuronal cell migration. *Curr Opin Neurobiol* 18, 237–244.
41. Stumm, R. and Hollt, V. (2007) CXC chemokine receptor 4 regulates neuronal migration and axonal pathfinding in the developing nervous system: implications for neuronal regeneration in the adult brain. *J Mol Endocrinol* 38, 377–382.

42. Ben-Hur, T., Ben-Menachem, O., Furer, V., Einstein, O., Mizrachi-Kol, R., and Grigoriadis, N. (2003) Effects of proinflammatory cytokines on the growth, fate, and motility of multipotential neural precursor cells. *Mol Cell Neurosci* 24, 623–631.
43. Forsberg-Nilsson, K., Behar, T.N., Afrakhte, M., Barker, J.L., and McKay, R.D. (1998) Platelet-derived growth factor induces chemotaxis of neuroepithelial stem cells. *J Neurosci Res* 53, 521–530.
44. Kendall, S.E., Najbauer, J., Johnston, H.F., Metz, M.Z., Li, S., Bowers, M., Garcia, E., Kim, S.U., Barish, M.E., Aboody, K.S., and Glackin, C.A. (2008) Neural stem cell targeting of glioma is dependent on phosphoinositide 3-kinase signaling. *Stem Cells* 26, 1575–1586.
45. Ponte, A.L., Marais, E., Gallay, N., Langonne, A., Delorme, B., Herault, O., Charbord, P., and Domenech, J. (2007) The in vitro migration capacity of human bone marrow mesenchymal stem cells: comparison of chemokine and growth factor chemotactic activities. *Stem Cells* 25, 1737–1745.
46. Ozaki, Y., Nishimura, M., Sekiya, K., Suehiro, F., Kanawa, M., Nikawa, H., Hamada, T., and Kato, Y. (2007) Comprehensive analysis of chemotactic factors for bone marrow mesenchymal stem cells. *Stem Cells Dev* 16, 119–129.
47. Ankeny, D.P., McTigue, D.M., and Jakeman, L.B. (2004) Bone marrow transplants provide tissue protection and directional guidance for axons after contusive spinal cord injury in rats. *Exp Neurol* 190, 17–31.
48. Pearse, D.D. and Bunge, M.B. (2006) Designing cell- and gene-based regeneration strategies to repair the injured spinal cord. *J Neurotrauma* 23, 438–452.
49. Donnelly, D.J. and Popovich, P.G. (2008) Inflammation and its role in neuroprotection, axonal regeneration and functional recovery after spinal cord injury. *Exp Neurol* 209, 378–388.
50. Shi, E., Kazui, T., Jiang, X., Washiyama, N., Yamashita, K., Terada, H., and Bashar, A.H. (2007) Therapeutic benefit of intrathecal injection of marrow stromal cells on ischemia-injured spinal cord. *Ann Thorac Surg* 83, 1484–1490.
51. Saito, F., Nakatani, T., Iwase, M., Maeda, Y., Hirakawa, A., Murao, Y., Suzuki, Y., Onodera, R., Fukushima, M., and Ide, C. (2008) Spinal cord injury treatment with intrathecal autologous bone marrow stromal cell transplantation: the first clinical trial case report. *J Trauma* 64, 53–59.
52. Callera, F. and do Nascimento, R.X. (2006) Delivery of autologous bone marrow precursor cells into the spinal cord via lumbar puncture technique in patients with spinal cord injury: a preliminary safety study. *Exp Hematol* 34, 130–131.
53. Mehta, T., Feroz, A., Thakkar, U., Vanikar, A., Shah, V., and Trivedi, H. (2008) Subarachnoid placement of stem cells in neurological disorders. *Transplant Proc* 40, 1145–1147.
54. Callera, F. and de Melo, C.M. (2007) Magnetic resonance tracking of magnetically labeled autologous bone marrow CD34+ cells transplanted into the spinal cord via lumbar puncture technique in patients with chronic spinal cord injury: CD34+ cells' migration into the injured site. *Stem Cells Dev* 16, 461–466.
55. Bunge, M.B. and Wood, P.M. (2006) Transplantation of Schwann cells and olfactory ensheathing cells to promote regeneration in the CNS. in *Textbook of neural repair and rehablilitation* (Selzer, M., Clarke, S., Cohen, L., Duncan, P., and Gage, F.H., Eds.), pp. 513–531, Cambridge University Press, Cambridge.
56. Nandoe Tewarie, R.D., Hurtado, A., Levi, A.D., Grotenhuis, J.A., and Oudega, M. (2006) Bone marrow stromal cells for repair of the spinal cord: towards clinical application. *Cell Transplant* 15, 563–577.
57. Murray, M., Kim, D., Liu, Y., Tobias, C., Tessler, A., and Fischer, I. (2002) Transplantation of genetically modified cells contributes to repair and recovery from spinal injury. *Brain Res Brain Res Rev* 40, 292–300.
58. Schwartz, M. and Yoles, E. (2006) Immune-based therapy for spinal cord repair: autologous macrophages and beyond. *J Neurotrauma* 23, 360–370.

Chapter 11

CNS Delivery of Peptides Across the BBB Using the Dual-Artery In Situ Brain Perfusion Model

Ken A. Witt, Patrick T. Ronaldson, Karin E. Sandoval, and Thomas P. Davis

Abstract

Peptides have been shown to be potent regulators of central nervous system (CNS) activity. As such, peptide-based drugs can serve as a highly effective means to treat diseases of the CNS. Despite this potential, peptides have limited capacity to permeate across the blood–brain barrier (BBB). Nevertheless, advancing strategies in peptide drug design, delivery systems, and transporter targeting are emerging, with such potential to overcome the restrictive nature of the BBB. This chapter addresses peptide drug delivery into the CNS, and in vivo evaluation of peptide permeability across the BBB using the dual-artery in situ brain perfusion model coupled with the technique of capillary depletion.

Key words: Blood–brain barrier, peptide drug delivery, in situ brain perfusion, capillary depletion.

1. Introduction

The understanding that endogenous peptides of the central nervous system (CNS) can induce potent biological effects has brought about significant research focused on enhancing the delivery of exogenously administered peptides to the brain. Assuming that a peptide drug is sufficiently able to overcome peripheral obstacles (i.e., absorption from the chosen route of administration, protein binding, peripheral organ uptake, and liver metabolism), the drug must still gain access to the brain. In this regard, both physicochemical attributes of the given peptide and physiological characteristics of the blood–brain barrier (BBB) must be considered.

The BBB comprises both a physical and enzymatic barrier segregating the brain from the systemic circulation. This dynamic structure regulates the critical processes of nutrient uptake and

K.K. Jain (ed.), *Drug Delivery to the Central Nervous System*, Neuromethods 45,
DOI 10.1007/978-1-60761-529-3_11, © Humana Press, a part of Springer Science+Business Media, LLC 2010

waste removal, thus enabling the BBB to maintain the fragile homeostasis of the brain. As paracellular permeability is significantly limited due to epithelial-like tight junctions of the BBB, peptide drug design is most often focused on enhancing diffusion capacity or affinity for specific uptake transporters (i.e., mimetic design). In addition, peptide drug design must also take into consideration the functional expression of BBB efflux transporters, due to the broad affinity of such transporters for lipophilic drug molecules. As such, rational approaches to peptide drug design and targeting must also be coupled with appropriate animal modeling and evaluation to ensure successful entry into the CNS.

2. Peptide Diffusion

Lipid solubility, charge, and molecular size have long been identified as dependent factors for the diffusion of molecules across cellular membranes (1). Generally, compounds that are non-ionized at physiological pH are lipophilic and have a MW < 500 Da have the greatest potential to passively diffuse across the BBB. Peptides generally do not share these physicochemical properties and, therefore, are not generally favored for diffusion. Many peptides are highly water soluble (i.e., polar). The polar functional groups of a peptide impart a degree of dipolarity and hydrogen bonding capability, which reduces the ability of the peptide to partition into a non-polar environment (i.e., lipid membrane). Shifting the lipophilicity of a peptide from polar to non-polar is a viable method of diffusion enhancement. This can be brought about by the addition of non-polar groups or removal of polar groups. Localization of respective polar and non-polar amino acids (i.e., whether respective groups are encapsulated within the center of a peptide or exist on the perimeter) can also impact a peptide's ability to diffuse across a lipid membrane. In this regard, addition of non-polar groups to the surface of a peptide may enhance lipophilicity if they are able to reduce or mask the hydrogen bonding effects. Various methods used to increase lipid solubility of peptides include methylation, halogenation, acylation, amino acid substitution, and/or amino acid addition (2, 3). Nevertheless, enhancement of lipid solubility does have limitations. Lipid soluble compounds may be extensively bound to plasma proteins, thereby, reducing the free concentration of drug available for brain diffusion. Furthermore, lipid soluble compounds will have a greater capacity for peripheral tissue uptake. Alterations in lipophilicity may also alter conformational flexibility, molecular size, and affinity for receptor targets. Lastly, increasing the lipid solubility of a compound may also enhance affinity for efflux transporters, particularly those that are expressed at the BBB.

Beyond enhancing lipid solubility, other peptide modification and delivery strategies to enhance brain uptake include glycosylation, structural modification for enhance enzymatic stability, cyclization, cationization, vector-based models, prodrugs, polymer conjugation, nanoparticles, and increasing the affinity for nutrient transporters at the BBB. Each of these approaches comes with its own advantages and disadvantages and has been described elsewhere (3, 4). Ideally, the appropriate combination of such techniques will enhance BBB passage and increase the end-biological effect.

3. Transport Mechanisms

Blood–brain barrier transport mechanisms exemplify both a challenge and an opportunity for peptide drug delivery. For many peptides, uptake into and extrusion from the brain is determined by the functional expression of both influx and efflux transport proteins. While transport uptake mechanisms at the BBB exist for several endogenous peptides/proteins, examples of designed peptide-drugs that use such carriers remain relatively few. This is due to both the emphasis on drug action-design versus delivery-design and the specificity of transport carriers. In theory, higher capacity carriers (e.g., hexose, large-neutral amino acid carriers) should have the greatest ability to transport a therapeutic dose of a designed peptide. Some established drugs, either derived from amino acids or having amino acid constituents, have shown uptake via the large-neutral amino acid transporter into the brain (e.g., baclofen, L-DOPA) (5). The analgesic enkephalin peptide analog biphalin has also been shown to use the large-neutral amino acid transporter (6). Nevertheless, determinations of those drugs shown to use carrier mediated uptake into the brain have only come after their viability as functional agents was identified.

Efflux transporters at the BBB are a significant obstacle to CNS drug permeation for many drugs, including peptides. The specific chemical attributes that render a drug selective for an individual efflux transporter is complex. Thus, evaluating the role of these transporters in restricting uptake and regulating distribution of therapeutics in the CNS is a critical step in drug development. The ATP-binding cassette (ABC) transporters are the primary superfamily of proteins involved in BBB efflux. Several ABC superfamily members such as P-glycoprotein (P-gp), multidrug resistance-associated proteins (MRP in humans; Mrp in rodents), and breast cancer resistance protein (BCRP; also known as ABCG2) have been localized to the luminal and/or abluminal surface of the BBB (7). These transporters are known to be involved in cellular extrusion of a vast array of structurally and functionally unrelated compounds (8). In general, P-gp transports cationic, basic, and neutral compounds, while MRPs/Mrps are involved in cellular efflux of anionic drugs as well as their glucuronidated, sulfated, and glutathione-conjugated metabolites. For example, P-gp has been shown to be involved in BBB efflux transport of opioid peptides, such as DPDPE, DADLE, and Met-enkephalin (9–11); while the tripeptide glutathione (GSH) along with its oxidized form glutathione disulfide (GSSG) are established substrates for multiple MRP/Mrp isoforms including MRPs/Mrps 1, 2, and 4 (12–14). BCRP has significant overlap in substrate specificity profile with P-gp and has been shown to recognize a vast array of sulfoconjugated organic anions, hydrophobic, and amphiphilic compounds (15).

Apart from ABC transporters, organic anion transporting polypeptides (OATP in humans; Oatp in rodents) may also be involved in peptide transport at the BBB (7, 16). OATPs/Oatps are a group of membrane solute carriers involved in bidirectional and sodium-independent transport of amphipathic substrates. Previous studies suggest that OATPs/Oatps have overlapping and partially distinct substrate preferences for various solutes, including pharmacological agents and anionic peptides (17). For example, the opioid peptides DPDPE and deltorphin II are known substrates for OATP1A2, an OATP isoform known to be expressed at the human BBB (16). Additionally, organic anion transporter 3 (OAT3), which is known to transport peptides such as GSH (18), has been identified at the basolateral surface of brain microvascular endothelial cells (7); however, its role in the transport of peptides and/or peptidomimetic drugs at the BBB has not been elucidated. Other peptide transporters such as peptide transporter 2 (PEPT2) and peptide histidine transporters 1 and 2 (PHT1, PHT2) have also been identified in the CNS; however, they have not been localized to the BBB (19, 20).

4. Method of Brain Uptake Measurement

Several methods are available for evaluation of peptide delivery across the BBB, both in vivo and in vitro. Although in vitro methods allow for high throughput screening, they often lack the critical multidimensional characteristics and physiological conditions of in vivo modeling. Specifically, in vivo models lend themselves to evaluations of cerebral blood-flow and pressure, cerebrospinal fluid, and regional transporter dynamics. Moreover, BBB function and structural integrity varies with age, disease, and brain region, which may be critical when evaluating a drug designed for regional effect in a specific disease state. With this understanding, the in situ brain perfusion technique is ideal for evaluation of peptide-drug entry into the brain.

The in situ brain perfusion method, which has gone through multiple inceptions since its original development by Takasato et al. (21), is a potent technique. This approach maintains many of the advantages of the brain uptake index method (22), while bypassing the systemic circulation. It permits direct brain perfusion, eliminating peripheral metabolism and organ uptake. Control of perfusate flow also allows for the maintenance of a constant drug concentration, simplifying kinetic analyses. Furthermore, the perfusate medium can be adjusted by adding inhibitors (both pharmacological and mechanistic), binding proteins, or simply altering the ionic nature. Lastly, perfusions can be performed at various set periods (5 s to 30 min), permitting the investigator to adjust parameters to further delineate transport mechanisms, as well as measure more slowly penetrating substances.

While the original Takasato method requires significant training and practice, a straightforward alternative is to perform the perfusion directly via the common carotids (23). Common carotid cannulation is mechanistically simple and can be performed on both arteries in the same animal. Such dual-artery cannulation allows for evaluation of both brain hemispheres, not only providing more tissue for assessment but reducing the potential of cross-over distribution via the circle of Willis. The power of dual-artery in situ perfusion is further enhanced when combined with the capillary depletion method (24, 25). Capillary depletion measures association of a drug to the capillary component, thereby delineating the true transport of a peptide into brain parenchyma from simple binding and/or endocytosis by the brain microvasculature.

5. Materials

5.1. Dual-Artery In Situ Perfusion

5.1.1. Perfusion Medium

The optimal medium for an in situ perfusion depends on the experimental goals. The chosen medium should supply the essential nutrients to maintain structural and metabolic integrity of the brain.

1. One of the most commonly used perfusion media is a buffered Krebs–Henseleit physiological saline or Ringer which is composed of 24 mM $NaHCO_3$, 4.2 mM KCl, 2.4 mM NaH_2PO_4, 128 mM NaCl, 1.5 mM $CaCl_2$, 10 mM D-glucose, and 0.9 mM $MgCl_2$, pH = 7.4 (26) (Sigma-Aldrich). Dextran

(MW 60 kDa, 39 g/l) and albumin (bovine serum albumin type-V, 1 g/l) can be added to maintain osmotic levels and to serve as effective means for binding of drugs, thus mimicking typical blood proteins (Sigma-Aldrich) (*see* **Note 1**).

5.1.2. Perfusion Equipment

1. Oxygen (95% O_2/CO_2) tank with appropriate regulator (local distributor).
2. Peristaltic pump (Harvard Apparatus or World Precision Instruments) (*see* **Note 2**).
3. Syringe pump with capacity for two 60-ml syringes (Harvard Apparatus or World Precision Instruments).
4. Circulating water with heat control (Fisher).
5. Pressure transducer (World Precision Instruments), with chart recorder optional.
6. Perfusion line polyethylene tubing, pump tubing (Fisher Scientific), and cannula tubing (Plastics-1) (*see* **Note 3**).
7. 60-ml syringes and 18-gauge needles (Fisher Scientific).
8. Stop-cocks, T-pieces, and tubing connectors (Fisher Scientific).
9. In-line bubble-traps (*see* **Note 4**).

5.1.3. Surgical Supplies

1. Heating lap (Fisher Scientific).
2. Surgical board (Fisher Scientific).
3. Surgical silk suture (5-0) (Harvard Apparatus).
4. Plastic syringes (1 ml) (Fisher Scientific).
5. Surgical instruments: Rongeurs, micro-dissecting and standard tweezers and forceps, micro-dissecting spring scissors, scalpel handle and blades, utility scissors, serrated scissors, hemostats, spatulas (Fine Science Tools, Roboz Surgical Instruments, or World Precision Instruments).
6. Anesthetic (e.g., sodium pentobarbital) (Sigma-Aldrich).
7. Heparin (Sigma-Aldrich).
8. Absorbent pads (Fisher Scientific).

5.2. Capillary Depletion

5.2.1. Buffers/Stock Solutions required for Capillary Depletion

Below are recipes for the physiological buffer and dextran stock solutions required to successfully perform capillary depletion. All solutions should be chilled to 4°C prior to use and maintained on ice for the duration of the procedure.

1. The physiological buffer recommended for the capillary depletion technique is comprised of the following: 10 mM HEPES, 141 mM NaCl, 4 mM KCl, 1 mM $MgSO_4$, 1 mM NaH_2PO_4, 10 mM D-glucose, and 2.8 mM $CaCl_2$, pH = 7.4.
2. A 26% clinical grade dextran (MW 60 kDa) solution is also required (26 g dextran/100 ml of ddI-H_2O).

5.2.2. Supplies

1. Glass vessel and Teflon pestle for tissue homogenization (Wheaton, Millville, NJ).
2. Rongeurs, spatula, and iris forceps to remove and separate brain tissue (Fine Science Tools, Foster City, CA).
3. Polypropylene tubes (14 ml) (Fisher Scientific).

5.3. Measurement and Auxiliary Supplies

1. Analytical balance (Mettler Toledo).
2. Microcentrifuge tubes (1.5 and 2.0 ml) (Fisher Scientific).
3. Pipettes: repeat pipette, 10–200 μl, and 200–1000 μl (Eppendorf or Rainin).
4. Pipette tips (Eppendorf or Rainin).
5. Microcentrifuge (Fisher Scientific).
6. Flasks and/or beakers (500 ml and 1000 ml) (Fisher Scientific).
7. Timer (Fisher Scientific).
8. Vials for sample collection (dependent on choice of radioisotope and/or instrument).
9. Graduated cylinders (Fisher Scientific).
10. Compound to be evaluated (*see* **Note 5**).

6. Methods

6.1. In Situ Brain Perfusion

The perfusion system set-up requires ~6 × 4 ft area (*see* **Fig. 11.1**). A stable shelf above the work for the peristaltic pump and pressure transducer provides an efficient surgical area.

6.1.1. Perfusion Preparation

1. The perfusion medium should be filtered (0.2 μm) and warmed to 37°C.
2. The solution is bubbled with 95% O_2/CO_2.
3. Remove an appropriate amount of perfusion medium necessary for the perfusion period desired and number of animals to be evaluated. The amount of radioactivity is dependent on the isotope chosen (5–20 μCi), with the chosen ratio of perfusate-to-radioactivity kept consistent throughout experimental sets. Once the drug and medium are mixed, place equal amounts into syringes. The syringes are placed in the syringe pump (0.5 ml/min or 1.0 ml/min) and hooked into the perfusion lines. Once the syringe pump is primed, it should, be isolated from the main perfusion line (by using a valve or occluding the lines using hemostat) (*see* **Note 6**).
4. Place the remaining perfusion medium in a flask for continued bubbling with 95% O_2/CO_2 (a slow gradual rate), and connect perfusion tubes (*see* **Note 7**).

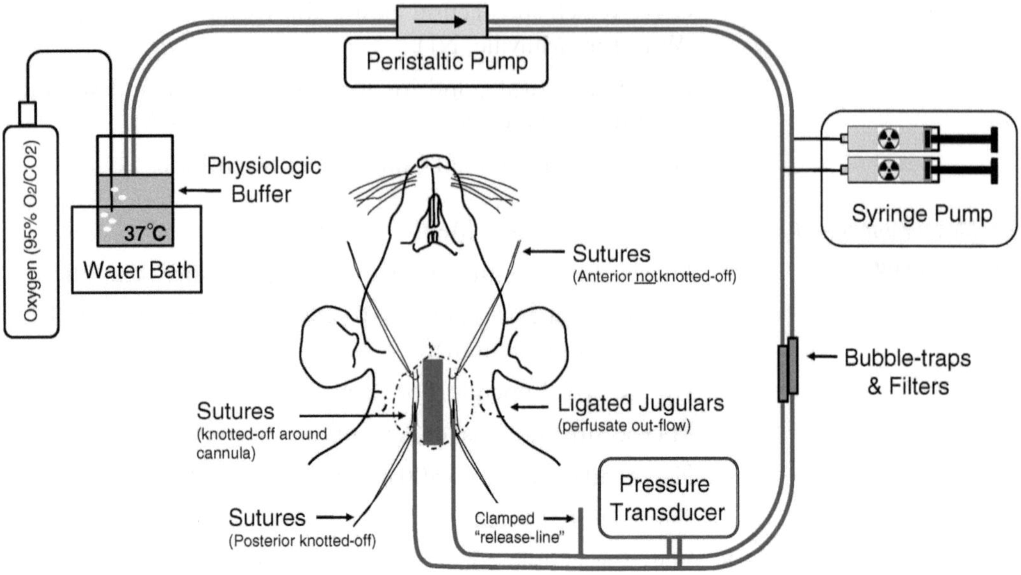

Fig. 11.1. Schematic of in situ brain perfusion set-up.

5. Connect all tubing of the perfusion system, using T-pieces where appropriate to 60-ml syringe lines, bubble traps, and transducer. Run the system to free bubbles and identify leaks or clogs, as well as baseline pressure readings (*see* **Note 8**).

6.1.2. Surgery and Perfusion

1. Anesthetize the rat with suitable anesthetic (e.g., sodium pentobarbital 64.8 mg/ml or "rat cocktail" consisting of acepromazine (0.6 mg/ml), ketamine (3.1 mg/ml), and xylazine (78.3 mg/ml)) and inject a dose of heparin (~100 μl of 12,500 U/ml, i.p.) (*see* **Note 9**).
2. Secure the fully anesthetized rat to surgical board under heating lap.
3. An incision is made along the ventral midline of the throat, exposing muscle underneath. Separate the Sternohyoid muscle with tweezers, exposing the trachea.
4. Tease out the respective common carotid arteries (they are quite pliable), which lie along the sides of the trachea. Separate any connective tissue and the vagus nerve with forceps.
5. To each carotid, loosely lace a piece of 5-0 silk (~2–3 inches) along the upper (anterior) and lower (posterior) portion of the exposed artery.
6. Starting with the right artery, tie off the posterior piece of silk suture, isolating the artery from the peripheral circulation. Then use a hemostat and clip the anterior piece of suture to prevent cerebral blood back-flow, but DO NOT tie-off.

7. Use micro-dissecting scissors to carefully cut an opening in the artery, making sure not to cut deeply.
8. Insert the primed cannula into the artery and loop another piece of suture over the portion of the artery with cannula inserted and tie-off. Releasing the hemostat on the anterior suture blood should backflow into the perfusion line.
9. Start the peristaltic pump (the syringe pump is NOT turned on at this point).
10. Immediately cut the jugular vein along the side that was cannulated, allowing for venous outflow.
11. In a similar manner, cannulate and tie-off the left carotid followed by cutting the jugular vein. Note that the left line must have an extra piece of tubing, connected via T-piece, to act as a release for the perfusate that is now running through the line. Once the cannulation is complete the "release-line" can be shut or clamped allowing perfusate to then flow through the cannula into the artery.
12. Once both cannulas are tied in and the pump is running, unclamp the syringe pump line and turn pump on for set period of time (*see* **Note 10**).
13. Upon completion of the time period, remove the cannulas and place them into respectively marked vials to collect perfusate/radioactive-marker (~1 ml) (*see* **Note 11**).
14. Turn off the respective pumps and decapitate the animal, removing the brain and placing it on a piece of wax weighing paper.
15. The brain should have its choroid plexus removed at this point, which can be easily identified if Evans blue was added to the perfusion medium by the deep blue staining.
16. The brain is then dissected into desired regions using a surgical scalpel. If taking multiple small regions that may require greater time, it is best to place the brain/wax-paper on top of a Petri dish and then setting it on top of ice in a bucket. Alternatively, this would be the point in which the brain tissue can be removed using capillary depletion analyses.
17. Respective tissue sections can then be place in tared vials, making sure to record the weight before and after the brain tissue is place in the vial.
18. Measurement is then determined by scintillation counting or other analytic means (*see* **Notes 5** and **12**).

6.2. Capillary Depletion

1. After completion of in situ perfusion (step 15 above, 6.1.2), the brain is removed and the choroid plexus excised.
2. Brain tissue (~0.5 g) is homogenized (13-strokes), using a glass homogenizer with 4 ml of ice-cold capillary depletion buffer.

3. After initial homogenization, 3.5 ml of ice-cold 26% dextran (MW 60 kDa) solution is added to the homogenate to achieve a final dextran concentration of 13%. The sample is homogenized again (three strokes).
4. Two aliquots of brain homogenate are taken, one for homogenate assessment (~2 mL; "whole brain" portion) with the remainder centrifuged at 5400 × *g* for 15 min at 4°C.
5. After centrifugation, the capillary-depleted supernatant and vasculature-containing pellet are carefully separated. Two aliguots of capillary-depleted supernatant are collected for assesment. The vasculature-containing pellet is then resuspended in PBS and two aliquots are collected for measurement of radioactivity.
6. Homogenate, vascular pellet, and supernatant are solubilized using tissue solubilizer and radioactivity is counted as described under **Section 6.3.1** (*see* **Note 13**).

6.3. Calculations

6.3.1. In Situ Perfusion

1. The most direct approach it to evaluate the "unidirectional" uptake of a drug into the brain (J_{in}) is given by:

$$J_{in} = dC_{br}/dt = K_{in}C_{pf}$$

where C_{br} is the quantity of drug in the brain (in units of mass drug/g brain), C_{pf} is the concentration of drug in perfusate (drug mass/volume), K_{in} is the blood-to-brain transfer coefficient, and t is the perfusion time. K_{in} is a proportionality constant that measures the ability of a drug to cross into the brain, and is determined by the permeability of the drug (P), the cerebral perfusion fluid flow-rate (F), the surface area of the crebral microvasculature available for drug transport (A), and amount of unbound/unionized drug in the perfusate (f) (27). Assuming minimal drug binding ($f \approx 1.0$) and $F >> K_{in}$, then K_{in} is essentially flow dependent. If K_{in} is relatively large to F (i.e., $K_{in} > 0.3F$), then uptake will be dependent on flow and permeability (27, 28) (*see* **Note 14**).
2. Additionally, K_{in} can be determined by plotting C_{tot}/C_{pf} versus T (time) over a series of time points, and fitting the linear portion of the uptake curve using least-squares regression (29).

$$C_{tot}/C_{pf} = K_{in}T + (C_{vas}/C_{pf})$$

where C_{tot} is the total amount of drug in the brain at the end of the set perfusion time (C_{br}) and that which remains in the vessels (C_{vas}).

3. Radiolabeled drug uptake into the brain is also commonly plotted as the amount of radioactivity in the brain (C_{br}; dpm/g^{-1}) presented as a percentage of that in the perfusate (C_{pf}; dpm/ml^{-1}) and termed R_{Br}% (*see* **Fig. 11.2**).

$$R_{Br}\% = C_{br}/C_{pl} \times 100$$

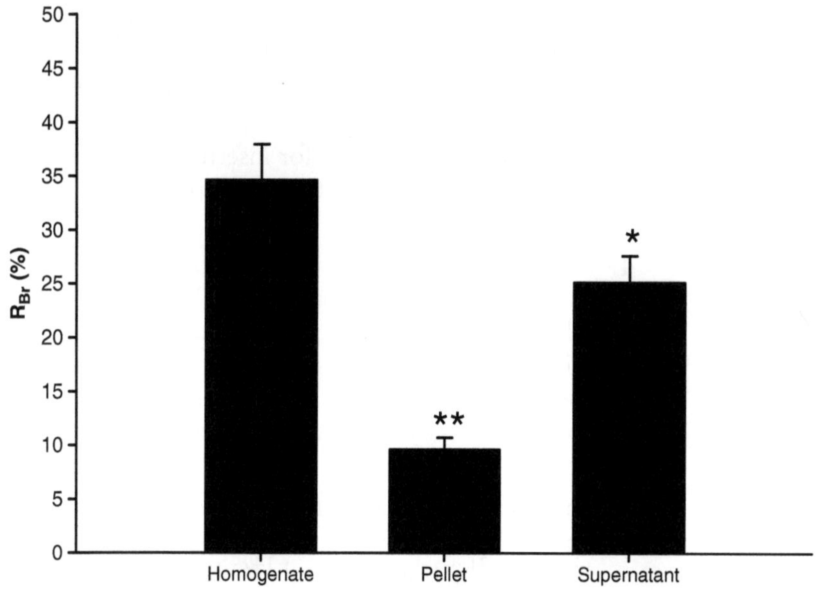

Fig. 11.2. Brain-to-perfusate ratio percentage (R_{Br} %) of [^{125}I]insulin detected in whole brain homogenate, pellet (i.e., capillary fraction), and supernatant (i.e., brain parenchyma fraction) as determined by capillary depletion analysis following 20 min in situ brain perfusion. Insulin, an endogenous hormone comprised of two peptide chains, is known to cross the BBB via receptor-mediated endocytosis. Capillary depletion analysis shows that, for insulin, the total brain uptake (i.e., R_{Br} % of the homogenate) overestimates the amount of compound that actually accumulates within the brain parenchyma. These data show that approximately 28% of [^{125}I]insulin is detected in the capillary fraction and 72% of [^{125}I]insulin is present in brain parenchyma following a 20-min perfusion experiment. Therefore, a more accurate assessment of insulin transport across the BBB is achieved by combining capillary depletion with the technique of in situ brain perfusion. Data are expressed as mean ± SEM, *$p < 0.05$, **$p < 0.01$ as compared to homogenate. Adapted from Hom et al. (31).

6.3.2. Capillary Depletion

1. The amount of radioactivity in the brain homogenate, supernatant, and pellet (C_{br}; dpm/g) and the amount of radioactivity in the perfusate (C_{pf}; dpm/ml) are measured as described under **Section 6.3.1** (*see* **Note 15**). The volume of distribution (V_D; μl/g) of the tracer compound is expressed as the ratio of tissue to perfusate:

$$V_D = (C_{br}/C_{pf}) = (\text{dpm/g of tissue})/(\text{dpm/ml perfusate})$$

7. Notes

1. Evans blue (Sigma-Aldrich) can also be added (50 μg/l), which binds irreversibly to serum albumin and can be used to help identify significant alterations in paracellular permeability that may occur due to clot or incident surgical issues.

2. Pulsatile flow is considered more physiologic than non-pulsatile flow; however, non-pulsatile pumps are valid for drug uptake evaluations given the relatively short perfusions and terminal end-point.
3. The actual cannula used for insertion into the artery may be metal or a relatively stiff plastic capable of repeated use and pressure.
4. Although various bubble-traps for perfusion can be purchased with and without filters (Terumo), a simple valve inlet connected to a sealable tube can be rigged with minimal effort. Care should be taken with filters used and placement along the perfusion lines, as respective compounds examined may adhere to filter if placed after syringe pump inlet.
5. Based on individual preference and peptide investigated the typical isotopes used for these examinations are ^{3}H, ^{14}C, ^{125}I, or ^{131}I. Both ^{3}H- and ^{14}C-labeled compounds require tissue digestion (TS-2, Research Products International) and scintillation fluid (Beckman-Coulter) for measurement (beta-counter), whereas ^{125}I or ^{131}I can be measured directly (gamma-counter). Although various companies will custom radiolabel peptides, this can be extremely costly. A simple alternative is to iodinate a peptide directly via the chloromine-T procedure (30), assuming the peptide has an open benzene group (e.g., tyrosine). Alternately, GC-MS, HPLC or other suitable analytical technique can be used.
6. One may also add other compounds/drugs into the perfusate before placing it into the syringes with the radiolabeled peptide. Excess "cold" peptide can be added to identify soluble transport (6) or inhibitors of specific transport mechanisms can also be used (11). Such compounds/drugs may also be added directly to the non-target drug containing perfusate, or alternatively additional syringe inlets into perfusion line can be set up.
7. Although the alternative methods may incorporate erythrocytes for additional oxygen carrying capacity just prior to perfusion, the integrity of the BBB is maintained throughout the dual-artery perfusion (23). Due to variability in erythrocyte viability and oxygen carrying capacity over the course of a perfusion set, as well as their impact on binding to the drug, it is not used in this procedure.
8. The perfusion medium can be maintained at 37° with either temperature regulated heat plate or simply set in a regulated water bath. The lines may also be run through a heating source/coil. Generally, for short perfusions the radioactive

perfusate in the 60-ml syringes on the syringe drive will maintain enough of its heat not to significantly affect the overall temperature of the perfusate it enters into.

9. Consistent use of specific anesthetic is necessary, as changes in anesthetic may alter results. Additionally, male rats are preferred to females to avoid alterations caused by variable estrous cycling. Alternately, this procedure is functional for use in other animals with appropriate modifications to perfusion rates and pressure.
10. A chosen perfusion rate should give adequate pressure during perfusion in the range of 80–130 mmHg. Ideally the infusion rate should be the minimum necessary to obtain a complete perfusion of the brain. A constant rate of 3–5 ml/min generally works well using the physiologic Ringer listed; however, alterations in Ringer content may significantly alter perfusion pressure. Pressures above 200 mmHg should be avoided as they may cause damage to the BBB.
11. It is often desirable to further flush the capillaries without radioactive marker at the end of the perfusion, removing excess radioactive compound from the intraluminal capillaries. In this case, simply shut off the syringe drive at the end of the experiment, while the peristaltic pump is still running. After ~20 s most of the radioactive containing perfusate in the brain is flushed, although this may vary depending on distance traveled (i.e., length of perfusion tubing). Consistency in time should always be maintained within any experimental set.
12. If metabolism is a factor, either with the peptide itself or with the radiolabeled tag, it is often necessary perform additional chromatograph assessment. (i.e., HPLC analysis with radiomatic detection).
13. In some instances, all blood vessels will not pellet and will be present in the supernatant as contamination. To assess the percent contamination of the supernatant, the relative reactivity of vascular enzyme markers, such as alkaline phosphatase or γ-glutamyl transpeptidase, can be measured (24).
14. Unidirectional uptake of a drug makes the assumption that back-flux and metabolism of the compound are negligible. For this reason early time points, before significant back-flux or metabolism occurs is preferential, which may be of issue when evaluating drugs with low permeability rates. Use of multiple time uptake analyses will identify back-flux issues, via observed loss of linearity.
15. If a flushing period is not performed (*see* **Note 11**) it is necessary to correct for radioactive protein contained within the lumen of the brain vasculature that leaks following homogenization (24).

References

1. Habgood, M.D., Begley, D.J., and Abbott, N.J. (2000) Determinants of passive drug entry into the central nervous system. *Cell Mol Neurobiol* 20, 231–253.
2. Witt, K.A. and Davis, T.P. (2006) CNS drug delivery: opioid peptides and the blood-brain barrier. *AAPS J* 8, E76–E88.
3. Witt, K.A., Gillespie, T.J., Huber, J.D., Egleton, R.D., and Davis, T.P. (2001) Peptide drug modifications to enhance bioavailability and blood-brain barrier permeability. *Peptides* 22, 2329–2343.
4. Patel, M.M., Goyal, B.R., Bhadada, S.V., Bhatt, J.S., and Amin, A.F. (2009) Getting into the brain: approaches to enhance brain drug delivery. *CNS Drugs* 23, 35–58.
5. del Amo, E.M., Urtti, A., and Yliperttula, M. (2008) Pharmacokinetic role of L-type amino acid transporters LAT1 and LAT2. *Eur J Pharm Sci* 35, 161–174.
6. Abbruscato, T.J., Williams, S.A., Misicka, A., Lipkowski, A.W., Hruby, V.J., and Davis, T.P. (1996) Blood-to-central nervous system entry and stability of biphalin, a unique double-enkephalin analog, and its halogenated derivatives. *J Pharm Exp Ther* 276, 1049–1057.
7. Loscher, W. and Potschka, H. (2005) Drug resistance in brain diseases and the role of drug efflux transporters. *Nat Rev* 6, 591–602.
8. Leslie, E.M., Deeley, R.G., and Cole, S.P. (2005) Multidrug resistance proteins: role of P-glycoprotein, MRP1, MRP2, and BCRP (ABCG2) in tissue defense. *Toxicol Appl Pharmacol* 204, 216–237.
9. Chen, C. and Pollack, G.M. (1998) Altered disposition and antinociception of [D-penicillamine(2,5)] enkephalin in mdr1a-gene-deficient mice. *J Pharm Exp Ther* 16, 545–552.
10. Kastin, A.J., Fasold, M.B., and Zadina, J.E. (2002) Endomorphins, Met-enkephalin, Tyr-MIF-1, and the P-glycoprotein efflux system. *Drug Metab Dispos Biol Fate Chem* 30, 231–234.
11. Ouyang, H., Andersen, T.E., Chen, W., Nofsinger, R., Steffansen, B., and Borchardt, R.T. (2009) A comparison of the effects of p-glycoprotein inhibitors on the blood-brain barrier permeation of cyclic prodrugs of an opioid peptide (DADLE). *J Pharm Sci* 98, 2227–2236.
12. Hirrlinger, J. and Dringen, R. (2005) Multidrug resistance protein 1-mediated export of glutathione and glutathione disulfide from brain astrocytes. *Methods Enzymol* 400, 395–409.
13. Paulusma, C.C., van Geer, M.A., Evers, R., Heijn, M., Ottenhoff, R., Borst, P., and Oude Elferink, R.P. (1999) Canalicular multispecific organic anion transporter/multidrug resistance protein 2 mediates low-affinity transport of reduced glutathione. *Biochem J* 338, 393–401.
14. Rius, M., Hummel-Eisenbeiss, J., Hofmann, A.F., and Keppler, D. (2006) Substrate specificity of human ABCC4 (MRP4)-mediated cotransport of bile acids and reduced glutathione. *Am J Physiol Gastrointest Liver Physiol* 290, G640–G649.
15. Staud, F. and Pavek, P. (2005) Breast cancer resistance protein (BCRP/ABCG2). *Int J Biochem Cell Biol* 37, 720–725.
16. Gao, B., Hagenbuch, B., Kullak-Ublick, G.A., Benke, D., Aguzzi, A., and Meier, P.J. (2000) Organic anion-transporting polypeptides mediate transport of opioid peptides across blood-brain barrier. *J Pharm Exp Ther* 294, 73–79.
17. Hagenbuch, B. and Meier, P.J. (2004) Organic anion transporting polypeptides of the OATP/ SLC21 family: phylogenetic classification as OATP/ SLCO superfamily, new nomenclature and molecular/functional properties. *Pflugers Arch* 477, 653–665.
18. Lash, L.H., Putt, D.A., Xu, F., and Matherly, L.H. (2007) Role of rat organic anion transporter 3 (Oat3) in the renal basolateral transport of glutathione. *Chem Biol Interact* 170, 124–134.
19. Shen, H., Smith, D.E., Keep, R.F., and Brosius, F.C. III (2004) Immunolocalization of the proton-coupled oligopeptide transporter PEPT2 in developing rat brain. *Mol Pharm* 1, 248–256.
20. Yamashita, T., Shimada, S., Guo, W., Sato, K., Kohmura, E., Hayakawa, T., Takagi, T., and Tohyama, M. (1997) Cloning and functional expression of a brain peptide/histidine transporter. *J Biol Chem* 272, 10205–10211.
21. Takasato, Y., Rapoport, S.I., and Smith, Q.R. (1984) An in situ brain perfusion technique to study cerebrovascular transport in the rat. *Am J Physiol* 247, H484–H493.

22. Oldendorf, W.H. (1970) Measurement of brain uptake of radiolabeled substances using a tritiated water internal standard. *Brain Res* 24, 372–376.
23. Preston, J.E., al Sarraf, H., and Segal, M.B. (1995) Permeability of the developing blood-brain barrier to 14C-mannitol using the rat in situ brain perfusion technique. *Brain Res Dev Brain Res* 87, 69–76.
24. Triguero, D., Buciak, J., and Pardridge, W.M. (1990) Capillary depletion method for quantification of blood-brain barrier transport of circulating peptides and plasma proteins. *J Neurochem* 54, 1882–1888.
25. Zlokovic, B.V., Banks, W.A., el Kadi, H., Erchegyi, J., Mackic, J.B., McComb, J.G., and Kastin, A.J. (1992) Transport, uptake, and metabolism of blood-borne vasopressin by the blood-brain barrier. *Brain Res* 590, 213–218.
26. Momma, S., Aoyagi, M., Rapoport, S.I., and Smith, Q.R. (1987) Phenylalanine transport across the blood-brain barrier as studied with the in situ brain perfusion technique. *J Neurochem* 48, 1291–1300.
27. Smith, Q.R. (1989) Quantitation of blood-brain barrier permeability, in *Implications of the Blood-Brain Barrier and its Manipulation* (Neuwelt E.A., Ed.) pp. 85–118, Plenum Press, New York.
28. Takasato, Y., Momma, S., and Smith, Q.R. (1985) Kinetic analysis of cerebrovascular isoleucine transport from saline and plasma. *J Neurochem* 45, 1013–1020.
29. Patlak, C.S., Blasberg, R.G., and Fenstermacher, J.D. (1983) Graphical evaluation of blood-to-brain transfer constants from multiple-time uptake data. *J Cereb Blood Flow Metab* 3, 1–7.
30. Bolton, A.E. (1986) Comparative methods for the radiolabeling of peptides. *Methods Enzymol* 124, 18–29.
31. Hom, S., Egleton, R.D., Huber, J.D., and Davis, T.P. (2001) Effect of reduced flow on blood-brain barrier transport systems. *Brain Res* 890, 38–48.

Chapter 12

An Engineered Peptide Compound Platform Technology Incorporating Angiopep for Crossing the BBB

Reinhard Gabathuler

Abstract

The blood–brain barrier (BBB), formed by the endothelial cells of the brain capillaries, restricts access to brain cells of blood-borne compounds and allows only nutrients essential for normal metabolism to reach brain cells. This results in the inability of both small and large therapeutic compounds to cross the BBB. Therefore, various strategies need to be developed to enhance the amount and concentration of therapeutic compounds in the brain. A new family of peptides called Angiopeps derived from proteins expressing the Kunitz domain is transported very efficiently across the BBB using a physiological mechanism. Angiopep-2 transport across the BBB is, in part, mediated by the low density lipoprotein receptor related protein 1 (LRP1). This peptide family is the base of the engineered peptide compound (EPiC) platform technology allowing the synthesis of compounds which enter the central nervous system (CNS) to treat brain disorders. Using the EPiC platform, active drugs are modified by incorporation of Angiopep-2 allowing entry in the brain parenchyma. Many anti-cancer drugs, including paclitaxel, show poor delivery to brain tumors due to low permeability and active efflux transport at the BBB. As a proof of concept we have shown that Angiopep-2, a 19 amino acid peptide, is able to carry a payload composed of three molecules of an anti-cancer agent paclitaxel across the BBB resulting in a therapeutic concentration in the brain parenchyma. This first product called ANG1005 is now being evaluated in two phase I/II clinical trials. The use of this EPiC platform technology allows access to the brain parenchyma of small drugs, peptides, and larger hydrophilic agents for the treatment of CNS diseases.

Key words: Blood–brain barrier, transcytosis, Angiopep, drug targeting, brain metastasis, glioblastomas, central nervous system.

1. Introduction

The BBB is mainly formed by brain capillary endothelial cells, which are closely sealed by tight junctions (1). In addition, brain capillaries possess few fenestrae and few endocytic vesicles,

K.K. Jain (ed.), *Drug Delivery to the Central Nervous System*, Neuromethods 45,
DOI 10.1007/978-1-60761-529-3_12, © Humana Press, a part of Springer Science+Business Media, LLC 2010

compared to the capillaries of other organs (2). Because of the presence of these tight junctions and the lack of fenestrae, the BBB acts as a physical barrier and is frequently the rate-limiting factor for the penetration of proteins, pharmacological agents, or peptides into the CNS (1, 3). Thus, many potential drugs that could be effective at their site of action have failed during development because they do not reach therapeutic concentrations within the CNS (4).

Several approaches have been described for drug delivery to the brain such as local invasive delivery by direct injection or infusion, induction of enhanced permeability as well as various physiological targeting strategies (5–7).

Demeule et al. have identified a family of peptides, named Angiopeps, which are derived from the Kunitz domains of aprotinin and other human proteins (8). Involvement of the Kunitz protease inhibitory domain on certain proteins such as the secreted ß-amyloid precursor protein (APP) has been shown to be important in the interaction with both LRP1 and megalin (LRP2) and have been shown to be responsible for their endocytosis (9, 10). The very efficient transport of these peptides into the brain is the main property for their use as vehicle to transport drugs across the BBB. This new family of peptides allow access to the brain parenchyma of modified pharmacological agents that do not readily enter the brain (8). Angiopep-2, a 19 amino acid peptide, is one of the peptide designed to target the low density lipoprotein receptor-related protein (LRP) in order to mediate transcytosis across the BBB.

The property of this new family of peptides, Angiopeps, has been used in the development of a platform technology called the engineered peptide compound (EPiC) platform technology which is used to design new drugs for the treatment of brain diseases. This peptide sequence is incorporated in new engineered peptide compounds which have now access to the brain parenchyma for the treatment of brain disorders.

In the present article, some applications of the EPiC platform technology are presented. Anti-cancer drug such as the chemotherapeutic agent paclitaxel is modified by incorporation of the Angiopep amino acid sequence for brain distribution to treat brain cancers. The preclinical proof of concept on rodents and human data validate this technology. Additional data for other application will be summarized for small compounds such as fluorescent markers, small anti-cancer agent such as chemotherapeutics, to neuropeptides and to larger hydrophilic therapeutic molecules.

2. Engineered Peptide Compounds (EPiC) Platform Technology

2.1. Angiopep-2 Is Efficiently Transported into the Brain Parenchyma

A family of 96 peptides derived from the Kunitz domain of human proteins were tested using the BBB in vitro model and in situ brain perfusion. Among them, Angiopep-2 was selected and further characterized. Data published previously (8, 11) suggest that Angiopep-2 is transcytosed across the endothelial cells forming the BBB by a receptor-mediated transport mechanism involving LRP1. Decrease by approximately 50% in the transcytosis rate of ^{125}I-Angiopep-2 in the presence of excess unlabeled Angiopep-2 (10 μM), partial inhibition of transcytosis by the receptor associated protein (RAP) which was previously shown to inhibit the binding of all ligands to all members of the LDL-family receptor and specific inhibition of the transcytosis of activated alpha-2 macroglobulin, which is the most specific known LRP1 ligand (12), by Angiopep-2 suggest that part of the transport of this peptide is associated with LRP1 (11, 13).

The apparent distribution volume of Angiopep-2 measured by in situ brain perfusion indicated that the brain and parenchymal volume of distribution for Angiopep-2 is much higher than the volume distribution of other proteins and peptides (**Table 12.1**) measured by in situ brain perfusion in mice and rats. Ten to one hundred times higher rate of uptake was measured with Angiopep-2 than with any other peptides as reported in the literature using the same techniques. Furthermore, the apparent distribution

Table 12.1
Blood–brain barrier influx rate constants for selected molecules, antibody, peptides, and growth factors

Peptides and proteins	Brain K_{in} (mL/s/g)	References
Angiopep-2	$8.8 \pm 1.3 \times 10^{-4}$	(14)
Insulin Rec mAb	$1 \pm 0.3 \times 10^{-4}$	(15)
Leu-Enkephalin	$6.0 \pm 0.2 \times 10^{-5}$	(16)
RAP	$1.0 \pm 0.1 \times 10^{-5}$	(17)
Beta Amyloid	$6.5 \pm 0.1 \times 10^{-6}$	(18)
DADLE	$6.5 \pm 1.3 \times 10^{-6}$	(19)
TNF-α	$4.3 \pm 0.1 \times 10^{-6}$	(20)
Arg Vasopressin	$2.5 \pm 0.1 \times 10^{-6}$	(21)

volume in brain parenchyma of Angiopep-2 increased linearly for at least 15 min of perfusion and was much higher than for the vascular marker [^{14}C]inulin. Moreover, the addition of Angiopep-2 (up to 500 nM) did not affect the volume of distribution for inulin, indicating that the BBB integrity was unaffected under these conditions by the peptide.

These data confirm that Angiopep-2 is transcytosed very efficiently at a magnitude higher than any other peptides or proteins investigated so far. Angiopep-2 is transported across the BBB by a receptor medicated mechanism suggesting a potential use to increase brain penetration of centrally active therapeutics.

2.2. A New Compound Cy5.5-Angiopep-2 Is Detected in the Mice Brain Parenchyma by Fluorescence Microscopy

Different peptides were generated presenting different affinities to interact with LRP. In Angiopep-7 the lysine present at positions 10 and 15 on Angiopep-2 were replaced by arginine residues. Parenchymal uptake of Angiopep-7 was decreased by about 85%. To visualize and localize these peptides new conjugate were generated. Cy5.5, a 1-kD dye, was conjugated to the N-terminal of Angiopep-2 and Angiopep-7 which allowed detection by fluorescence microscopy in mice brain sections. Fluorescence microscopy confirmed that Cy5.5-Angiopep-2 crossed the BBB and accumulated in brain parenchyma 24 h after IV administration (**Fig. 12.1**). The fluorescence pattern of Cy5.5-Angiopep-2 was diffuse and homogenous throughout brain parenchyma. At

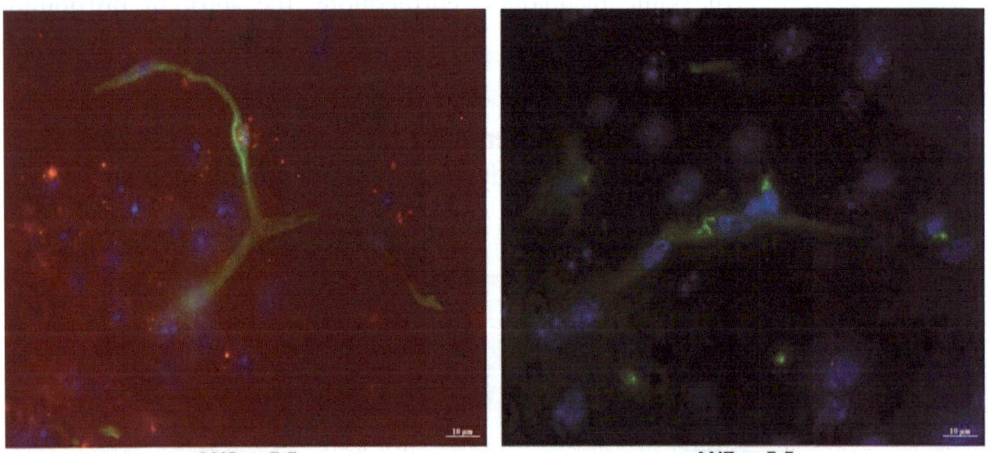

Fig. 12.1. Parenchymal distribution of Cy5.5-Angiopep-2. Fluorescent microscopy was done on brain sections 24 h after IV bolus injection of Cy5.5-Angiopep-2 or Cy5.5-Angiopep-7 (100 μg), brains were fixed, and fluorescence microscopy was performed on 50 μm thick brain sections and were viewed in the near-infrared mode (*red*; a 660- to 680-nm excitation and a 700-nm longpass emission filter) using Zeiss Axiovert 200 fluorescent microscope (Carl Zeiss). Cy5.5-Angiopep-2 was detected as white spots in the brain parenchyma. Nuclei were labeled with DAPI and detected as grey regions in the brain parenchyma and brain capillaries were labeled with tomato lectin, the brain endothelial cells are seen forming the brain capillaries. Cy5.5-Angiopep-7 is not transported across the BBB and is considered to be a negative control only brain cell nuclei and capillaries are detected.

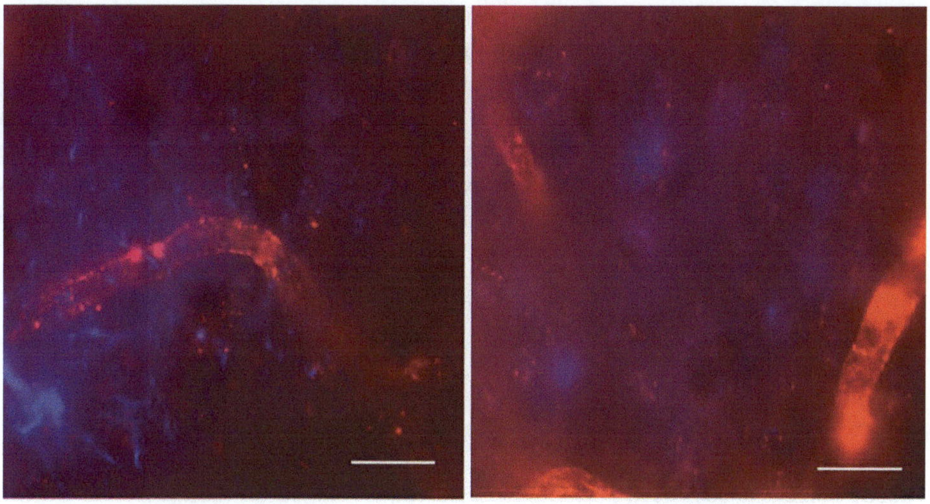

Fig. 12.2. Co-localization of Cy5.5-Angiopep-2 with astrocyte and neuron markers in the brain parenchyma. Fluorescent microscopy was done on brain sections after 10 min in situ brain perfusion of Cy5.5-Angiopep-2 in mice. Brains were fixed and fluorescence microscopy was performed on 50-μm-thick brain sections and was viewed in the near-infrared mode (red; a 660- to 680-nm excitation and a 700-nm longpass emission filter) using Zeiss Axiovert 200 fluorescent microscope (Carl Zeiss). Cy5.5-Angiopep-2 was detected as white spots and regions in the brain endothelial cells forming the BBB and in the brain parenchyma co-localized with a grey staining which corresponds to the GFAP staining for astrocytes (*left*) or the grey staining which corresponds to the NeuN staining for the neuron nuclei (*right*).

the cellular level, the fluorescence co-localized with neurons and astrocytes in brain sections as detected after 10 min in situ brain perfusion (**Fig. 12.2**). The observed localization results are consistent with the distribution of the receptor LRP-1 which is present on astrocytes and neurons. Thus, the broad distribution pattern likely reflects the high capillary density in the cerebrum, with virtually every neuron perfused by its own vessel. Regarding the co-localization results, the data suggest that, after passing through the BBB, Angiopep-2 agents pass into neurons and astrocytes by receptor-mediated transcytosis via LRP-1, which is expressed on both of these parenchymal cell types.

2.3. First Proof of Concept with a Small Anti-cancer Agent: Paclitaxel Chemically Conjugated to Angiopep-2, ANG1005

2.3.1. Description and Brain Distribution of ANG1005

Paclitaxel acts by promoting microtubule assembly and stabilizes microtubule dynamics, preventing them from depolymerizing, leading to inhibition of cell proliferation and induction of apoptosis (7). It is commonly used to treat ovarian, breast, and non-small cell lung cancer (22). In vitro, paclitaxel has been shown to be active against gliomas and brain metastases (23). However, in patients with recurrent primary brain tumors, paclitaxel has shown only a modest activity (24). Lack of paclitaxel brain uptake is associated with low blood–brain barrier permeability due to the expression of the multidrug resistance efflux pump P-glycoprotein (P-gp) at the apical membrane of the brain endothelial cells (25).

A new compound for the treatment of brain cancer was synthesized by conjugation of three molecules of paclitaxel to the amino groups of Angiopep-2 using a cleavable ester linkage, to produce ANG1005. The chemical structure of ANG1005 conjugate is shown in **Fig. 12.3**. Each ANG1005 molecule consists of three paclitaxel molecules attached to one Angiopep-2 molecule by cleavable ester linkage. A molecular weight of 5109.14 g mol^{-1} was defined by mass spectrometry.

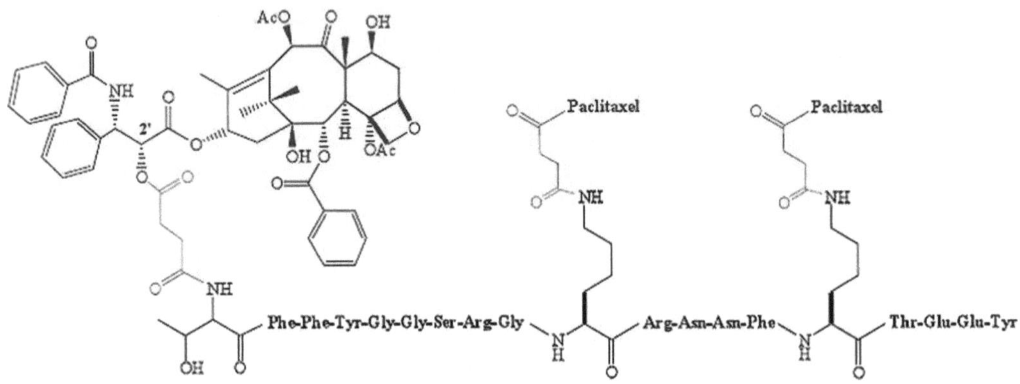

Fig. 12.3. Schematic representation of ANG1005. The molecule contains the proprietary amino acid sequence responsible for receptor-mediated transcytosis and three paclitaxel molecules, only the first of which is presented completely.

ANG1005 crosses the BBB and distributes into the brain parenchyma at a much higher level than free paclitaxel (26). ANG1005 exerts cytotoxic effects on tumor cell lines comparable to that of free paclitaxel, and acts as free paclitaxel at the G2/M phase junction and induces microtubule stabilization in tumor cells (26). Paclitaxel associated to ANG1005 bypass the P-gp (26).

We evaluated the ability of ^{125}I-ANG1005, ^{3}H-paclitaxel, and other chemotherapeutics to be taken up into brain by in situ perfusion in rats. The data confirm that the rate of transport of ANG1005 across the BBB is 10 to over 100 times higher than unconjugated molecules (**Table 12.2**) (27). ANG1005 rate of uptake is similar to the rate of uptake of glucose which uses the Glut-1 transporters expressed at the luminal and abluminal side of the endothelial cells of the BBB.

When considering that three molecules of paclitaxel are attached to ANG1005, 300 times more paclitaxel is entering in the brain parenchyma when conjugated to Angiopep-2 (27). This has been confirm by direct analysis of the brain distribution of ^{125}I-ANG1005 compared to that of free ^{14}C-paclitaxel following intravenous injection in animals with brain metastases of breast cancer using quantitative autoradiography (14). Overall our results show that ANG1005, a paclitaxel-peptide conjugate, represents a new drug with high potential for the treatment of primary or metastatic brain tumors.

Table 12.2
Transport rate (Kin) of anti-cancer drugs in the mice brain parenchyma

Molecules	Brain K_{in} (mL/s/g)
ANG1005	$8.8 \pm 0.6 \times 10^{-3}$
Temolozomide (TMZ)	$1 \pm 0.1 \times 10^{-3}$
Paclitaxel	$8.5 \pm 0.5 \times 10^{-5}$
Doxorubicin	$\sim 5 \times 10^{-5}$
Etoposide	$\sim 4 \times 10^{-6}$

Initial transport rate measured by in situ brain perfusion in mice demonstrates that Angiopep-2 modified drugs are better transported than unconjugated drugs

2.3.2. Antitumor Effect in Rodents

ANG1005 has shown in vivo antitumoral efficacy in intracranial mice tumor model (26). ANG1005 treatments were highly active in brain and lung carcinoma tumor xenograft models and resulted in significant increased mice survival in orthotopic brain tumor model in nude mice (26).

To examine and follow the antitumor efficacy of ANG1005, nude rats were implanted intracranially with U87 human glioblastoma cells, and after a 10-day period to allow for tumor growth, animals were injected with ANG1005 (75 mg/kg IP, days 10, 13, and 17), paclitaxel (5 mg/kg IV, days 10, 13, 17, and 20), or vehicle (25 mL/kg IP, days 10, 13, 17, and 20). Tumor size was examined on the first day of ANG1005 injection (day 10) and on days 17 and 24 using magnetic resonance imaging (**Fig. 12.4**). After treatment with either paclitaxel, which was administered at the maximum tolerated dose (MTD), or vehicle, tumors grew significantly over the 2-week follow-up period, engulfing an entire cerebral lobe by day 24. By contrast, ANG1005 prevented any further tumor growth beyond that present at day 10, the first day of study treatment. These results support and extend previous studies, which demonstrated that ANG1005 induced potent inhibition of human tumor xenografts in rats compared to paclitaxel.

2.3.3. Antitumor Effect and ANG1005 Uptake in Glioblastomas in Humans

ANG1005 is currently undergoing two parallel, multi-centre, open-label, phase I/II, dose-escalation studies in human patients with either recurrent primary glioma or secondary brain metastases. The primary objectives of the studies are to characterize the safety and tolerability of ANG1005 and to identify the MTD in each patient population. ANG1005 is administered by IV infusion once every 21 days (1 treatment cycle) for up to six cycles. Disease status is measured by either magnetic resonance imaging (MRI) or

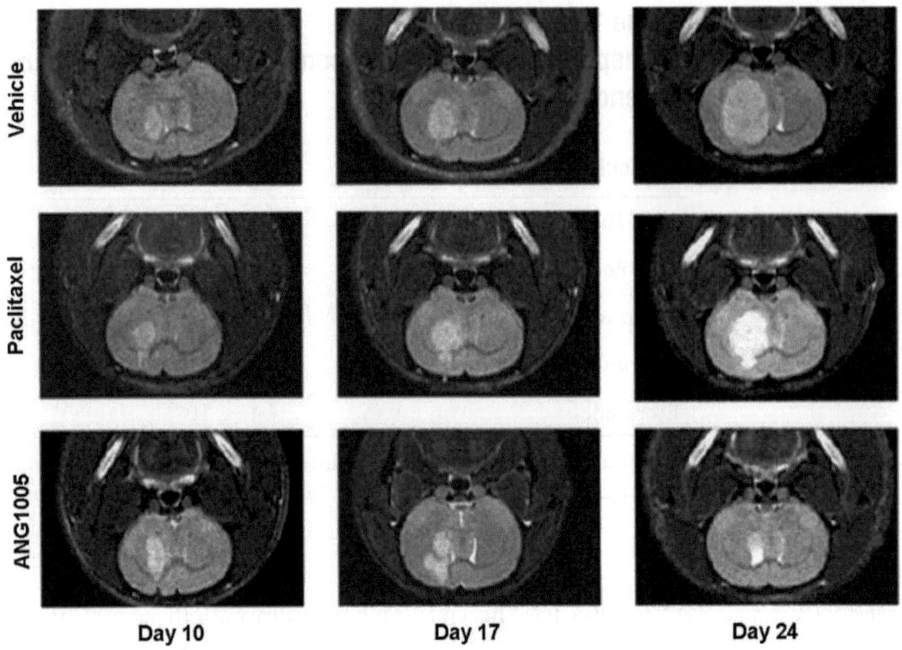

Fig. 12.4. Antitumor activity of ANG1005 in rat glioblastoma xenografts compared to paclitaxel and vehicle.

computed tomography (CT) performed at 6-week intervals during therapy (i.e., following every two cycles of ANG1005 administration). Patients will remain on the study until they demonstrate progressive disease, exhibit unacceptable toxicities, or withdraw consent. At present, the current doses of ANG1005 are 420 and 650 mg/m^2, respectively. In a sub-study of the recurrent primary glioma protocol, a patient with recurrent malignant glioma scheduled for debulking surgery provided informed consent to participate in a study analysing the transport of ANG1005 through the BBB into tumor tissue. The patient received an infusion of ANG1005 at a dose of 200 mg/m^2, and the tumor was removed 4 h following the end of the infusion. The concentration of ANG1005 in blood plasma was 11 μM immediately after tumor extraction. However, because ANG1005 contains three molecules of paclitaxel, the equivalent concentration of paclitaxel molecules was higher by 3-fold, i.e., 33 μM. ANG1005 appeared to be relatively stable in plasma, as 96% of the total circulating paclitaxel molecules remained as ANG1005 during the first 4 h. The concentration of ANG1005 in total brain tumor tissue was corrected for vascular content, which resulted in a concentration of 0.67 μM in parenchymal glioma tissue, a paclitaxel equivalent concentration of 2.02 μM. After transcytosis, approximately 28% of total paclitaxel in glioma parenchyma was in the free form (0.78 μM). In humans, ANG1005 showed significant uptake into glioma tumor tissue 4 h after infusion. These data confirm that when associated to the Angiopep-2 peptides, in the form of a new chemical entity ANG1005 a chemotherapeutic such as paclitaxel can reach therapeutic concentration in the brain parenchyma and especially in tumor tissue and finally validates the EPiC platform technology in humans.

Furthermore, one patient with secondary brain metastases demonstrated significant tumor shrinkage and other benefits after ANG1005 infusions. This patient had ovarian cancer and metastases in lung, lymph, and brain. Prior therapy included ascending doses of docetaxel/carboplatin, gemcitabine/topotecan, and a single dose of doxorubicin. The patient was initiated on ANG1005 infusions at 650 mg/m^2. After 6 weeks of therapy (two cycles), an impressive partial response was observed characterized by significant shrinkage of brain tumors (**Fig. 12.5**). These data are consistent with uptake of ANG1005 into brain parenchyma via receptor-mediated transcytosis, the first such demonstration for a pharmaceutical in humans.

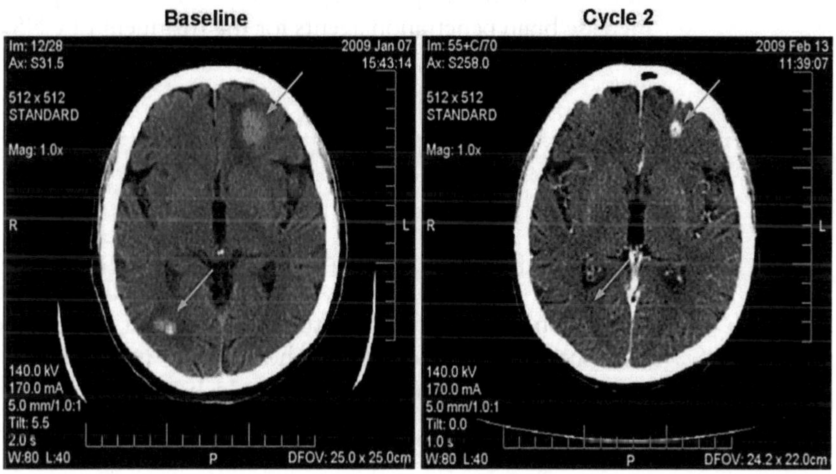

Fig. 12.5. CT scans of two patients with secondary brain metastases before and after treatment with ANG1005. The brain metastases are identified by a grey arrow. The primary cancer was ovarian cancer for this patient.

2.4. Future Application of the New Engineered Peptide Compound (EPiC) Platform for the Transport of Small and Large Therapeutics to the CNS

Other applications of the EPiC platform technology are under investigation. Despite the clinical effectiveness of doxorubicin and etoposide in the treatment of many malignant tumors, clinical trials involving systemic administration have demonstrated very limited efficacy in the treatment of gliomas. This limited efficacy can be explained by the poor penetration of the drug through the BBB and by the effect of the multidrug resistance pump, P-gp. The brain uptake of new chemical entities synthesized by conjugation of the peptide Angiopep-2 (An2) with three molecules of the anticancer drugs doxorubicin or etoposide are tested using the in situ mice brain perfusion technique. Preliminary data obtained by mice in situ brain perfusion suggest and confirm that the etoposide and doxorubicin conjugates are transported very efficiently across the BBB at a rate comparable to ANG1005.

Early development suggests the possibility of the application of this platform technology to neuroactive peptides, larger proteins, for example, mAbs and nucleic acid-based agents such as siRNA. The new chemical derivatives have been synthesized using different approaches and characterized after purification by mass spectrometry. The activity of the modified compounds was measured and was unaffected by the incorporation of Angiopep-2. Furthermore, we show by mice in situ brain perfusion that these new derivatives are able to cross the BBB at a higher rate than the unmodified molecules which normally do not cross the BBB efficiently. Higher brain distribution is measured for the peptides, mAbs, and siRNA when modified with our peptide technology. Finally, in vivo studies using mice models demonstrate that these derivatives can reach therapeutic amount in the brain.

By using our EPiC platform we are creating new chemical entities for small drugs, peptides, and larger hydrophilic proteins with increase brain penetration agents for the treatment of CNS diseases.

3. Conclusion

The EPiC platform provides a number of important advantages over these prior approaches. First, because LRP-1 is distributed widely on brain capillaries, the BBB transfer appears to occur efficiently and broadly throughout brain parenchyma. Finally, receptor-mediated transcytosis should be compatible with small and large ligands, as well as ligands of different charge or lipophilicity. The results presented here and in earlier studies suggest that the EPiC platform may have significant utility in overcoming this constraint on the development of novel neurotherapeutics. It has been argued that one of the primary roadblocks preventing the development of more neurotherapeutics has been the BBB (14, 28–30). The EPiC platform described in this report has led to the rapid development of other new potential therapeutics including ANG1005 and other small drug and peptide conjugates and is applicable to the development of still more (10). Recently, for instance, NCEs consisting of monoclonal antibodies and siRNAs have been generated and shown to transport efficiently across the BBB (Demeule et al., unpublished data). In addition to the EPiC platform, a variety of other strategies to breach the BBB are currently under development (31). In concert, the application of these approaches can be expected to usher in a particularly fruitful era of neurotherapeutic development in the near future.

References

1. Abbott, N.J., Ronnback, L., and Hansson, E. (2006) Astrocyte-endothelial interactions at the blood-brain barrier. *Nat Rev Neurosci* 7, 41–53.
2. Pardridge, W.M. (1999) Blood–brain barrier biology and methodology. *J Neurovirol* 5, 556–569.
3. Begley, D.J. (2004) ABC transporters and the blood–brain barrier. *Curr Pharm Des* 10, 1295–1312.
4. de Boer, A.G. and Gaillard, P.J. (2007) Drug targeting to the brain. *Annu Rev Pharmacol Toxicol* 47, 323–355.
5. Begley, D.J. (2004) Delivery of therapeutic agents to the central nervous system: the problems and the possibilities. *Pharmacol Ther* 104, 29–45.
6. Gaillard, P.J., Brink, A., and de Boer, A.G. (2005) Diphteria toxin receptor-targeted brain drug delivery. *Int Cong Series* 1277, 185–195.
7. Gupta, M.L. Jr., Bode, C.J., Georg, G.I., and Himes, R.H. (2003) Understanding tubulin–Taxol interactions: mutations that impart Taxol binding to yeast tubulin. *Proc Natl Acad Sci USA* 100, 6394–6397.
8. Demeule, M., Regina, A., Che, C., Poirier, J., Nguyen, T., Gabathuler, R., Castaigne, J.-P., and Beliveau, R. (2008) Identification and design of peptides as a new drug delivery system for the brain. *J Pharmacol Exp Ther* 324, 1064–1072.
9. Cam, J.A. and Bu, G. (2006) Modulation of beta-amyloid precursor protein trafficking and processing by the low density lipoprotein receptor family. *Mol Neurodegener* 1, 8.
10. Kounnas, M.Z., Moir, R.D., Rebeck, G.W., Bush, A.I., Argraves, W.S., Tanzi, R.E., Hyman, B.T., and Strickland, D.K. (1995) LDL receptor-related protein, a multifunctional ApoE receptor, binds secreted beta-amyloid precursor protein and mediates its degradation. *Cell* 82, 331–340.
11. Demeule, M., Currie, J.-C., Bertrand, Y., Che, C., Nguyen, T., Regina, A., Gabathuler, R., Castaigne, J.-P., and Beliveau, R. (2008) Involvement of the low-density lipoprotein receptor-related protein in the transcytosis of the brain delivery vector Angiopep-2. *J Neurochem* 106, 1534–1544.
12. Hussain, M.M., Strickland, D.K., and Bakillah, A. (1999) The mammalian low-density lipoprotein receptor family. *Annu Rev Nutr* 19, 141–172.
13. Moestrup, S.K., Cui, S., Vorum, H., Bregengard, C., Bjorn, S.E., Norris, K., Gliemann, J., and Christensen, E.I. (1995) Evidence that epithelial glycoprotein 330/megalin mediates uptake of polybasic drugs. *J Clin Invest* 96, 1404–1413.
14. Pardridge, W.M. (2003) Blood-brain barrier drug targeting: the future of brain drug development. *Mol Interv* 3, 90–105.
15. Wu, D., Yang, J., and Pardridge, W.M. (1997) Drug targeting of a peptide radiopharmaceutical through the primate blood-brain barrier in vivo with a monoclonal antibody to the human insulin receptor. *J Clin Invest* 100, 1804–1812.
16. Zloković, B.V., Lipovac, M.N., Begley, D.J., Davson, H., and Rakić, L. (1987) Transport of leucine-enkephalin across the blood-brain barrier in the perfused guinea pig brain. *J Neurochem* 49, 310–315.
17. Pan, W., Kastin, A.J., Zankel, T.C., van Kerkhof, P., Terasaki, T., and Bu, G. (2004) Efficient transfer of receptor-associated protein (RAP) across the blood-brain barrier. *J Cell Sci* 117, 5071–5078.
18. Banks, W.A., Kastin, A.J., Barrera, C.M., and Maness, L.M. (1991) Lack of saturable transport across the blood-brain barrier in either direction for beta-amyloid1-28 (Alzheimer's disease protein). *Brain Res Bull* 27, 819–823.
19. Chen, W., Yang, J.Z., Andersen, R., Nielsen, L.H., and Borchardt, R.T. (2002) Evaluation of the permeation characteristics of a model opioid peptide, H-Tyr-D-Ala-Gly-Phe-D-Leu-OH (DADLE), and its cyclic prodrugs across the blood-brain barrier using an in situ perfused rat brain model. *J Pharmacol Exp Ther* 303, 849–857.
20. Pan, W. and Kastin, A.J. (2002) TNFalpha transport across the blood-brain barrier is abolished in receptor knockout mice. *Exp Neurol* 174, 193–200.
21. Tanabe, S., Shimohigashi, Y., Nakayama, Y., Makino, T., Fujita, T., Nose, T., Tsujimoto, G., Yokokura, T., Naito, M., Tsuruo, T., and Terasaki, T. (1999) In vivo and in vitro evidence of blood-brain barrier transport of a novel cationic arginine-vasopressin fragment 4–9 analog. *J Pharmacol Exp Ther* 290, 561–568.
22. McGrogan, B.T., Gilmartin, B., Carney, D.N., and McCann, A. (2007) Taxanes, microtubules and chemoresistant breast cancer. *Biochim Biophys Acta* 1785, 96–132.

23. Tseng, S.H., Bobola, M.S., Berger, M.S., and Silber, J.R. (1999) Characterization of paclitaxel (Taxol) sensitivity in human glioma- and medulloblastoma-derived cell lines. *Neuro Oncol* 1, 101–108.
24. Chamberlain, M.C. and Kormanik, P. (1995) Salvage chemotherapy with paclitaxel for recurrent primary brain tumors. *J Clin Oncol* 13, 2066–2071.
25. Breedveld, P., Beijnen, J.H., and Schellens, J.H. (2006) Use of P-glycoprotein and BCRP inhibitors to improve oral bioavailability and CNS penetration of anticancer drugs. *Trends Pharmacol Sci* 27, 17–24.
26. Regina, A., Demeule, M., Che, C., Lavallee, I., Poirier, J., Gabathuler, R., Beliveau, R., and Castaigne, J.-P. (2008) Antitumor activity of ANG1005, a conjugate between paclitaxel and the new brain delivery vector Angiopep-2. *Br J Pharmacol* 155, 185–197.
27. Thomas, F.C., Taskar, K., Rudraraju, V., Goda, S., Thorsheim, H.R., Gaasch, J.A., Mittapalli, R.K., Palmieri, D., Steeg, P.S., Lockman, P.R., and Smith, Q.R. (2009) Uptake of ANG1005 - A Novel Paclitaxel Derivative, Through the Blood-Brain Barrier in to Brain and Experimental Brain Metastases of Breast Cancer. Pharm. Res. (on line, ahead of print)
28. Cecchelli, R., Berezowski, V., Lundquist, S., Culot, M., Renftel, M., Dehouck, M.P. et al. (2007) Modelling of the blood-brain barrier in drug discovery and development. *Nat Rev Drug Discov* 6, 650–661.
29. Pardridge, W.M. (2007) Drug targeting to the brain. *Pharm Res* 24, 1733–1744.
30. Pardridge, W.M. (2005) The blood-brain barrier: bottleneck in brain drug development. *NeuroRx* 2, 3–14.
31. Dove, A. (2008) Breaching the barrier. *Nat Biotechnol* 26, 1213–1215.

Chapter 13

Adeno-Associated Virus Mediated Gene Therapy in Ischemic Stroke

Yongting Wang, Falei Yuan, and Guo-Yuan Yang

Abstract

Gene therapy offers a novel approach for the treatment of experimental stroke. The adeno-associated virus (AAV) mediated vascular endothelial growth factor (VEGF) gene transfer into the ischemic brain is described in detail in this chapter. Other methods are also illustrated here, including the generation of mouse middle cerebral artery occlusion (MCAO) model, injection of viral vector into mouse brain, and standard assays for determining the successes of brain ischemia and gene transfer.

Key words: Gene therapy, ischemia, adeno-associated virus, VEGF, middle cerebral artery occlusion, neurons.

1. Introduction

1.1. Stroke and Gene Therapy

Stroke is the leading cause of death in China and the United States. Stroke is a multigenic complex disorder; it can be classified into two major categories, hemorrhagic stroke and ischemic stroke. Of the two types, ischemic stroke results from vascular occlusion and accounts for about 80% of all stroke (1). Stroke induces rapid neuronal death and loss of brain function due to the lack of oxygen and energy. Currently, apart from t-PA, the only medication approved by FDA, there are no other effective medications for the treatment of ischemic stroke.

Focal cerebral ischemia-injured brain tissue can be divided into three zones: ischemic core, ischemic penumbra, and extra penumbral cortical zone. Since neuronal injury in the ischemic penumbra may be reversible, inducing angiogenesis and neurogenesis in this zone may provide a unique approach in rescuing ischemic brain

K.K. Jain (ed.), *Drug Delivery to the Central Nervous System*, Neuromethods 45,
DOI 10.1007/978-1-60761-529-3_13, © Humana Press, a part of Springer Science+Business Media, LLC 2010

injury (2). The gene therapy technique in cerebral ischemic stroke was first applied by Betz's group in 1995 and showed exciting effect (3). Since then, different genes have been delivered through gene therapy in experimental studies with positive results. Hence, further study to improve transduction efficiency is necessary. There are mainly two types of experimental animal models in ischemic stroke research including focal (4–7) and global ischemic stroke (8–10). They can further be divided into transient or permanent ischemia based on whether reperfusion is followed. Focal ischemia, especially middle cerebral artery occlusion (MCAO) in rodents is widely used as an experimental focal cerebral ischemia model because this model mimics acute focal ischemia in humans.

Several gene therapy strategies have been introduced to rescue the damaged brain after ischemic stroke such as providing neuroprotection, promoting neurogenesis and angiogenesis, anti-apoptosis, and anti-inflammation. Growth factors are the most selected genes to be transduced in the ischemic brain because they are capable of inducing neurogenesis and angiogenesis in the brain. Such growth factors include vascular endothelial growth factor (VEGF), brain-derived neurotrophic factor (BDNF), glial cell line-derived neurotrophic factor (GDNF), nerve growth factor (NGF), fibroblast growth factor -2 (FGF-2), and epidermal growth factor (EGF). Among them, the role of BDNF and GDNF is of special interest because both exert potent neurotrophic effects.

Angiogenesis is a process involving the growth of neo-microvessels from pre-existing vessels. Several factors have been demonstrated to activate angiogenesis such as VEGF, PDGF, and TGF- β. It is noteworthy that some factors are not only simply angiogenic but also play roles in promoting neurogenesis. The shortage of oxygen and nutrients supply is followed by neuronal death. The increase of ROS and breakdown of cytoskeleton further induces apoptosis and necrosis. Gene delivery of B-cell lymphoma 2(Bcl-2) and heat shock protein (HSP) may attenuate neuronal death (11).

Inflammatory factors play important roles in ischemic-induced brain injury (12). When stroke occurs, various cells such as leukocytes, microglia/macrophages, and lymphocytes are able to migrate to the damaged site and release a variety of inflammatory mediators such as cytokines (Interleukin-1/IL1, IL-6, IL-8, and TNF-α), chemokines (monocyte chemotactic protein-1/MCP-1 and macrophage inflammatory protein-1/MIP-1), and adhesion molecules (intracellular adhesion molecule-1/ICAM-1 and platelet endothelial cell adhesion molecule/PECAM-1). Acute inflammatory response induces further brain tissue damage. Therefore, in vivo introduction of the antagonist of those factors through gene transfer will provide appropriated approaches to protect tissue from further injury. Recently, a number of experiments using anti-inflammatory factor such as IL-1 receptor antagonist (IL-1ra) or matrix metalloproteinase (MMP) inhibitors have been studied in ischemic animal models; the results are encouraging (13, 14).

Currently, several novel strategies have been developed in the stroke treatment using gene delivery. One of them is the delivery of microRNAs (miRNAs). miRNAs are small, endogenous, conserved, and non-coding RNAs that modulate gene expression by either promoting the degradation of mRNA or down regulating the protein production through translational repression. miRNAs are pervasive in human and animals. More than 400 miRNAs have been identified in human. Estimated 1000 miRNA genes have been predicted by computational programs (15). Discovery of miRNAs in brain tissue provides a glimpse of undiscovered regulatory mechanisms underlying cerebrovascular diseases. Current study demonstrates that miRNA can be delivered via gene transfer approach (16). Key technical aspects such as optimization of selectivity, stability, in vivo delivery, efficacy, and safety need to be extensively investigated before RNA silencing can become a successful therapeutic strategy (17, 18).

1.2. Vector Types and Delivery

In general, vectors can be divided into three groups: viral, non-viral, and cellular forms. Viral vectors include adenovirus (Ad), adeno-associated virus (AAV), herpes simplex virus (HSV), Sendai virus, hemagglutinating virus of Japan (HVJ), lentivirus, etc. (8, 19 22). Ad, AAV and HSV are the most widely used vectors in brain tissue since they are more efficient, safer, and relatively easy to be delivered. Multiple brain cell types including neurons, glial cells, vascular smooth muscle cells, and endothelial cells can produce target genes via recombinant Ad, AAV, or HSV systems. However, viral vectors have immunogenic toxicities, which should be overcome before they can be clinically applied (23).

Employing cellular vectors is a novel approach to circumvent immune response to viral vectors. Potential cellular vehicles may be used include fibroblast cells, mesencyhmal stem cells (MSCs), endothelial progenitor cells (EPCs), and astrocytes. This approach presents the potential benefit because these cells may be acquired from the patient's own body, which potentially reduces the incidence of immuno-rejection. For example, ex vivo gene delivery of NGF in fibroblasts has exhibited success in Alzheimer's disease and animal ischemic stroke models (24, 25). Another benefit of cellular vectors is the potential use of stem cells. Circulating EPCs could be attracted by SDF-1, a chemokine which is released by injured brain cells. Neuro-progenitor cells (NPCs) could migrate from the subventricular zone (SVZ) to the ischemic brain region. EPCs and NPCs have been shown to play an important role in the remodeling, repairing, and regeneration of brain function following ischemic brain injury (26, 27).

Non-viral vectors have lower transfection efficiency but safer compared to their viral counterparts (28). Modes for non-viral gene delivery include ligand–DNA conjugate system, cationic conjugate, direct injection of naked DNA, and calcium phosphate precipitation. Many of these methods demonstrate promises for future gene therapy studies (29–32). The use of DNA-cationic liposome complexes in a gene therapy study for ischemic stroke has afforded successful results (31). Another possibility is to inject naked DNA directly into the tissue. Introducing naked DNA of hypoxia inducible factor-1 alpha (HIF-1) into cerebral ischemia model was reported to promote angiogenesis in the target tissue (29).

The available routes of administration to stroke animal model include injection into cerebrospinal fluid parenchyma, and cerebral vasculature. An alternative route is to inject a viral vector at a peripheral site that has the potential to retroactively transport into the central nervous system (33). Direct introduction of vectors into the cerebrospinal fluid (CSF), within either the ventricular or perivascular system, has been utilized in rodent stroke models (32, 34, 35). Intravascular injection is an optional approach to different neurological disease processes and cerebrovascular diseases (4, 36, 37). However, the blood-brain barrier (BBB) is an obstacle to the delivery of vectors through intravascular route. Hypertonic sugar solutions such as mannitol may induce the transient osmotic opening of the BBB, and allow vectors to travel through the BBB and reach the target regions (38). Injection of Ad or AAV vectors into the perivascular system is often used to deliver into the cortex and basal ganglia. This route is especially important in cerebral vasospasm model (39–41).

The viral vectors mostly infect epidermal cells, choroids plexus, and superficial cell layers of the ventricular wall after intraventricular injection (3, 34, 42, 43). Administration of vector containing fibroblast growth factor-2 (FGF-2) improved neurological function in a mouse transient MCAO model, despite the fact that infected cells were limited to the ependymal cell layer (34). Although the brain tissue that is around 2 mm away from the injecting point is believed to not been transfected, therapeutic gene products, especially cytokines and chemokines, could be released into the cerebral spinal fluid and may defuse appropriately to deeper locations in the brain (3).

1.3. VEGF and AAV

VEGF, a 34- to 45-kDa dimeric glycosylated protein, is usually upregulated in ischemic tissues. VEGF is highly expressed in brain tissue and especially in the penumbra after cerebral ischemia (44). VEGF is expressed in different brain cells including endothelial cells, smooth muscle cells, astrocytes, and neurons (19). VEGF is explored as a therapeutic reagent in cerebral ischemia models because it not only induces cerebral angiogenesis but also enhances neuroprotection and promotes neurogenesis following cerebral ischemia. Pretreatment with VEGF through recombinant AAV-mediated gene transfer reduces infarct volume in mouse MCAO model (20). Intraventricular administration of VEGF activates angiogenesis in the ischemic penumbra and increases neuron survival, subsequently reducing infarct volume and improving neurological performance (20). Therefore, VEGF exerts its neuroprotective effect in the acute phase of cerebral ischemia with the longer latency effects on angiogenesis and on neo-neuron survival (45).

Adeno-associated virus (AAV) is a small single-stranded DNA virus, which infects human and primates. AAV based vector system is increasingly considered as a promising gene transfer vehicle because of its many advantages over adenovirus, retrovirus, or lentivirus. First, AAV is nonpathogenic and elicits no or less inflammatory response than other viral vectors (46). Second, recombinant AAV (rAAV) can infect both dividing and non-dividing cells and mediate lone-term gene expression up to months or several years in vivo (47–51). Third, rAAV probably maintains in vivo mostly as episomal genomes extrachromosomally and occasionally as integrated concatemers, which diminishes the insertional mutagenesis associated with random integration (52). Many kinds of serotypes of AAV have been characterized and different types show distinct tissue and cellular tropism transduction. AAV-2 and AAV-5 were always used in gene transduction into central nervous system (CNS), while current report demonstrates rAAV-1 also had much higher transduction efficiency in CNS (53). Distribution of rAAV1 transduced cells in the striatum is much wider than that of rAAV-2 (54).

Since uncontrolled VEGF expression can cause unwanted side effects such as hemangioma formation and synovium, it is important to control the expression of VEGF protein (55, 56). Examples of inducible promoters that have been used to control gene expression include the tetracycline operons, RU 486, edyasone, and other inducible systems (56–64). For the treatment of ischemic tissue, an ideal control would be for the expression of angiogenic factors to respond to hypoxia. Hypoxia inducible protein-1 (HIF-1) is a protein that accumulates in tissues under hypoxic condition. HIF-1α is greatly increased in ischemic myocardium from 48h after left anterior descending coronary artery (LAD) occlusion (65). A method has been developed to achieve hypoxia-induced VEGF gene expression in vitro and in ischemic heart in vivo by controlling AAV-mediated VEGF expression using nine copies of hypoxia response elements (HRE) isolated from Epo gene enhancer and a minimum SV40 promoter (65). The AAVH9VEGF gene has been successfully transduced in the ischemic mouse brain and was shown to increase VEGF expression and consequently, reduce brain injury and promote local angiogenesis.

Although there is not yet experiment performed on human after stroke, AAV vectors has been used in treating Parkinson's disease (66). AAV-GAD was injected subthalamically into 11 men and 1 woman with Parkinson's disease. The results are encouraging; all patients improved their movement function. This is a promising start for using AAV vectors in gene therapy.

In the following paragraphs, we describe the delivery of VEGF through AAV-mediated pathway in detail.

2. Materials

2.1. AAV Construction, Production, and Purification

1. AAV Helper-Free Gene Delivery and Expression System (Strategene Inc., La Jolla, CA)
2. Restriction enzymes and T4 DNA ligase
3. Agarose gel equipment
4. Competent cells: *E. coli* strains XL1 blue, DH5α
5. Dulbecco's modified Eagle's medium (DMEM) with 4.5 g/L glucose and L-glutamine (Cellgro, Herndon, VA)
6. Fetal bovine serum (FBS) (Hyclone, Logan, UT)
7. HEK293 cells (American Type Culture Collection, Gaithersburg, MD)
8. 2 × HBS (280 mM NaCl, 10 mM KCl, 1.5 mM $Na_2HPO_4 \cdot 7H_2O$, 12 mM dextrose, and 50 mM HEPES)
9. Dialysis cassettes (Slide-A-Lyzer 7 K, Pierce, Rockford, IL)
10. High speed centrifuge and ultra speed centrifuge
11. Cell culture room equipped with incubators and biological hood
12. Quantikine human VEGF ELISA kit (R&D Systems, Minneapolis, MN)
13. 2 × SSC (300 mM NaCl, 30 mM sodium citrate, pH 7)

2.2. Middle Cerebral Artery Occlusion (MCAO) Model and Viral Vector Injection

The optimal small animal surgery room needs to be kept quiet and pathogen-free with the room temperature maintained at 25°C. The following items are recommended for small animal surgery.

1. Operation microscope (M651, Leica, Germany)
2. Bipolar coagulator (Malis™ Precision-Control bipolar coagulator, CMC™-II-PC, Codman and Shurtleff Inc., Randolph, MA)
3. Exercise Physiological System (AD Instruments, PowerLab/4SP, Castle Hill, Australia)
4. Temperature controller (Homeothermic Blanket Control Unit, Harvard Apparatus, Cambridge, MA)

5. Laser Doppler Flowmeter (VMS-LDF2, Moor Instrument Ltd., Millwey,UK)
6. Dry sterilizer (Simon Keller AG, Fine Science Tools, CH 3400, Foster City, CA)
7. Vaporizer (Isoflurane Matrx, Midmark International, Orchard Park, NY)
8. Stereotaxic frame (World Precision Instruments Inc., Sarasota, FL)
9. pH/blood gas analyzer (Bayer, Radiolab 248, Tarrytown, NY)
10. High-speed micro drill (Fine Science Tools, Foster City, CA)
11. Surgical equipment kit (nylon suture, bipolar forceps, micro-scissors for small vessels, surgical scissors for animal skin and tissue use, microforceps, needle holder, skin hook for the exposure, ruler)
12. Hamilton syringe with replaceable beveled needle (WPI, Sarasota, FL)

2.3. Histology and Immunohistochemistry

2.3.1. X-Gal Staining

1. 2% paraformaldehyde in 0.1 M PIPES, pH 6.9
2. 0.5% glutaraldehyde
3. X-gal staining solution 1 (5 mM $K_3[Fe(CN)_6]$, 5 mM K4[Fe (CN)$_6$], 2 mM $MgCl_2$, 0.01% sodium deoxycholate, 0.02% NP-40)
4. X-gal staining solution 2 (1 mg/ml 5-bromo-4-chloro-3-indolyl-β-D-galactoside in PBS)

2.3.2. Cresyl Violet Staining

1. 0.1% cresyl violet acetate (Sigma, St. Louis, MO)
2. Distilled water
3. 10% acetic acid (Sigma, St. Louis, MO)

2.3.3. Lectin Staining for Capillary Density

1. Luorescein-lycopersicin esculentum lectin (Vector Lab, Burlingame, CA)

2.3.4. Fluoro-Jade B Staining

1. 0.06% potassium permanganate (KMnO4) (Sigma, St. Louis, MO).
2. A 0.001% Fluoro-Jade staining solution (Chemicon, Temecula, CA).
3. 0.1% acetic acid (Sigma, St. Louis, MO).

3. Methods

3.1. AAV Vector Construction

Plasmid AAV-LacZ is included in the AAV Helper-Free Gene Delivery and Expression System (Strategen Inc.). The pCMV-VEGF vector can be generated by inserting the human $VEGF_{165}$ cDNA to multiple cloning sites in pCMV-MCS vector. The expression cassette, including CMV promoter, VEGF cDNA and HGH polyadenylation(poly A) signal, can be isolated from pCMV-VEGF vector by Not I digestion. pAAV-VEGF vector can be generated by replacing the LacZ expression cassette with VEGF expression cassette. VEGF expression mediated by pAAV-VEGF can be confirmed by infecting 293 cells and analyzing VEGF concentration in supernatant using a Quantikine human VEGF ELISA kit (R&D Systems).

3.2. AAV Virus Production

3.2.1. Transfection

1. First, 4.5×10^6 HEK 293 cells are seeded in 15-cm tissue culture dishes 2 days before transfection. Cells in one 90% confluent dish can be passed to eight dishes. The cells are maintained in 25 ml DMEM with 10% fetal bovine serum (FBS), and 25 mM HEPES.
2. The AAV vector containing the gene of interest is co-transfected with pAAV-RC and pHelper vectors into 293 cells by using the calcium phosphate precipitation method. pAAV-RC contains AAV *rep* and *cap* genes (*see* **Note 1**). pHelper has the adenoviral VA, E2A, and E4 regions that mediate AAV vector replication. To transfect one 15-cm dish, a total of 50 μg DNA (17 μg DNA of each plasmid) is mixed with 1 ml of 300 mM $CaCl_2$. A mixture for transfection of four dishes can be prepared in one 50-ml Corning tube.
3. A sterile pipet is placed into the tube. Air is gently bubbled while 1 ml (4 ml for four plates) of 2 × HBS is added to the tube drop by drop.
4. The transfection mix is distributed into each plate.
5. The plates are incubated in 37°C for 6h.
6. The medium is changed to DMEM containing 2% FBS and 25 mM HEPES. Cells are cultured at 37°C for 54h.

3.2.2. Purification (see Note 2)

1. Cells are dislodged from the dishes by gentle pipeting and transferred into 50-ml Corning tubes.
2. Media is removed by centrifugation (1000 × *g* for 5 min at 4°C).
3. Cells are resuspended in 100 mM Tris–HCl, 150 mM NaCl, pH 8.0 (1×10^7 cells/ml) and lysed by three freeze and thaw cycles (alternating between dry ice-ethanol and 37°C water baths).
4. The cell lysate is centrifuged at 10,000 × *g* for 15 min to remove the cell debris.
5. The cleared supernatant is precipitated with 25 mM $CaCl_2$ at 0°C for 1h.
6. Precipitate is removed by centrifugation (10,000 × *g*, for 15 min at 4°C).

7. NaCl and PEG (8000) are added into the supernatant to make a final concentration of 620 mM NaCl and 8% PEG.
8. The supernatant is incubated on ice for 3h.
9. AAV vector containing precipitate is collected by centrifugation (300 × *g*, for 30 min at 4°C) and resuspended in 5 ml of 50 mM HEPES, 150 mM NaCl, 25 mM EDTA, pH 8.0 (Adjust pH with NaOH).
10. Insoluble material is removed by centrifugation (10,000 × *g* for 15 min at 4°C).
11. The AAV vector is purified by $CsCl_2$ gradient centrifugation. Solid $CsCl_2$ is added to the supernatant to produce a density of 1.4 g/ml. Samples are centrifuged at 15°C for 16h at 22,300 × *g*.
12. The gradient is fractionated (0.5–1 ml/fraction) and assayed by dot blot to detect the viral particles (*see* **Section 3.2.3**).
13. The AAV vector-containing fractions are pooled, put in dialysis cassettes (Slide-A-Lyzer 7 K, Pierce) and dialyzed against 1 l buffer containing 10 mM HEPES (pH 7.4), 140 mM NaCl, 0.1% Tween-80, 5% Sorbital three times at 4°C. The dialysis buffer is changed every 2h.

3.2.3. Titer Determination

Viral titers are determined by dot blot analysis of the DNA content (*see* **Note 3**).

1. First, 2 μl of each fraction collected after $CsCl_2$ centrifugation or dialyzed viral stock are added to 20 μl of 500 mM NaOH and dotted on Hybond H^+ membrane (Strategene).
2. Then 1 ng of gene fragment (about 1 kb) is serially diluted and dotted on the same membrane as standards.
3. The membrane is air-dried and neutralized in 500 mM Sodium phosphate buffer, pH 7.4, for 10 min, washed in 2 × SSC briefly, and air-dried.
4. The membrane is hybridized with isotope or digoxin-labeled (Roche Diagnostics, Indianapolis, IN) probes for genes in the AAV vectors using a standard protocol.
5. The density of each fraction or viral stock is compared with that of the standards. The relative copy number can be calculated. Copies of a 1-ng gene fragment can be calculated as follows (copies/ml): $(1 \times 10^{-9}/\text{length of the fragment (bp)} \times 660) \times 6 \times 10^{23} \times 2 \times 10^{3}$.

3.2.4. Toxicity

The viral stocks' toxicity to cells is checked by infection of 293 cells.

1. 293 cells are seeded in 48-well tissue culture dishes, 1×10^5 cells/well. Cells are cultured in 0.5 ml DMEM with 10% FBS for 24h at 37°C.

2. Then 50 μl of viral stock is added to each well and cultured with the cells for 24h.
3. The medium is changed after 24h of incubation.
4. The status of the infected cells is monitored every day for 3–5 days and compared with uninfected control cells. The viral stock will be considered toxic if cell death or slower growth is observed. More dialysis steps can be used to reduce toxicity, if a viral stock is found to be toxic to cells.

3.3. Animal Procedure

3.3.1. Rat Middle Cerebral Artery Occlusion (see Note 4 for Mouse Model)

1. Adult rats are anesthetized using 1.5–2.0% isoflurane inhalation.
2. A PE-50 catheter is inserted into the femoral artery for continuous monitoring of arterial blood pressure, and sampling for analysis of blood gases and blood pH.
3. The temperature of both the temporal muscle and the rectum is measured by thermocouples (E type, OMEGA Engineering, Inc.). The body temperature of the animal is kept in normal physiological range with a feedback-enabled heating unit.
4. Drill a burr hole in the pericranium 3.5 mm lateral to the sagittal suture and 1 mm posterior to the coronal suture (keep the inner layer intact).
5. Measure baseline of cerebral blood flow using a Laser Doppler Flowmeter; monitor and record surface blood flow before, during, and after occlusion.
6. Make a midline incision on the front of the neck, and expose left common carotid artery (CCA).
7. Isolate left internal carotid artery (ICA) and coagulate other small branches.
8. Ligate the pterygopalatine artery from the bifurcation of ICA and the pterygopalatine artery.
9. A 3-cm length of 3-0 nylon suture with a slightly enlarged and rounded tip is inserted into the transected lumen of the external carotid artery (ECA) and gently advanced from the ICA across to the opening of the MCA. The distance from the tip of the suture to the bifurcation of CCA is 17–19 mm.
10. Measure surface blood flow to make sure the blood flow drop to below 20% of baseline.
11. For reperfusion, the inserted suture is withdrawn into the ECA to restore ICA-MCA blood flow.
12. The skin is sutured and the animal is returned to the cage when it recovers from the anesthesia.

3.3.2. AAV-Mediated Gene Transfer (see Note 5 for Mouse Model)

1. Rat is anesthetized with 50 mg/kg ketamine and 10 mg/kg xylazine (Sigma, San Louis, MO) body weight intraperitoneally.
2. Rat is placed in a stereotactic frame with a mouth holder (David Kopf Instruments, Tujunga, CA).
3. Cut a skin incision in the midline of skull, drill a burr hole in the pericranium 3.5 mm lateral to the sagittal suture and 1 mm posterior to the coronal suture.
4. A 50-μl Hamilton syringe is stereotactically inserted 5.5 mm into the lateral caudate.
5. Ten μl AAV viral suspension containing 2×10^{10} genome copies of virus is injected into the right caudate putamen at a rate of 0.2 μl/min (13). The volume and concentration can be adjusted based on different purpose. The needle is withdrawn after 15 min of injection.
6. Seal the bone hole and the suture the wound.
7. Rat is returned to their home cages after it recovers from anesthesia.

3.4. Histology and Immunohistochemistry

3.4.1. X-Gal Staining

1. The brains are cut into 20-μm coronal sections and fixed in 0.5% glutaraldehyde for 10 min.
2. The sections are incubated in X-gal staining solution for 2h and photographed.
3. The transduction volumes are calculated by multiplying the transduction areas by the thickness of the sections using NIH Image 1.63 software.

3.4.2. Cresyl Violet Staining for Infarct Volume

1. Brain sections are incubated in xylene for 3 min.
2. The sections are incubated in 100, 95, 90, and 70% ethanol for 3 min each and washed in distilled water for 3 min.
3. Sections are incubated in cresyl violet solution for 5–10 min depending on the freshness of the staining solution.
4. Sections are differentiated by incubation in 70, 90, and 100% ethanol sequentially, 2×3 min each. Incubation time can vary slightly according to degree of staining and thickness of the sections.
5. Sections are then incubated in xylene for 2×5 min.
6. Sections are mounted with Permount.
7. Color assignment: nissl granules: violet; nuclei: pale violet; and background: colorless.

3.4.3. Fluoro-Jade B Staining

1. The frozen sections are fixed with 2% paraformaldehyde in 0.1 M PIPES, pH 6.9, for 20 min.
2. After being washed, the sections are stained with 0.06% potassium permanganate ($KMnO_4$) for 30 min at room temperature.

3. The sections are immersed a 0.001% fluoro-Jade B staining solution (Chemicon, Temecula, CA) in 0.1% acetic acid for 20 min.
4. The staining sections are evaluated using a fluorescence microscope.

3.4.4. Determination of Capillary Density

1. Five minutes before sacrificing the animal, 100 μl of fluorescein-lycopersicon esculentum lectin (100 μg, Vector Labs, CA) is injected intravenously (19, 67).
2. After the animals are sacrificed, their brains are removed, and 20-μm-thick frozen sections are fixed with 100% ethanol at 20°C for 20 min.
3. Sections are incubated with 2 g/ml fluorescein-lycopersicin esculentum lectin (Vector Lab, Burlingame, CA) at 4°C overnight.
4. Two coronal sections from the lectin staining brain, 1 mm anterior and 1 mm posterior from the needle track, are chosen for capillary density analysis. The microvessel density is quantified by counting the total number of three areas of microscopic fields per tissue section immediately adjacent (left, right, and bottom) to the needle track.
5. As a surrogate of vessel counting, vessel density was determined by lectin optical density measurements using NIH Image 1.63 software.
6. The procedure needs to be analyzed blindly.

4. Notes

1. The AAV Helper-Free Gene Delivery and Expression System was generated based on AAV serotype 2 virus. Many new AAV serotypes have been cloned in recent years (68–73). Recombinant cross-packaging of the AAV genome of one serotype into capsids of other AAV serotypes to achieve optimal tissue-specific gene transduction is now possible. Studies by our group and others have shown that AAV serotype 1 results in more efficient transduction of genes in the murine and human adult heart compared with serotypes 2, 3, 4, and 5. AAV serotype 1 also mediated more efficient gene transduction in the brain the AAV serotype 2 (74, 75). The AAV packaged in serotypes other than 2 can be made by replacing the serotype 2 *CAP* sequence in pAAV-RC with the *CAP* sequence of the corresponding serotype.
2. AAV titer can also be determined by quantitative PCR assay (76). Infectious particles can be determined by using serial

diluted viral stocks to infect 293 or other cells and counting the infected cells or transgenic expressions. DNase I digestion can be used to eliminate unpackaged DNA in viral stocks (77).

3. AAV can be purified by using other methods, such as nonionic iodixanol gradients, ion exchange, or heparin affinity chromatography by either conventional or high-performance liquid chromatography columns (77, 78). Ion exchange or heparin affinity chromatography can also be used in combination with $CsCl_2$ gradients or nonionic iodixanol gradients.
4. The method for mouse MCAO is similar to the rat model described above. The only difference is the suture size (2-cm length of 5-0 nylon). The distance from the tip of the suture to the bifurcation of CCA is 10–11 mm.
5. The method for mouse AAV-mediated gene transfer is similar to the rat model described above. The differences are (1) the bone hole location is pericranium 2 mm lateral to the sagittal suture and 1 mm posterior to the coronal suture, and (2) the total inject volume of AAV vector is 2 μl.

References

1. Gunel, M., Awad, I. A., Finberg, K. et al (1996) A founder mutation as a cause of cerebral cavernous malformation in Hispanic Americans. *New Engl J Med* 334(15), 946–951.
2. Astrup, J., Siesjo, B. K., and Symon, L. (1981) Thresholds in cerebral ischemia – the ischemic penumbra *Stroke*; a Journal of Cerebral Circulation, 12(6), 723–725.
3. Betz, A. L., Yang, G. Y., and Davidson, B. L. (1995) Attenuation of stroke size in rats using an adenoviral vector to induce overexpression of interleukin-1 receptor antagonist in brain. *J Cereb Blood Flow Metab* 15(4), 547–551.
4. Baker, A. H., Sica, V., Work, L. M. et al (2007) Brain protection using autologous bone marrow cell, metalloproteinase inhibitors, and metabolic treatment in cerebral ischemia. *Proc Natl Acad Sci USA* 104(9), 3597–3602.
5. Davis, A. S., Zhao, H., Sun, G. H., Sapolsky, R. M., and Steinberg, G. K. (2007) Gene Therapy using SOD1 protects striatal neurons from experimental stroke. *Neurosci Lett* 411(1), 32–36.
6. Zhu, W., Khachi, S., Hao, Q. et al (2008) Upregulation of EMMPRIN after permanent focal cerebral ischemia. *Neurochem Int* 52(6), 1086–1091.
7. Wang, Z. Q., Chen, X. C., Yang, G. Y., and Zhou, L. F. (2004) Inflammatory reaction after focal cerebral ischemia in mouse. *Chin Med J (Engl)* 117(4), 586–591.
8. Shirakura, M., Fukumura, M., Inoue, M. et al (2003) Sendai virus vector-mediated gene transfer of glial cell line-derived neurotrophic factor prevents delayed neuronal death after transient global ischemia in gerbils. *Exp Anim* 52(2), 119–127.
9. Yonekura, I., Kawahara, N., Nakatomi, H., Furuya, K., and Kirino, T. (2004) A model of global cerebral ischemia in C57 BL/6 mice. *J Cereb Blood Flow Metab* 24(2), 151–158.
10. Wellons, J. C. 3rd, Sheng, H., Laskowitz, D. T., Burkhard Mackensen, G., Pearlstein, R. D., and Warner, D. S. (2000) A comparison of strain-related susceptibility in two murine recovery models of global cerebral ischemia. *Brain Res* 868(1), 14–21.
11. Yenari, M. A., Liu, J., Zheng, Z., Vexler, Z. S., Lee, J. E., and Giffard, R. G. (2005) Antiapoptotic and anti-inflammatory mechanisms of heat-shock protein protection. *Ann N Y Acad Sci* 1053, 74–83.
12. Wang, Q., Tang, X. N., and Yenari, M. A. (2007) The inflammatory response in stroke. *J Neuroimmunol* 184(1–2), 53–68.

13. Yang, G. Y., Zhao, Y., Davidson, B. L., and Betz, A. L. (1997) Overexpression of interleukin-1 receptor antagonist in the mouse brain reduces ischemic brain injury. *Brain Res* 751, 181–188.
14. Zhao, B. Q., Wang, S., Kim, H. Y. et al (2006) Role of matrix metalloproteinases in delayed cortical responses after stroke. *Nat Med* 12(4), 441–445.
15. Berezikov, E., Guryev, V., van de Belt, J., Wienholds, E., Plasterk, R. H., and Cuppen, E. (2005) Phylogenetic shadowing and computational identification of human microRNA genes. *Cell* 120(1), 21–24.
16. Xia, H., Mao, Q., Paulson, H. L., and Davidson, B. L. (2002) siRNA-mediated gene silencing in vitro and in vivo. *Nat Biotechnol* 20(10), 1006–1010.
17. Pushparaj, P. N., Aarthi, J. J., Manikandan, J., and Kumar, S. D. (2008) siRNA, miRNA, and shRNA: in vivo applications. *J Dental Res* 87(11), 992–1003.
18. Kurreck, J. (2009) RNA interference: from basic research to therapeutic applications. *Angewandte Chemie (International ed)* 48(8), 1378–1398.
19. Yang, G. Y., Xu, B., Hashimoto, T. et al (2003) Induction of focal angiogenesis through adenoviral vector mediated vascular endothelial cell growth factor gene transfer in the mature mouse brain. *Angiogenesis* 6(2), 151–158.
20. Shen, F., Su, H., Fan, Y. et al (2006) Adeno-associated viral vector-mediated hypoxia-inducible vascular endothelial growth factor gene expression attenuates ischemic brain injury after focal cerebral ischemia in mice. *Stroke*; a Journal of Cerebral Circulation, 37(10), 2601–2606.
21. Gabriel, R. A. and Yang, G. Y. (2007) Gene therapy in cerebrovascular diseases. *Curr Gene Ther* 7(6), 421–433.
22. Zheng, Z. and Yenari, M. A. (2006) The application of HSP70 as a target for gene therapy. *Front Biosci* 11, 699–707.
23. Xu, Z. L., Mizuguchi, H., Sakurai, F. et al (2005) Approaches to improving the kinetics of adenovirus-delivered genes and gene products. *Adv Drug Deliv Rev* 57(5), 781–802.
24. Tuszynski, M. H., Thal, L., Pay, M. et al (2005) A phase 1 clinical trial of nerve growth factor gene therapy for Alzheimer disease. *Nat Med* 11(5), 551–555.
25. Pechan, P. A., Yoshida, T., Panahian, N., Moskowitz, M. A., and Breakefield, X. O. (1995) Genetically modified fibroblasts producing NGF protect hippocampal neurons after ischemia in the rat. *Neuroreport* 6(4), 669–672.
26. Lee, S. P., Youn, S. W., Cho, H. J. et al (2006) Integrin-linked kinase, a hypoxia-responsive molecule, controls postnatal vasculogenesis by recruitment of endothelial progenitor cells to ischemic tissue. *Circulation* 114(2), 150–159.
27. Jin, K., Minami, M., Lan, J. Q. et al (2001) Neurogenesis in dentate subgranular zone and rostral subventricular zone after focal cerebral ischemia in the rat. *Proc Natl Acad Sci USA* 98(8), 4710–4715.
28. Heistad, D. D. (2006) Gene Therapy for vascular disease. *Vascul Pharmacol* 45(5), 331–333.
29. Matsuda, T., Abe, T., Wu, J. L., Fujiki, M., and Kobayashi, H. (2005) Hypoxia-inducible factor-1alpha DNA induced angiogenesis in a rat cerebral ischemia model. *Neurol Res* 27(5), 503–508.
30. Quissell, D. O., Barzen, K. A., Gruenert, D. C., Redman, R. S., Camden, J. M., and Turner, J. T. (1997) Development and characterization of SV40 immortalized rat submandibular acinar cell lines. *In Vitro Cell Dev Biol Anim* 33(3), 164–173.
31. Sun, Y., Ouyang, Y. B., Xu, L. et al (2006) The carboxyl-terminal domain of inducible Hsp70 protects from ischemic injury in vivo and in vitro. *J Cereb Blood Flow Metab* 26(7), 937–950.
32. Li, H. and Qian, Z. M. (2002) Transferrin/transferrin receptor-mediated drug delivery. *Med Res Rev* 22(3), 225–250.
33. Frampton, A. R. Jr., Goins, W. F., Nakano, K., Burton, E. A., and Glorioso, J. C. (2005) HSV trafficking and development of gene therapy vectors with applications in the nervous system. *Gene Therapy* 12(11), 891–901.
34. Watanabe, T., Okuda, Y., Nonoguchi, N. et al (2004) Postischemic intraventricular administration of FGF-2 expressing adenoviral vectors improves neurologic outcome and reduces infarct volume after transient focal cerebral ischemia in rats. *J Cereb Blood Flow Metab* 24(11), 1205–1213.
35. Pang, L., Ye, W., Che, X. M., Roessler, B. J., Betz, A. L., and Yang, G. Y. (2001) Reduction of inflammatory response in the mouse brain with adenoviral- mediated transforming growth factor-ss1 expression. *Stroke*; a Journal of Cerebral Circulation, 32(2), 544–552.
36. Dickson, P. V., Hamner, J. B., Burger, R. A. et al (2007) Intravascular administration of

tumor tropic neural progenitor cells permits targeted delivery of interferon-beta and restricts tumor growth in a murine model of disseminated neuroblastoma. *J Pediatr Surg* 42(1), 48–53.

37. Thomas, J. E. and Rosenwasser, R. H. (1999) Reversal of severe cerebral vasospasm in three patients after aneurysmal subarachnoid hemorrhage: initial observations regarding the use of intraventricular sodium nitroprusside in humans. *Neurosurgery* 44(1), 48–57, discussion-8.
38. Rapoport, S. I. (2000) Osmotic opening of the blood-brain barrier: principles, mechanism, and therapeutic applications. *Cell Mol Neurobiol* 20(2), 217–230.
39. Khurana, V. G., Smith, L. A., Baker, T. A., Eguchi, D., O'Brien , T. ,, and Katusic , Z. S. (2002) Protective vasomotor effects of in vivo recombinant endothelial nitric oxide synthase gene expression in a canine model of cerebral vasospasm *Stroke*; a Journal of Cerebral Circulation, 33(3), 782–789.
40. Ono, S., Komuro, T., and Macdonald, R. L. (2002) Heme oxygenase-1 gene therapy for prevention of vasospasm in rats. *J Neurosurg* 96(6), 1094–1102.
41. Yamaguchi, M., Zhou, C., Heistad, D. D., Watanabe, Y., and Zhang, J. H. (2004) Gene transfer of extracellular superoxide dismutase failed to prevent cerebral vasospasm after experimental subarachnoid hemorrhage *Stroke* ; a Journal of Cerebral Circulation, 35(11), 2512–2517.
42. Akli, S., Caillaud, C., Vigne, E. et al (1993) Transfer of a foreign gene into the brain using adenovirus vectors. *Nat Genet* 3, 224–228.
43. Bajocchi, G., Feldman, S. H., Crystal, R. G., and Mastrangeli, A. (1993) Direct in vivo gene transfer to ependymal cells in the central nervous system using recombinant adenovirus vectors. *Nat Genet* 3, 229–234.
44. Leker, R. R., Teichner, A., Ovadia, H., Keshet, E., Reinherz, E., and Ben-Hur, T. (2001) Expression of endothelial nitric oxide synthase in the ischemic penumbra: relationship to expression of neuronal nitric oxide synthase and vascular endothelial growth factor. *Brain Res* 909(1–2), 1–7.
45. Sun, Y., Jin, K., Xie, L. et al (2003) VEGF-induced neuroprotection, neurogenesis, and angiogenesis after focal cerebral ischemia. *J Clin Invest* 111(12), 1843–1851.
46. Vincent-Lacaze, N., Snyder, R. O., Gluzman, R., Bohl, D., Lagarde, C., and Danos, O. (1999) Structure of adeno-associated virus vector DNA following transduction of the skeletal muscle. *J Virol* 73(3), 1949–1955.
47. Flotte, T. R., Afione, S. A., Conrad, C. et al (1993) Stable in vivo expression of the cystic fibrosis transmembrane conductance regulator with an adeno-associated virus vector. *Proc Natl Acad Sci USA* 90(22), 10613–10617.
48. Kaplitt, M. G., Leone, P., Samulski, R. J. et al (1994) Long-term gene expression and phenotypic correction using adeno-associated virus vectors in the mammalian brain. *Nat Genet* 8(2), 148–154.
49. McCormick, W. F. and Schochet, S. S. Jr. (1976) Vascular malformations. In: Atlas of Cerebrovascular Disease. W. B. Saunders Company, Philadelphia, 72–105.
50. Snyder, R. O., Spratt, S. K., Lagarde, C. et al (1997) Efficient and stable adeno-associated virus-mediated transduction in the skeletal muscle of adult immunocompetent mice. *Hum Gene Ther* 8(16), 1891–1900.
51. Xiao, X., Li, J., and Samulski, R. J. (1996) Efficient long-term gene transfer into muscle tissue of immunocompetent mice by adeno-associated virus vector. *J Virol* 70(11), 8098–8108.
52. Nakai, H., Yant, S. R., Storm, T. A., Fuess, S., Meuse, L., and Kay, M. A. (2001) Extrachromosomal recombinant adeno-associated virus vector genomes are primarily responsible for stable liver transduction in vivo. *J Virol* 75(15), 6969–6976.
53. Davidson, B. L., Stein, C. S., Heth, J. A. et al (2000) Recombinant adeno-associated virus type 2, 4, and 5 vectors: transduction of variant cell types and regions in the mammalian central nervous system. *Proc Natl Acad Sci USA* 97(7), 3428 3432.
54. Wang, C., Wang, C. M., Clark, K. R., and Sferra, T. J. (2003) Recombinant AAV serotype 1 transduction efficiency and tropism in the murine brain. *Gene Therapy* 10(17), 1528–1534.
55. Lee, R. J., Springer, M. L., Blanco-Bose, W. E., Shaw, R., Ursell, P. C., and Blau, H. M. (2000) VEGF gene delivery to myocardium: deleterious effects of unregulated expression. *Circulation* 102(8), 898–901.
56. Springer, M. L., Chen, A. S., Kraft, P. E., Bednarski, M., and Blau, H. M. (1998) VEGF gene delivery to muscle: potential role for vasculogenesis in adults. *Molecular cell* 2(5), 549–558.
57. Bohl, D., Salvetti, A., Moullier, P., and Heard, J. M. (1998) Control of erythropoietin delivery by doxycycline in mice after

intramuscular injection of adeno-associated vector. *Blood* 92(5), 1512–1517.

58. Hofmann, A., Nolan, G. P., and Blau, H. M. (1996) Rapid retroviral delivery of tetracycline-inducible genes in a single autoregulatory cassette. *Proc Natl Acad Sci USA* 93(11), 5185–5190.
59. Bohl, D., Naffakh, N., and Heard, J. M. (1997) Long-term control of erythropoietin secretion by doxycycline in mice transplanted with engineered primary myoblasts. *Nat Med* 3(3), 299–305.
60. Gossen, M. and Bujard, H. (1992) Tight control of gene expression in mammalian cells by tetracycline-responsive promoters. *Proc Natl Acad Sci USA* 89(12), 5547–5551.
61. Wang, Y., O'Malley, B. W. Jr., Tsai , S. Y., and O'Malley, B. W. (1994) A regulatory system for use in gene transfer. *Proc Natl Acad Sci USA* 91(17), 8180–8184.
62. Blau, H. M. and Rossi, F. M. (1999) Tet B or not tet B: advances in tetracycline-inducible gene expression. *Proc Natl Acad Sci USA* 96(3), 797–799.
63. Rossi, F. M., Guicherit, O. M., Spicher, A. et al (1998) Tetracycline-regulatable factors with distinct dimerization domains allow reversible growth inhibition by p16. *Nat Genet* 20(4), 389–393.
64. Kringstein, A. M., Rossi, F. M., Hofmann, A., and Blau, H. M. (1998) Graded transcriptional response to different concentrations of a single transactivator. *Proc Natl Acad Sci USA* 95(23), 13670–13675.
65. Su, H., Arakawa-Hoyt, J., and Kan, Y. W. (2002) Adeno-associated viral vector-mediated hypoxia response element-regulated gene expression in mouse ischemic heart model. *Proc Natl Acad Sci USA* 99(14), 9480–9485.
66. Kaplitt, M. G., Feigin, A., Tang, C. et al (2007) Safety and tolerability of gene therapy with an adeno-associated virus (AAV) borne GAD gene for Parkinson's disease: an open label, phase I trial *Lancet*; (Feb 13), 369(9579), 2097–2105.
67. Lee, C. Z., Xu, B., Hashimoto, T., McCulloch, C. E., Yang, G. Y., and Young, W. L. (2004) Doxycycline suppresses cerebral matrix metalloproteinase-9 and angiogenesis induced by focal hyperstimulation of vascular endothelial growth factor in a mouse model *Stroke*; a Journal of Cerebral Circulation, 35(7), 1715–1719.
68. Gao, G. P., Alvira, M. R., Wang, L., Calcedo, R., Johnston, J., and Wilson, J. M. (2002) Novel adeno-associated viruses from rhesus monkeys as vectors for human gene therapy. *Proc Natl Acad Sci USA* 99(18), 11854–11859.
69. Xiao, W., Chirmule, N., Berta, S. C., McCullough, B., Gao, G., and Wilson, J. M. (1999) Gene Therapy vectors based on adeno-associated virus type 1. *J Virol* 73(5), 3994–4003.
70. Rutledge, E. A., Halbert, C. L., and Russell, D. W. (1998) Infectious clones and vectors derived from adeno-associated virus (AAV) serotypes other than AAV type 2. *J Virol* 72(1), 309–319.
71. Muramatsu, S., Mizukami, H., Young, N. S., and Brown, K. E. (1996) Nucleotide sequencing and generation of an infectious clone of adeno-associated virus 3. *Virology* 221(1), 208–217.
72. Chiorini, J. A., Yang, L., Liu, Y., Safer, B., and Kotin, R. M. (1997) Cloning of adeno-associated virus type 4 (AAV4) and generation of recombinant AAV4 particles. *J Virol* 71(9), 6823–6833.
73. Chiorini, J. A., Kim, F., Yang, L., and Kotin, R. M. (1999) Cloning and characterization of adeno-associated virus type 5. *J Virol* 73(2), 1309–1319.
74. Du, L., Kido, M., Lee, D. V. et al (2004) Differential myocardial gene delivery by recombinant serotype-specific adeno-associated viral vectors. *Mol Ther* 10(3), 604–608.
75. Shen, F., Su, H., Liu, W., Kan, Y. W., Young, W. L., and Yang, G. Y. (2006) Recombinant adeno-associated viral vector encoding human VEGF165 induces neomicrovessel formation in the adult mouse brain. *Front Biosci* 11, 3190–3198.
76. Clark, K. R., Sferra, T. J., Lo, W., Qu, G., Chen, R., and Johnson, P. R. (1999) Gene transfer into the CNS using recombinant adeno-associated virus: analysis of vector DNA forms resulting in sustained expression. *J Drug Target* 7(4), 269–283.
77. Zolotukhin, S., Potter, M., Zolotukhin, I. et al (2002) Production and purification of serotype 1, 2, and 5 recombinant adeno-associated viral vectors. *Methods (San Diego, Calif)* 28(2), 158–167.
78. Zolotukhin, S., Byrne, B. J., Mason, E. et al (1999) Recombinant adeno-associated virus purification using novel methods improves infectious titer and yield. *Gene Therapy* 6(6), 973–985.

Chapter 14

Use of Mesenchymal Stem Cells for Gene Delivery to Intracranial Glioma

Sin-Soo Jeun, Seong Muk Kim, Jung Yeon Lim, and Chung Heon Ryu

Abstract

Virus-mediated gene therapies against brain tumors have been limited by the difficulty in tracking glioma cells infiltrating the brain parenchyma. Mesenchymal stem cells (MSCs) are particularly attractive cells for clinical use in cell-based therapies because they have tumor-targeting properties, can be easily isolated and expanded to the numbers required for use, and can be genetically manipulated with viral vectors. In addition, most of the replication-deficient adenoviral vectors that have been used to transduce MSCs are based on human Ad serotype 5 (Ad5). However, transduction of MSCs by conventional Ad5 vectors is inefficient, even when very high multiplicities of infection are used because MSCs do not express the cellular coxsackie-adenovirus receptor. This chapter describes in detail how such adenoviral transduction of MSCs mediated by cell-permeable peptides should be prepared and handled, and applied for the use of targeted therapeutic gene delivery into glioma by an in vitro migration assay and by in vivo injection of MSCs into the tumor mass or the opposite hemisphere of established human glioma in nude mice.

Key words: Glioma, umbilical cord blood, mesenchymal stem cells, cell-permeable peptides, adenovirus.

1. Introduction

Glioblastomas are the most common primary malignant brain tumors in humans. Therapeutic gene delivery with viral vectors by direct injection into the primary brain tumor or postoperative tumor cavity has failed to reach outgrowing tumor islands because of the tumor cells' infiltration into the normal brain parenchyma (1). New effective therapeutic tools are needed that specifically target tumor cells. Recent studies suggest that stem cells can be used as vehicles for delivering therapeutic genes to treat brain tumors (2, 3). Because the clinical application of neural stem

K.K. Jain (ed.), *Drug Delivery to the Central Nervous System*, Neuromethods 45,
DOI 10.1007/978-1-60761-529-3_14, © Humana Press, a part of Springer Science+Business Media, LLC 2010

cells is limited by ethical problems associated with their isolation and immunologic incompatibility in allogenic transplantation, a new therapeutic strategy has been developed that uses mesenchymal stem cells (MSCs) for the targeted delivery and local production of biologic agents in tumors (4, 5).

Human umbilical cord blood (UCB) is an alternative source of adult stem cells. In a previous study, we isolated multipotent UCB-MSCs that have the capacity to differentiate into several mesodermal tissues (bone, cartilage, tendon, muscle, and adipose), endodermal tissue (hepatocyte), and ectodermal tissue (neurons) (6). Recently, UCB-MSCs were proven to be more advantageous in cell procurement, storage, and transplantation than BM-derived MSCs (BM-MSCs) (7). These characteristics make UCB-MSCs potent candidates for the clinical application of allogenic MSC-based therapies. However, transduction of MSCs by conventional Ad5 vectors is inefficient, even when very high multiplicities of infection (MOIs) are used because MSCs do not express CAR (8). One way to increase cell entry of Ad vectors into MSCs may be through cell-permeable peptides [protein transduction domain (PTD)], small polybasic peptides derived from the transduction domains of certain proteins that cross the cell membrane through a receptor-independent mechanism (9). These cell-permeable peptides have been used to introduce biologically active cargo molecules, such as DNA, peptides or proteins, and viruses into cells. Recently, we developed a novel PTD, which could significantly enhance adenoviral transduction into MSCs (10, 11).

These results suggest that human UCB-MSCs have potential use as effective delivery vehicles for therapeutic genes in the treatment of intracranial glioma.

2. Materials

2.1. MSC Isolation and Characterization

2.1.1. Human UCB or Rat Bone Marrow Harvest

1. A 16-gauge needle from a UCB collection bag containing 23 ml of CPDA-1 anticoagulant is obtained from Greencross Co., Ltd. (Yongin, Kyungki-do, Korea).
2. For rat bone marrow harvest, 150–200 g (4 weeks old) male Sprague-Dawley (SD) rats are used, and fine scissors and forceps (Fine Science Tools, Foster City, CA, USA, Cat. 18000-17) are prepared for the harvest of femurs and tibias from rat.
3. Phosphate-buffered saline (PBS) without Ca^{2+} and Mg^{2+} (CMF-PBS; GIBCO-Invitrogen, Carlsbad, CA, USA, Cat. 20012) and 2% (v/v) heat-inactivated (30 min at 56°C) fetal bovine serum (FBS; GIBCO-Invitrogen, Cat. 16000) is prepared.

4. Sterile dissecting scissors, insulin syringe or 5-ml syringe with a 22-G needle, and 15- and 50-ml polypropylene tubes are prepared.

2.1.2. Isolation and Expansion in Culture of MSC

1. Ficoll–Paque is obtained from Amersham Biosciences (Uppsala, Sweden, Cat. 17-1440-03).
2. Minimum essential medium-α (α-MEM; GIBCO-Invitrogen, Cat. 12571), supplemented with 10% (v/v) heat-inactivated FBS and 1 × pen/strep (final concentrations of 100 U/ml and 100 μg/ml, respectively; 10,000 U/ml penicillin/10,000 μg/ml streptomycin; GIBCO-Invitrogen, Cat. 15140) is prepared.
3. All disposable cell culture plastics are from Corning Incorporated (Corning, NY, USA) and 15 ml or 50 ml polypropylene conical centrifuge tube are prepared from SPL Lifesciences (Gyeonggi-do, Korea, Cat. 50015 or 50050).
4. Dulbecco's Modified Eagle's Medium (DMEM; HyClone Laboratories, Logan, UT, USA, Cat. SH30243.01), supplemented with 20% (v/v) heat-inactivated FBS. The DMEM, formulated with high D-glucose (4.5 g/l), $NaHCO_3$ (3.7 g/l), and HEPES (25 mM), contained 1 × pen/strep (final concentrations of 100 U/ml and 100 μg/ml, respectively; 10,000 U/ml penicillin/10,000 μg/ml streptomycin; GIBCO-Invitrogen).
5. 0.25% (w/v) trypsin-1 mM EDTA solution is prepared from GIBCO-Invitrogen (Cat. 25200).
6. Neither recombinant cytokine nor growth factor is used in the culture medium.

2.1.3. Immunophenotyping of UCB-Derived MSPC

1. The anti-human Ab CD45-FITC, CD34-FITC, CD14-FITC, HLA-DR-FITC, CD13-PE, CD29-PE, CD44-PE, CD31-FITC, CD106-PE, CD90-FITC, CD51/61-FITC, and CD64-FITC from BD Bioscience Pharmingen (San Jose, CA, USA) are prepared, kept at 4°C, light protected, and can be used for no longer than 6 months.
2. Isotypic controls, mouse IgG_1-FITC, IgG_1-PE, IgG_{2a}-FITC, and IgG_{2b}-PE are used (BD Bioscience Pharmingen).
3. Anti-human Ab SH2, SH3 and SH4 are made from each hybridoma cell line (American Type Culture Collection, Rockville, MD, USA).
4. Anti-mouse IgG-FITC (Jackson ImmunoResearch, West Grove, PA, USA, Cat. 115-096-146) is prepared as a secondary Ab.
5. PBS or PBS supplemented with 0.5% bovine serum albumin (BSA; Amresco, Solon, OH, Cat. 0332) is prepared for washing step.

2.1.4. Osteogenic Differentiation

1. Osteogenic induction medium is composed of α-MEM supplemented with 10% FBS, 1 × pen/strep (final concentrations of 100 U/ml and 100 μg/ml, respectively; 10,000 U/ml penicillin/10,000 μg/ml streptomycin; GIBCO-Invitrogen, Cat. 15140), 10 nM dexamethasone from a 1 mM dexamethasone (Sigma, Cat. G9891) stock solution, 10 mM β-glycerolphosphate from a 0.5 M β-glycerolphosphate (Sigma, Cat. G9891) stock solution, and 50 μM ascorbic acid from a 200 mM ascorbic acid (L-ascorbic acid-2-phosphate sesquimagnesium salt; Sigma, Cat. A8960) stock solution. The medium can be used up to 1 month at 4°C.
2. The onset of osteoblasts is evaluated by the expression of alkaline phosphatase (ALP), using an ALP staining kit (Sigma, Cat. 85L-3R). Mineralized matrix is also stained with silver nitrate (Sigma, Cat. S7276) by the method of von Kossa.

2.1.5. Chondrogenic Differentiation

1. Chondrogenic induction medium is composed of serum-free high-glucose DMEM (GIBCO-Invitrogen, Cat. 21063-029) supplemented with 100 nM dexamethasone from a 1 mM dexamethasone (Sigma, Cat. G9891) stock solution, 50 μM ascorbic acid from a 200 mM ascorbic acid (L-ascorbic acid-2-phosphate sesquimagnesium salt; Sigma, Cat. A8960) stock solution, 100 μg/ml sodium pyruvate (Sigma, Cat. P5280), 40 μg/ml L-proline (Sigma, Cat. P1428), 10 ng/ml of transforming growth factor (TGFβ3; Sigma, Cat. T5425), 500 ng/ml bone morphogenic protein 6 (BMP-6, R&D Systems, Minneapolis, MN, USA, Cat. 507-BP-020), and 50 mg/ml ITS^{+} premix (6.25 μg/ml insulin, 6.25 μg/ml transferrin, 6.25 μg/ml selenious acid, 1.25 μg/ml BSA and 5.35 μg/ml linoleic acid, BD Biosciences, Cat. 354352).
2. For histologic and immunohistochemical analyses, OCT compound (Tissue-Tek, Torrance, CA, USA, Cat. 4583) is prepared.
3. 0.1% (w/v) Safranin O (Sigma, Cat. 84120) staining solution is prepared with filtering through filter paper before use.
4. Blocking solution is prepared in PBS with 5.0% BSA.
5. Type II collagen (Oncogene, San Diego, CA, USA, Cat. 234184) is diluted in 0.05 M Tris–HCl with 1.0% BSA.
6. DAKO EnVision^{TM+} System Peroxidase (DAB) kit (DAKO, Carpinteria, CA, USA, Cat. K4006) according to the manufacturer's instructions is used for detection.

2.1.6. Adipogenic Differentiation

1. Adipogenic induction medium is consisted of DMDM (GIBCO-Invitrogen) supplemented with 10% FBS, 1 × pen/strep (final concentrations of 100 U/ml and 100 μg/ml, respectively; 10,000 U/ml penicillin/10,000 μg/ml

streptomycin; GIBCO-Invitrogen, Cat. 15140), 0.5 mM IBMX from a 10 mM 3-isobutyl-1-methylxanthine (IBMX; Sigma, Cat. 17018) stock solution in methanol, 1 μM dexamethasone from a 1 mM dexamethasone (Sigma, Cat. G9891) stock solution, 0.2 mM indomethacin from a 40 mM indomethacin (Sigma, Cat. 17378) stock solution in methanol, and 10 μM h-insulin from a 10 mg/ml insulin (Sigma, Cat. 19278) stock solution.

2. 0.5% (w/v) Oil Red O (Sigma, Cat. O0625) stock solution is prepared in isopropanol and filtered the mixture with filter paper and can be used up to 6 months at 4°C.
3. Working solution is prepared with mixing three parts Oil Red O stock solution and two parts water for 10 min. The solution is filtered using filter paper for use.

2.2. Genetic Manipulation of MSC Mediated by PTD

2.2.1. Peptides

1. All peptides (HP1, HP2, HP3, HP4, HP5, HP6, HP4-1, HP4-2, Tat_{49-57}, Tat_{48-60}, Hph-1, $Antp_{43-58}$) are synthesized by PEPTRON Inc (Daejeon, Korea).

2.2.2. Construction of Replication-Defective Adenoviral Vectors

1. Recombinant replication-defective adenoviruses are generated using the AdEasy Vector System (QBioGene, Irvine, CA, USA) as per supplier's protocol.
2. The cDNAs of enhanced green fluorescent protein (EGFP) are subcloned into the adenoviral shuttle vector, pShuttleCMV. Two hundred and ninety-three cells (ATCC, Cat. CRL-1573) are prepared for generation and expansion of recombinant adenoviral vectors.
3. The tissue culture infectious dose 50 methods are used according to the supplier's protocol (QBioGene).

2.2.3. In Vitro Adenovirus Transduction

1. PTD stock solution (2 mM) is kept at –20°C and working solution is prepared in CMF-PBS with 100 μM concentration and then diluted with reduced serum medium (OPTI-MEM; GIBCO-Invitrogen, Cat.31985) for experiments.

2.3. Migration Studies

2.3.1. In Vitro Migration Assay

1. U-87MG, U-251MG, A172, and NIH3T3 cells are obtained from the American Type Culture Collection (ATCC).
2. Eagle's MEM (GIBCO-Invitrogen Cat. 10370-088) supplemented with 10% FBS are used in culture of glioma cells. Normal human astrocytes as a control cells are obtained from the Applied Cell Biology Research Institute (ACBRI; Kirkland, WA, Cat. ACBRI 371).
3. Transwell plates (Corning Incorporated, Cat. 3422) 6.5 mm in diameter with 8-μm pore filters are prepared.

4. BSA coating solution is prepared in a 10% (v/v) stock solution in CMF-PBS and sterilized over 0.22-μm filter (Millipore Corporation, Bedford, MA, USA, Cat. SLGP033RS). The stock solution can be store at −20°C no longer than 2 months. BSA working solution of 0.1% (v/v) is prepared by dilution in culture media for target cells.
5. Three-Step Stain Set (Diff-Quik; Sysmex, Kobe, Japan, Cat. 38721) is used for staining for migrated cells.

2.3.2. Implantation of Tumor Cells

1. Male athymic nude mice are obtained from Charles River Laboratories (6–8 weeks old).
2. Stereotactic frame (Harvard Apparatus, Cat. 726049) is used and fine scissors and forceps (Fine Science Tools, Cat. 18000-17) are prepared.
3. Sterile PBS, 1-ml syringe needle, insulin syringe with a 29-G needle, and bone wax (Ethicon, Belgium, Cat. W810) are used.
4. 10-μl Hamilton syringe with 26-G needle is obtained from Hamilton Company (Reno, NV, USA).
5. Microinfusion pump (KD Scientific Inc., Holliston, MA, USA) is used for transplantation of cells and surgical staples is prepared.

2.3.3. MSC Implantation and In Vivo Migration

1. PKH26 (Sigma, Cat. 078K1809) is prepared for MSC labeling.
2. Migration toward the tumor is assessed by direct visualization using fluorescence microscope or confocal microscope (LSM 510 Meta; Carl Zeiss).
3. 4% paraformaldehyde (Merck, Darmstadt, Germany, Cat. 1.04005.1000) solution is prepared in PBS and filtered with filter paper.
4. 30% sucrose (Amresco, Cat. 0335) is prepared in PBS.

3. Methods

3.1. MSC Isolation and Characterization

3.1.1. Human UCB or Rat Bone Marrow Harvest

1. The following procedures must be performed in a tissue culture hood and all solutions and equipment must be sterile and proper aseptic techniques should be used accordingly.
2. UCB samples ($n = 411$) are collected from the umbilical vein of deliveries with informed maternal consent.
3. A 16-gauge needle from a UCB collection bag containing 23 ml of CPDA-1 anticoagulant is inserted into the umbilical vein, and UCB is allowed to flow by gravity.

4. In all cases, UCB harvests are processed within 24 h of collection, with viability of more than 90%.
5. Rat bone marrow stromal cells (BMSC) is collected, after sacrifice of 150- to 200-g male SD rats with sterile scissors, slightly cut open both epiphysis areas of each bone from femurs and tibias, flush the marrow a few times with a 5-ml syringe coupled with a 22-G needle containing PBS supplemented with 2% FBS (CMF-PBS/2% FBS), and transfer to a 15-ml polypropyrene tube. Remove all connective tissue.

3.1.2. Isolation and Expansion in Culture of MSC

1. Mononuclear cells are isolated by flushing the shaft with 15 ml buffer (CMF-PBS/2% FBS) using a sterile hypodermic syringe. Cells are diluted with buffer up to 50 ml and centrifuged through a Ficoll–Hypaque gradient (density 1.077 g/cm^3) for 30 min at 1000 $\times g$.
2. The separated mononuclear cells from UCB are washed with PBS, suspended in culture medium, α-MEM supplemented with 10% FBS, and seeded at a concentration of 5 $\times$ 10^6 cells/cm^2.
3. The mononuclear cells from rat bone marrow are collected and suspended with culture medium, DMEM supplemented with 20% FBS. Cells are plated at 7 $\times$ 10^5 cells/cm^2 in coating-culture dish and maintained in culture medium (*see* **Note 1**).
4. Cultures are maintained at 37°C in a humidified atmosphere containing 5% CO_2 with a change of culture medium twice a week. Human UCB- or rat bone marow-derived mononuclear cells are set in culture, and the onset of fibroblast-like adherent cells is observed.
5. One to three weeks later, when the monolayer of MSC colonies reach 80% confluence, cells are trypsinized with 0.25% trypsin, washed, resuspended in culture medium, and subcultured at a concentration of 5 $\times$ 10^4 cells/cm^2. MSCs of each UCB harvest or rat bone marrow are expanded ex vivo by successive subcultivation under the same condition. Neither recombinant cytokine nor growth factor is used in the culture medium.
6. The 5th–8th passage cells of UCB harvests with more than 1000-fold expanding capacity ($n=9$) are examined for surface Ag phenotypes and in vitro differentiation potentials (*see* **Note 2**).

3.1.3. Immunophenotyping of UCB-Derived MSCs

1. For cell-surface Ag phenotyping, cells are detached and stained with FITC-or PE-coupled Ab. Labeled cells are analyzed by FACSCalibur flow cytometer (Becton Dickinson, San Jose, CA, USA) using the CELLQUEST software (Becton Dickinson).

2. To detect surface Ag, aliquots of the UCB-derived MSCs after treatment with 0.25% trypsin are washed three times with PBS supplemented with 0.5% BSA.
3. For direct assays, cells are immunolabeled with the anti-human Ab CD45-FITC, CD34-FITC, CD14-FITC, HLA-DR-FITC, CD13-PE, CD29-PE, CD44-PE, CD31-FITC, CD106-PE, CD90-FITC, CD51/61-FITC, and CD64-FITC. As isotypic controls, mouse IgG_{1-} FITC, IgG_1-PE, IgG_{2a}-FITC, and IgG_{2b}-PE are used (*see* **Note 3**).
4. For indirect assays, cells are immunolabeled with anti-human Ab SH2, SH3, and SH4 Ab made from each hybridoma cell line. As a secondary Ab, anti-mouse IgG-FITC is used.

3.1.4. Osteogenic Differentiation

1. To induce osteogenic differentiation, cells are cultured in osteogenic medium for 3 weeks.
2. The medium is composed of α-MEM supplemented with 10% FBS, 0.1 μm dexamethasone, 10 mM β-glycerol phosphate, and 50 μM L-ascorbic acid 2-phosphate. The medium is changed twice a week.
3. The onset of osteoblasts is evaluated by the expression of alkaline phosphatase (ALP), using an ALP staining kit. Mineralized matrix is also stained with silver nitrate by the method of von Kossa.

3.1.5. Chondrogenic Differentiation

1. To induce chondrogenic differentiation, approximately 2×10^5 cells are placed in a 15-ml polypropylene tube and centrifuged at $500 \times g$ for 5 min at room temperature, to form a pellet micromass. Then, the micromass is treated with chondrogenic medium for 6 weeks. Medium is replaced twice a week.
2. Chondrogenic medium consisted of high-glucose DMEM supplemented with 100 nM dexamethasone, 50 μg/ml L-ascorbic acid, 100 μg/ml sodium pyruvate, 40 μg/ml L-proline, 10 ng/ml TGF-β3, 500 ng/ml BMP-6, and 50 mg/ml ITS^+ premix (6.25 μg/ml insulin, 6.25 μg/ml transferrin, 6.25 ng/ml selenious acid, 1.25 mg/ml BSA, and 5.35 mg/ml linoleic acid).
3. For histologic and immunohistochemical analyses, the pellet aggregates are frozen in OCT compound and 5-μm sections are cut.
4. For histologic evaluation, sections are stained with Safranin-O. For immunohistochemistry, the frozen sections are fixed for 10 min in methanol after a brief immersion in distilled water to remove the OCT compound. Non-specific Ab binding sites are blocked by incubating the slides in PBS with 5.0% BSA for 1 h. Slides are then incubated for 1 h at room temperature with primary Ab, with epitope in type II collagen

diluted in 0.05 M Tris–HCl with 1.0% BSA (TBS). Reactivity is detected by using the DAKO EnVision^{TM+} System Peroxidase (DAB) kit according to the manufacturer's instructions.

3.1.6. Adipogenic Differentiation

1. To induce adipogenic differentiation, cells are treated with adipogenic medium for 4 weeks. Changing of medium is carried out twice a week.
2. Adipogenic medium consisted of DMDM supplemented with 10% FBS, 0.5 mM 3-isobutyl-1-methylxanthine, 1 μM dexamethasone, 0.2 mM indomethacin, and 10 μM h-insulin. The onset of adipocytes is evaluated with Oil-red O stain.

3.2. Genetic Manipulation of MSC Mediated by PTD

3.2.1. Peptides

1. All peptides (HP1, HP2, HP3, HP4, HP5, HP6, HP4-1, HP4-2, $Tat_{49–57}$, $Tat_{48–60}$, Hph-1, $Antp_{43–58}$) are synthesized by PEPTRON Inc (Daejeon, Korea).
2. $GP_{61–80}$ from lymphocytic choriomeningitis virus, Hph-1, $Tat_{49–57}$, $Antp_{43–58}$, and HP4 are labeled with FITC at the carboxyl terminus.
3. Peptides are dissolved in deionized water (stock solution, 2 mM) and sterile filtered before use.

3.2.2. Construction of Replication-Defective Adenoviral Vectors

1. Recombinant replication-defective adenoviruses are generated using the AdEasy Vector System.
2. Briefly, the cDNAs of enhanced green fluorescent protein (EGFP) are subcloned into the adenoviral shuttle vector, pShuttleCMV. After recombination with the adenoviral backbone vector, pAdEasy, in *Escherichia coli* BJ5183, the rAds are generated and expanded in 293 cells. Adenoviral titers are determined using the tissue culture infectious dose 50 (TCID50) methods according to the supplier's protocol.

3.2.3. In Vitro Adenovirus Transduction

1. Cells are seeded into 48-well plates at 3×10^4 cells per well or large scale and incubated overnight at 37°C. rAd/EGFP at various MOIs is diluted in OPTI-MEM to achieve a 2 × virus dilution.
2. PTD is similarly diluted into OPTI-MEM to yield 2 × dilutions (*see* **Note 4** and **5**).
3. Virus and PTD mixtures are combined at a 1:1 ratio and allowed to incubate for 30 min at room temperature. Cell monolayers are washed and overlaid with 150 μl of virus/PTD mixture.
4. After 30 min incubation in a CO_2 incubator at 37°C, the virus/PTD complex is removed and 300 μl of fresh serum-containing medium is added (*see* **Note 6**).
5. Cells are then incubated for an additional 46 h before green fluorescent protein (GFP) expression is analyzed. Gene transduction efficiency is analyzed using flow cytometry (*see* **Fig. 14.1**).

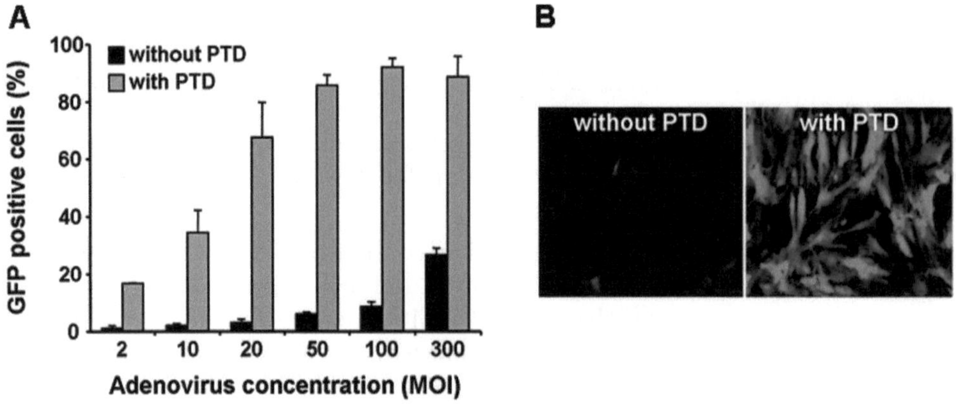

Fig. 14.1. Effect of PTD on adenoviral transduction into UCB-MSCs. A significant dose dependent increase in the number of GFP-positive cells was detected, and enhanced transduction efficiency was evident upon visual inspection when the Ad-EGFP infection was carried out in the presence of 0.01 μM PTD. Panel **A** shows the effect of increasing Ad-EGFP concentrations on proportions of GFP-expressing cells without or with 0.01 μM of PTD was revealed by flow cytometry analysis. Panel **B** shows the photomicrographs of MSC cultures transduced with Ad-EGFP (MOI 20) without or with 0.01 μM PTD 24 h after infection. Magnification, × 200. *Columns*, mean; *bars*, SEM.

3.3. Migration Studies

3.3.1. In Vitro Migration Assay

1. U-87MG cells are maintained in Eagle's MEM and U-251MG, A172, and NIH3T3 cells in DMEM. Normal human astrocytes as a control cells are cultured in DMEM.
2. Enhanced green fluorescent protein (EGFP)-expressing U-87MG cells (U87-EGFP) are derived from stable transfection of U-87MG cells with pEGFP-N1 vector and the stable transfectants are selected with neomycin analogue G418.
3. The migratory ability of MSCs is determined using Transwell plates containing 6.5 mm in diameter with 8 μm pore filters. Three different concentrations (1×10^5, 5×10^5, and 1×10^6) of glioma cells are incubated in serum-free medium (SFM) for 48 h and the resulting conditioned media are used as chemoattractants.
4. MSCs (2×10^4 cells) are suspended in SFM containing 0.1% BSA and seeded into the upper well, and 600 μl of conditioned medium (CM) is placed in the lower well of the Transwell plate.
5. Following incubation for 5 h at 37°C, cells that had not migrated from the upper side of the filters are scraped off with a cotton swab, and filters are stained with the Three-Step Stain Set. The number of cells that had migrated to the lower side of the filter is counted under a light microscope with five high-power fields (× 400).

3.3.2. Implantation of Tumor Cells

1. Male athymic nude mice are used in accordance with institutional guidelines under the approved protocols.

2. Animal on a stereotactic head frame is shaved dorsal surface of the animal's head and applied with two alternating coatings with betadine and 70% isopropyl alcohol for disinfection.
3. Using scissors and forceps remove the skin from the disinfected region and use a dry cotton swab to completely remove the periosteum membrane from the exposed skull surface. The skull of mouse is kept in moist by frequent application of sterile PBS following the removal of the periosteum.
4. For glioma cell implantations, a 1-ml syringe needle is used to drill through the bone at the location of the proposed implantation site.
5. Glioma cells are stereotactically inoculated with 1×10^5 U-87MG or U87-EGFP cells (in 3 μl PBS) into the right frontal lobe (2 mm lateral and 1 mm anterior to bregma, at 2.5 mm depth from the skull base) via a Hamilton syringe using a microinfusion pump.
6. Cells are implanted over a period of 5 min, and after waiting for 5 min the needle is removed slowly.
7. The burrow hole is sealed with bone wax and the wound is closed with surgical staples.

3.3.3. MSC Implantation and In Vivo Migration

1. The same animals implanted with glioma cells are anesthetized with the appropriate dose of ketamine and xylazine (*see* Step 2 in **Section 3.3.2**).
2. MSCs are labeled with PKH26. In brief, cells are suspended in Diluent C at a density of 2×10^6 cells/ml and mixed with an equal volume of the PKH26 dye in Diluent C. Labeling is carried out at room temperature for 5 min. The labeling reaction is terminated by the addition of an equal volume of MEM-α supplement with 10% FBS and then cells are centrifuged for 5 min at 3000 rpm. The pellet is suspended in an appropriate volume of CMF-PBS.
3. Seven days after tumor cell inoculation into the right frontal lobe, EGFP-expressing MSCs (MSCs-EGFP, 2×10^5 cells) are implanted into the opposite hemisphere (2 mm lateral and 1 mm anterior to bregma, at a depth of 2.5 mm from the skull base) or PKH26-labeled MSCs are implanted into the tumor mass (*see* **Note 7**).
4. Migration toward the tumor is assessed at 4, 7, or 10 days after MSCs inoculation by direct visualization using fluorescence microscope or confocal microscope.
5. Mouse brains are perfused with PBS followed by 4% paraformaldehyde under deep anesthesia at a specific time point after MSCs inoculation. The excised brains are postfixed overnight and then equilibrated in PBS containing 30% sucrose for

2 days. Fixed brains are embedded, snap frozen in liquid nitrogen, and stored at –70°C until use. Brains with therapeutic treatment at a specific time point after tumor inoculation are serially sectioned (20 μm, obtained every 200 μm into the tumor) and then stained with hematoxylin and eosin (H&E).

4. Notes

1. Normally, marrow cells collected from one SD rat are enough to seed five 100-mm tissue culture plates. The expected yield for one SD rat is $<5 \times 10^7$ cells. Cells derived from the same rat strain can sometimes vary from one laboratory to another.
2. Bone marrow stromal cells (BMSC) have the ability to self-renew, proliferate, and differentiate into a variety of cell types; however, questions may arise as to whether these traits differ between the cells that have different expansion times. In our previous study, we examined the stem cell potentiality of BMSC through their characterization, proliferative capacity, and the ability to differentiate into multiple lineages in the cultured 2nd passage cells and 10th passage cells. The results were as follows: First, the 10th passage cells had a larger and more flatted morphology than the 2nd passage cells and also exhibited a decreased labeling for BMSC-related antigens such as CD90, CD73. Second, the cell proliferative capacity was approximately two times greater in the 2nd passage cells, and the apoptosis phenomenon was detected in the 10th passage cells. Third, the ability to differentiate into mesodermal tissue (osteocytes, adipocytes) as well as into ectodermal tissue (neurons) was more effective in the 2nd passage cells. Taken together, early stage BMSC would be a valuable cell source for various in vitro applications, as well as cell therapy. Therefore, it must be considered early stage BMSC has a more powerful stem cell potentiality than late stage BMSC (12).
3. MSCs are negative for CD31, CD34, CD45, CD14, CD51/61, and HLA-DR and positive for CD13, CD29, CD44, and CD90 expressions, respectively. Although the 10th passage cells had a larger and more flatted morphology than the 2nd passage cells and also exhibited a little decreased labeling for BMSC-related antigens such as CD90 and CD73 (*see* **Note 2**), the phenotype of the cells from human UCB for these antigens is stable until passages 8–10.
4. Protamines are small and arginine-rich nuclear proteins that compact the genomic DNA structures by binding to them, and have a role in DNA delivery to the nucleus of the egg

during fertilization. Protamines have become one of the promising agents for intracellular delivery of not only exogenous DNA but also rAds. The main sources of protamines are sperms of various mature fish such as herring, salmon, and trout. Salmon and herring protamine showed similar transfection level, but higher than that of trout protamine. It was also reported that salmon protamine has a Tat-like PTD domain, which has the property of intracellular delivery of proteins. Previously, we hypothesized that herring protamine has an arginine-rich PTD domain and identified a novel PTD, HP4, that enhanced rAd delivery into several cancer cell lines and MSCs to a greater degree than herring PS itself, resulting in enhanced antitumor effect of therapeutic rAd in an ex vivo transduced mouse colon cancer model. Our results suggest that HP4 could strengthen the efficacy of rAd-mediated gene therapy by increasing the transduction efficiency of rAd to tumor cells or MSCs (10).

5. The concentration of PTD in the mixture of infectious medium is various to target cells through 0.01~0.1 μM. In our previous study, we found that transduction of UCB-MSCs could be greatly enhanced by incorporation of PTD (0.003–0.03 μM) into the Ad-EGFP infection medium. However, along with increased MOI and PTD concentration, cell death was also augmented, possibly reflecting adenovector cytopathogenic effects and/or cell damage due to excessive transgene expression (11). Taken together, a PTD concentration of 0.01 μM and relatively low doses of adenoviruses achieved high-level production of proteins without affecting cell viability and rendered optimal transduction efficiency of UCB-MSCs (*see* **Fig. 14.1**).
6. Normally, infectious medium of adenoviruses are incubated with target cells for 2 h. We evaluated the transduction efficiency of adenovirus into MSCs for various time (15 min ~ 2 h) of infection with virus and PTD mixtures. The transduction efficiency of EGFP expressing adenovirus into MSCs was similar from 30 min to 2 h of infection. The viability of infectious cells is much better than the cells incubated for long time, which means a short time (30 min) of incubation is enough to transduce MSCs.
7. Depending on the migrating ability and speed of migration, MSC can be placed at any distance from the gliomas in order to assess migration of MSC to gliomas in the brain. In our studies, we have implanted the MSC in the contralateral frontal lobe of the glioma-bearing mice or intratumoral injection in order to evaluate the migration of MSC toward gliomas (11).

Acknowledgments

This study was supported by a grant from the National R&D Program for Cancer Control (0820040) and a grant from the Korea Health 21 R&D Project (0405-DB01-0104-0006), Ministry of Health, Welfare, and Family affairs, Republic of Korea. The authors would like to acknowledge Prof. Young-Chul Sung (Pohang University of Science and Technology, Korea) for experimental contributions in this paper and Dr. Yoon-Sun Yang (MEDIPOST Co., Ltd., Korea) for providing the human UCB-MSC.

References

1. Kramm, C. M., Sena-Esteves, M., Barnett, F. H., Rainov, N. G., Schuback, D. E., Yu, J. S., Pechan, P. A., Paulus, W., Chiocca, E. A., and Breakefield, X. O. (1995) Gene therapy for brain tumors. *Brain Pathol* 5(4), 345–381.
2. Yip, S., Sabetrasekh, R., Sidman, R. L., and Snyder, E. Y. (2006) Neural stem cells as novel cancer therapeutic vehicles. *Eur J Cancer* 42(9), 1298–1308.
3. Hamada, H., Kobune, M., Nakamura, K., Kawano, Y., Kato, K., Honmou, O., Houkin, K., Matsunaga, T., and Niitsu, Y. (2005) Mesenchymal stem cells (MSC) as therapeutic cytoreagents for gene therapy. *Cancer Sci* 96(3), 149–156.
4. Nakamura, K., Ito, Y., Kawano, Y., Kurozumi, K., Kobune, M., Tsuda, H., Bizen, A., Honmou, O., Niitsu, Y., and Hamada, H. (2004) Antitumor effect of genetically engineered mesenchymal stem cells in a rat glioma model. *Gene Ther* 11(14), 1155–1164.
5. Nakamizo, A., Marini, F., Amano, T., Khan, A., Studeny, M., Gumin, J., Chen, J., Hentschel, S., Vecil, G., Dembinski, J., Andreeff, M., and Lang, F. F. (2005) Human bone marrow-derived mesenchymal stem cells in the treatment of gliomas. *Cancer Res* 65(8), 3307–3318.
6. Yang, S. E., Ha, C. W., Jung, M., Jin, H. J., Lee, M., Song, H., Choi, S., Oh, W., and Yang, Y. S. (2004) Mesenchymal stem/progenitor cells developed in cultures from UC blood. *Cytotherapy* 6(5), 476–486.
7. Jeong, J. A., Hong, S. H., Gang, E. J., Ahn, C., Hwang, S. H., Yang, I. H., Han, H., and Kim, H. (2005) Differential gene expression profiling of human umbilical cord blood-derived mesenchymal stem cells by DNA microarray. *Stem Cells* 23(4), 584–593.
8. Conget, P. A. and Minguell, J. J. (2000) Adenoviral-mediated gene transfer into ex vivo expanded human bone marrow mesenchymal progenitor cells. *Exp Hematol* 28(4), 382–390.
9. Gratton, J. P., Yu, J., Griffith, J. W., Babbitt, R. W., Scotland, R. S., Hickey, R., Giordano, F. J., and Sessa, W. C. (2003) Cell-permeable peptides improve cellular uptake and therapeutic gene delivery of replication-deficient viruses in cells and in vivo. *Nat Med* 9(3), 357–362.
10. Youn, J. I., Park, S. H., Jin, H. T., Lee, C. G., Seo, S. H., Song, M. Y., Lee, C. W., and Sung, Y. C. (2008) Enhanced delivery efficiency of recombinant adenovirus into tumor and mesenchymal stem cells by a novel PTD. *Cancer Gene Ther* 15(11), 703–712.
11. Kim, S. M., Lim, J. Y., Park, S. I., Jeong, C. H., Oh, J. H., Jeong, M., Oh, W., Park, S. H., Sung, Y. C., and Jeun, S. S. (2008) Gene therapy using TRAIL-secreting human umbilical cord blood-derived mesenchymal stem cells against intracranial glioma. *Cancer Res* 68(23), 9614–9623.
12. Lim, J. Y., Jeun, S. S., Lee, K. J., Oh, J. H., Kim, S. M., Park, S. I., Jeong, C. H., and Kang, S. G. (2006) Multiple stem cell traits of expanded rat bone marrow stromal cells. *Exp Neurol* 199(2), 416–426.

Chapter 15

Convection-Enhanced Drug Delivery to the Brain

Dale Ding, Charles W. Kanaly, Martin L. Brady, Stephan Mittermeyer, Raghu Raghavan, and John H. Sampson

Abstract

Convection-enhanced delivery (CED) is a method of direct intracerebral parenchymal infusion. It has been previously studied as a mechanism of drug delivery in glioma therapy, which is the focus of this review, and much work has gone into the utilization of this technique. CED can be modeled using several equations that describe the transport of fluid into porous tissue. More practically, variability in CED catheters has been studied and catheters are currently being designed to optimize drug delivery. While CED is a theoretically excellent way to bypass the blood–brain barrier (BBB), limitations of current approaches include excessive backflow along the catheter tract and leakage into subarachnoid and intraventricular spaces. Several drugs, including many recombinant cytotoxins, have been assessed for safety and efficacy in phase I and II clinical trials while many more are still in early preclinical stages. While drug development is important, we must also be able to assess the infusate's volume of distribution to verify adequate coverage of the target area. This can be evaluated through a number of different MRI sequencing techniques, or by co-infusing gadolinium as a tracer in a variety of different formulations, or even via software programs that can predict target distributions. This review summarizes the important advances that have been made to optimize the unique ability of CED to locally deliver high doses of powerful chemotherapeutics to gliomas for maximal tumor killing with minimal neurologic and systemic side effects.

Key words: Convection-enhanced delivery, glioblastoma, brain, drug delivery.

1. Introduction

The development of effective drug therapy for malignant gliomas remains a challenging task despite extensive research efforts. World Health Organization (WHO) grade III and IV malignant gliomas account for over half of all primary CNS gliomas, affecting approximately 20,000 people annually (1). Even after standard treatment, which involves gross total resection followed by concurrent radiation therapy and temozolomide, the prognosis for WHO grade IV glioblastoma multiforme (GBM) is still grim;

K.K. Jain (ed.), *Drug Delivery to the Central Nervous System*, Neuromethods 45,
DOI 10.1007/978-1-60761-529-3_15,

median survival from time of diagnosis is just over 14 months (2). Although temozolomide therapy was an important advance in sensitive patients, in patients who lack a methylated MGMT promoter combination radiotherapy and temozolomide did not show any statistically significant increase in survival when compared to patients who were treated with radiotherapy alone (3). Clearly, there is still a lot of room for improvement in the treatment of GBM.

While several new promising therapies are being developed, many potentially efficacious drugs are either partially or completely restricted from access to the CNS by the presence of the blood–brain barrier (BBB). Furthermore, systemic drug delivery can lead to significant side effects before an effective target drug concentration has been reached. In a landmark paper in 1994, Bobo et al. described convection-enhanced delivery (CED), a method of direct, pressure-driven infusion of molecules into specifically targeted areas of the brain in order to increase the local concentration and distribution of various drugs (4).

The principal advantages of CED over other commonly used methods of drug delivery to the CNS, including systemic delivery, direct delivery into the CSF via either intrathecal or intraventricular routes, and direct implantation at the surgical site (5) are as follows (1). CED can deliver molecules varying widely in weight and size across clinically significant intraparenchymal distances. This is in contrast to drug-impregnated polymer implants (i.e., Gliadel®) and intrathecal, intratumoral, and intraventricular drug injections, which yield small volumes of distribution with exponentially decaying concentrations up to a distance of usually only a few millimeters (2). CED circumvents the BBB, which may limit or prevent penetration of systemically delivered chemotherapies. Additionally, potentially effective systemic therapies with poor BBB penetration frequently generate dose-limiting toxicities before they reach tumoricidal levels in the brain (3). The CED target area can be modified by varying catheter placement and therefore possesses much more versatility in drug targeting than most other treatment modalities (4). CED results in a more homogeneous drug distribution over the area of drug delivery with a rapid decline in drug levels beyond this target area. This distribution pattern limits both neurotoxicity and systemic toxicity while maximizing drug levels to the target area. Alternatively, drug administration via other methods results in more heterogeneous drug distributions, which makes effective dosing to the entire target area difficult and increases the likelihood of adverse effects.

This review summarizes the principles that have been developed to describe CED, variations in CED methods and materials, some of the major experimental drugs that have been administered with this technique, the imaging techniques that have been used to monitor drug delivery, and the computer simulations that have been developed to assist with implementation of CED catheters in patients. It focuses on the use of CED in the treatment of malignant gliomas, but it is important to note that CED is also being studied as a treatment modality for other neurological diseases with varying degrees of success, including Parkinson's disease (6, 7), Gaucher disease (8, 9, 10), epilepsy (11, 12, 13, 14), and stroke (15).

2. Principles of CED

2.1. Diffusion Versus Pressure-Driven Delivery

Single bolus injection of a drug into the brain via a catheter delivers drug into the brain parenchyma via diffusion. This method causes the distribution of the infusate to have limited range, and the concentration of infusate drops off sharply as the distance from the catheter increases. However, application of constant pressure during infusion allows for a distribution of drug that travels a much greater distance from the catheter into the parenchyma with a steep boundary. The stark comparison between diffusion-driven delivery and pressure-driven CED is illustrated in **Fig. 15.1**. As this figure illustrates, the addition of constant pressure during infusion creates a more uniform distribution of infusate around the catheter and across the target area. The steep drop-off at the distribution boundary with pressure-driven infusion allows for decreased drug delivery outside of the target area, which should limit drug related adverse effects by sparing the surrounding normal brain tissue.

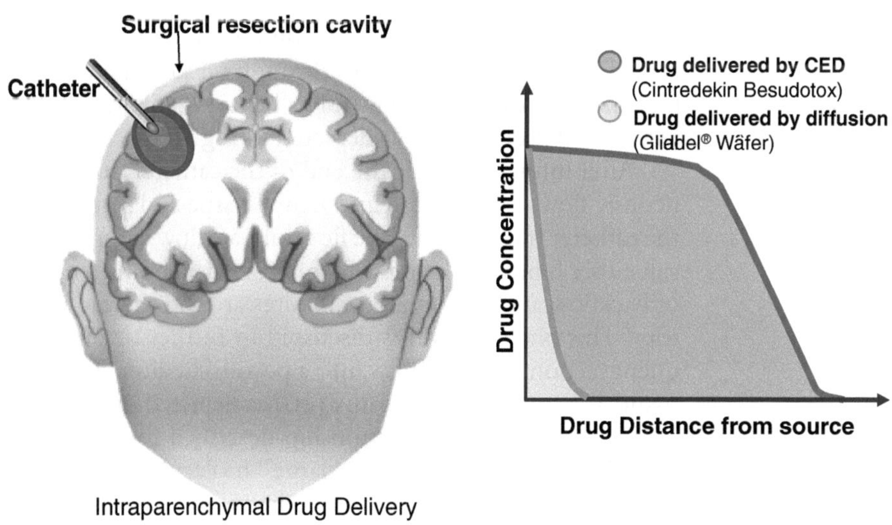

Fig. 15.1. Schematic showing concentration profiles for pressure-driven and diffusion-driven deliveries. Compared with diffusion-driven delivery, the convection-driven delivery results in a higher concentration extending farther from the delivery site.

2.2. Convection-Enhanced Delivery Equations

CED in the brain can be decomposed into three fundamental components – (1) flow of infusate, a viscous fluid, through the catheter, (2) flow of infused drug into the surrounding porous brain tissue, and (3) the reaction of parenchyma to the effective increase in volume induced by infusion.

Navier–Stokes equations mathematically describe the motion of fluids. Under a set of assumptions including steady state and boundary conditions defined by the catheter and infusate, Navier–Stokes equations can be simplified into more easily understood formulas that are called Poiseuille equations. The flow of viscous, incompressible fluid through the catheter is therefore known as Poiseuille flow. Equation [1] describes Poiseuille flow through the catheter into tissue in relation to catheter radius, fluid viscosity, catheter length, and the pressure difference between the two ends of the catheter which is the tissue pressure subtracted from the infusion pressure.

$$\mathrm{Flow}_{\mathrm{catheter}\rightarrow\mathrm{tissue}} = k\frac{(\Delta\mathrm{Pressure})(\mathrm{radius}_{\mathrm{catheter}})^4}{(\mathrm{viscosity}_{\mathrm{fluid}})(\mathrm{length}_{\mathrm{catheter}})} \qquad [1]$$

As shown in equation [1], the volumetric flow rate through the catheter into the surrounding tissue is directly proportional to the infusion pressure and the 4th power of the catheter radius, while it is inversely proportional to the fluid viscosity and catheter length. The term k is a constant. Solute molecules, such as drugs and imaging tracer molecules, are much larger than solvent molecules, like water, and therefore their flow can be thought of as logs in a river. The molecular size difference between "small" and "large" infused drugs is several orders of magnitude less than the difference in molecular size between the drugs and the solvent water molecules. Therefore, the difference between the flow of "small" and "large" molecules through CED catheters is negligible, and they can be thought of as behaving identically and both will obey Poiseuille flow as depicted in **Fig. 15.2**.

After infusate reaches the end of the catheter, it then simultaneously flows anterograde into neural tissue and retrograde along the catheter tract. Retrograde flow, or backflow, along the catheter can either hinder or help CED infusion, depending on the extent of backflow and anatomical structures around the catheter trajectory. This phenomenon is discussed in further detail in a subsequent section. Infusate flow into a porous medium, such as neural tissue, is known as Darcy's flux (16) as depicted in **Fig. 15.3**.

Similar to the size relationship described for Poiseuille flow, the relative size of "small" to "large" molecules pales in comparison to the relative size difference between the drugs and neural tissue. Therefore, if the drugs are considered to be logs traveling down a river in our analogy, then the porous tissue would be islands around which the infusate flows. Darcy's flux can be

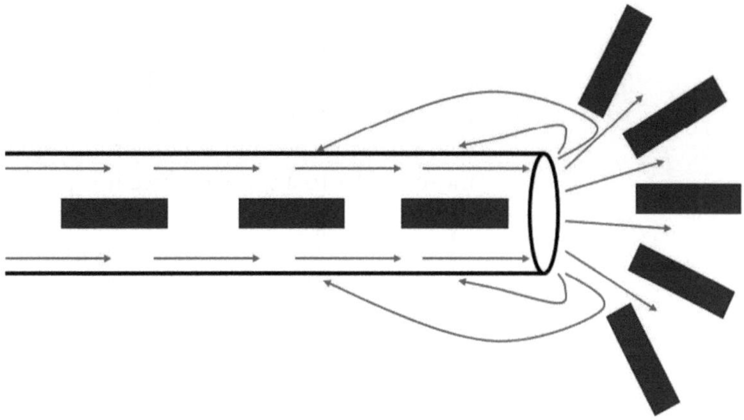

Fig. 15.2. A simplified depiction of infusate flow through a catheter. The rectangular "logs" represent the infused macromolecule and the *arrows* represent the infusate "river," or flow of solvent fluid in the catheter. After the infusate leaves the catheter tip, it flows into the surrounding neural tissue as well as retrograde along the catheter tract, a phenomenon known as backflow which is elaborated on later in more detail. Because this flow is described by Poiseuille equations, it is known as Poiseuille flow.

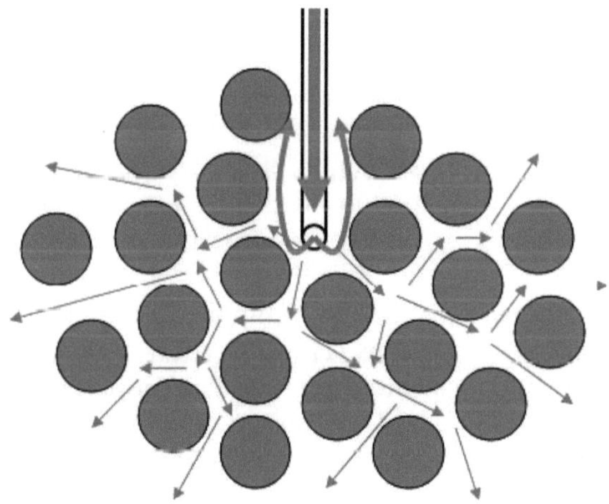

Fig. 15.3. A simplified depiction of Darcy's flux, showing anterograde infusate flow (*arrows*) out of the catheter tip and into a porous medium (*gray islands*) and simultaneous retrograde flow along the catheter tract. It should be noted that in addition to being porous, brain parenchyma is also elastic, and therefore is more completely described by Biot's theory of fluid flow into a poroelastic medium. The brain is more similar to a sponge as described by Biot's theory than to a rigid, porous material like soil, which would be best described by Darcy's flux.

mathematically described in equation [2], such that infusate flow through neural tissue varies according to the difference between infusion and tissue pressure, and inversely with tissue resistance.

$$\text{Flow}_{\text{tissue}} = \frac{\Delta\text{Pressure}}{\text{Resistance}_{\text{tissue}}} \qquad [2]$$

This simple relationship tells us that flow of the infusate through porous brain tissue is directly proportional to the pressure drop through the tissue and inversely proportional to the tissue resistance. Darcy's flux is akin to water flowing through sand or soil, since the medium is porous but rigid. Brain tissue is in reality more similar to a sponge, since it is porous but also very elastic. In order to properly describe the deformation of elastic tissue in response to a pressure resulting from fluid infusion, we analyze the linear relationship of stress to strain, as seen in equation [3]. Stress, or the difference between infusion pressure and tissue pressure, is related to strain, or the amount of linear tissue deformation, by a term that describes the intrinsic stiffness of the tissue in response to stress in different directions. In the case of fluid infusion, the deformation is perpendicular to the tissue surface and so the term relating the stress and strain is known as Young's modulus, or E.

$$\Delta \text{Pressure} = E(\text{Deformation}_{\text{tissue}}) \qquad [3]$$

Therefore, equations that describe the reaction of poroelastic media to fluid infusion are better models for CED infusion into brain parenchyma. The combination of equations that describe the elasticity of brain tissue, Darcy's flux equations for flow through rigid, porous media, and Navier–Stokes equations for fluid dynamics of a viscous infusate, results in Biot's theory for fluid saturation of poroelastic materials. The infused fluid expands and perfuses the tissue according to Biot's theory, which describes the reaction of poroelastic media, such as neural tissue, to fluid infusion.

2.3. Factors Affecting Drug Distribution

Effective drug distribution is critical to the success of CED infusions. The different factors that affect distribution can be divided into three broad categories, including influx, transport, and efflux (17). Influx is determined by flow rate of the infusate and duration of infusion. Transport is determined by convection, diffusion, conduction pathways (*i.e.*, white matter tracts), and cortical surfaces and sulci. Efflux is determined by binding properties and kinetics of the infused drug and capillary permeability. Saito et al. showed that alteration of the physical and chemical properties of an infused agent has a substantial effect on its distribution (18).

2.4. Backflow

Retrograde flow along the CED catheter track, or backflow, can be a major concern when determining the ability of CED to properly deliver drugs to the brain. Backflow forms a cylindrical fluid annulus around the CED catheter that can extend anywhere from a short distance back from the catheter tip to all the way along the catheter back to the brain surface. This phenomenon can actually increase the effective infusion area of the catheter because the backflow fluid annulus can also infuse drug into surrounding tissue as shown in **Fig. 15.4**. While a small amount of backflow is

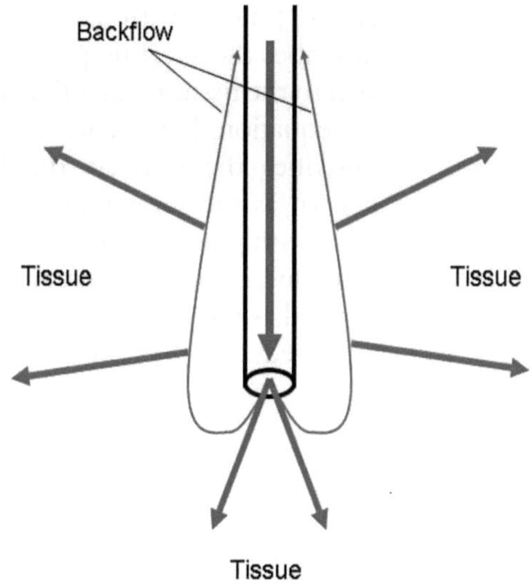

Fig. 15.4. Backflow along a CED catheter with drug infusion. At the catheter tip, anterograde flow into tissue occurs simultaneously with retrograde flow along the catheter. The retrograde flow, or backflow, creates a fluid annulus around the catheter, which itself can be a source of infusate for drug delivery to surrounding tissue. A certain degree of backflow is therefore beneficial by increasing the effective infusing surface of the catheter, which is now a cylinder formed by the backflow annulus rather than a single point at the catheter tip.

important and unavoidable, backflow that reaches a sulcus or the brain surface will result in fluid leakage and ineffective drug distribution in the tissue. As the fluid will follow the path of least resistance into the CSF compartment, no pressure gradient will be build up to drive the convective flow though the tissue.

Backflow can be separated into two general categories, depending on whether its presence is due to intrinsic or extrinsic causes. Extrinsic backflow arises from the artificially created void that can be created during catheter placement. If the catheter mechanically disrupts tissue integrity, a potential space can be created between the catheter and the surrounding brain tissue. In this situation, infusate will simply flow through this space along the catheter tract and tissue infusion will be severely compromised. Extrinsic backflow can be decreased by using a soft, flexible catheter rather than a rigid one (19). The flexible catheter will shift as the brain shifts, allowing for more predictable infusate distribution.

Intrinsic backflow occurs even when catheter insertion is technically perfect. This backflow arises because the difference between infusion-induced tissue hydrostatic pressure and intrinsic tissue shear forces is not zero, and the remaining difference manifests as intrinsic backflow (17). A theoretical model of intrinsic backflow was first described by Morrison et al., and his paper provides a technically detailed explanation of the mathematical modeling of backflow that will not be fully reiterated in this review (20). Backflow can be described at the most basic levels by equations [1], [2], and [3] along with an additional stress-strain relationship that describes the shearing pressure caused by backflow, which is parallel to the tissue (rather than the perpendicular pressure described by equation [3]). The term, akin to Young's modulus, that describes the intrinsic rigidity of tissue to shearing forces is known as the shear modulus, or *G*. Equation [4] describes the relationship between this additional shearing pressure to the shear modulus, catheter radius, and the thickness of the annulus which is defined as the perpendicular distance from the outer catheter wall to the edge of the cylindrical annulus (20).

$$\text{Pressure}_{\text{shear}} = 2G\left(\frac{\text{thickness}_{\text{annulus}}}{\text{radius}_{\text{catheter}}}\right) \qquad [4]$$

The shearing pressure is directly related to the shear modulus and annulus thickness while it is inversely related to the catheter radius. Chen et al. studied the effects of altering several variables on backflow. They concluded that increased rate of infusion and increased catheter diameter increase backflow while changes in infusate concentration and different pre-infusion sealing times had no effect on the amount of backflow (21).

2.5. Hardware Design

Catheters specifically designed for direct drug delivery into neural tissue are crucial to achieving reliable distribution patterns. Although the ideal drug delivery with CED is depicted in **Fig. 15.1** above, many factors can prevent this optimal distribution from occurring. In reality, CED does not occur from an ideal point source for the infusate to flow into solid tissue, since an end port catheter initially will show some backflow along the trajectory. Since the starting point for the convection is actually a cylinder around the catheter with the specific length of the backflow (22) the area of infusion can be larger than that obtained from an ideal point source. Rather than a disadvantage, it is therefore possible that backflow can be an advantage for covering the clinical target, depending on its geometry. However, such backflow length needs to be calculated and controlled for proper planning of catheter trajectories. This intrinsic backflow is mainly affected by the infusion flow rate and the outer diameter of the catheter (20). In order to increase the freedom to place catheters, a minimized outer diameter of catheters is mandatory, since decreasing this diameter will decrease the area of backflow. Reducing the backflow length increases the number of potential trajectories where the backflow area is not crossing any surface or sulcus, mandatory to avoid leakage. Skull-fixed catheters, which use burr hole caps to secure the catheter to the skull instead of the traditional loop stitch, have been used to decrease the likelihood of catheter dislocation, but they do not compensate for movement of the brain during infusion relative to the skull. Rigid catheters are likely to cause voids along the trajectory resulting in significant backflow and leakage of the drug into the subarachnoid space (19). To increase the volume of distribution, multiport catheters (with ports along the catheter axis) initially were considered to be the best design for dedicated CED catheters. However, in vitro studies showed that such a design leads to unpredictable distribution, since different ports may have differing amounts of tissue resistance and therefore only one of the ports may deliver the fluid as the fluid will follow the path of least resistance (17). In those in vitro studies, a disturbing effect on drug distribution was observed due to air bubbles within the infusion line; however, the degree of this effect remains to be investigated in an in vivo setting. To also reduce or eliminate backflow also for high infusion flow rates, a step design catheter was proposed to overcome limitations associated with CED based on in vitro and animal studies (23).

Design requirements in general depend on the application of the catheter. To further optimize CED, tailored catheters will have to be designed for acute infusion protocols (infusions over several days) versus chronic protocols (infusions over months or years) as well as for intratumoral versus peritumoral protocols in the oncology field. Balloon-tipped catheters are currently under investigation to be placed into resection cavities in order to treat peritumoral tissue or residual disease (24). In addition to these infusion protocol-dependent parameters, the delivered fluid has to be taken in account. Chemical or physical (*e.g.*, friction) reactions between catheter material and therapeutic agent have to be considered for proper catheter design. To successfully apply CED clinically, routine catheters designs will also have to comply with other common surgical techniques, such as stereotactic biopsies, so that potential human error due to unfamiliar procedures is minimized.

3. Drugs

A large number of therapeutics have been successfully administered with CED, but only a handful of these have been studied in phase I/II clinical trials. Instead of summarizing all of the CED studies that have been performed in animal tumor models, this review focuses on the anti-glioma therapeutics that have been tested in human clinical studies. Although not yet studied in humans, promising animal studies utilizing CED have tested the use of boron neutron capture therapy (25) (BCNT) with boronated epidermal growth factor (26), adeno-associated virus thymidine kinase (27) (AAV-tk), liposomal topotecan (28), liposomal irinotecan (29), tumor necrosis factor-related apoptosis-inducing ligand (TRAIL) with subsequent temozolomide (30), carboplatin (31), and gemcitabine (31).

3.1. Paclitaxel

The chemotherapy compound paclitaxel was tested by Lidar et al., who administered 20 cycles of CED paclitaxel over a period of 2–5 days to 15 patients with recurrent grade III or IV primary gliomas in a phase I/II clinical trial. Paclitaxel was infused through one or two catheters for 5 days at a rate of 0.3 mL/h up to a total of 7.2 mg of paclitaxel (diluted with 6 mL normal saline) daily in the first three patients and a total of 3.6 mg (diluted with 6.6 mL normal saline) of paclitaxel daily in the remaining patients. Patients were scanned using diffusion-weighted, T1-weighted, and T2-weighted MRI with tumor response categories defined as resolution of enhancing mass (complete response, CR), reduction by more than 50% of enhancing mass (partial response, PR), or no response. Five patients (33%) had complete responses and six patients (40%) had partial responses, yielding a 73% overall response rate. The overall median survival was 7.5 months. A significant number of treatment-related adverse events were documented, including six cases of transient chemical meningitis (30%), three cases of bacterial infections (15%), and four cases of transient neurological deterioration (20%), which were thought to be secondary to infusion-induced cerebral edema (32).

3.2. AP 12009

TGF-β has been strongly implicated in the biology of malignant gliomas and their progression (33, 34, 35). Schlingensiepen et al. developed AP 12009, an antisense oligonucleotide that binds to TGF-β2 mRNA and inhibits its translation (36). The results of three phase I/II dose-escalation clinical trials showed stable disease after 1 month in 7 out of 24 patients with treatment-refractory or recurrent grade III or IV malignant gliomas. AP 12009 was infused through a single catheter for 4 days (except in one patient in whom it was infused for 7 days), at concentrations ranging from 2.5 to 80 μM at a rate of 5.76 or 11.52 mL/day. Patients were given 1–12 treatment cycles of AP 12009, receiving 0.354–39.62 mg/cycle. Clinical evaluation was based on T1-weighted MRI. Two patients with anaplastic astrocytomas (AA) had complete responses, one after only one treatment cycle and another after 12 treatment cycles. The latter patient was still in complete remission at the time of publication, more than 4.5 years following AP 12009 therapy. Of the 29 treatment-related adverse events, there were no grade 4 toxicities and only two grade 3 toxicities as defined by the NCI common toxicity criteria. The median overall survival following treatment was 146.6 weeks for AA patients and 44.0 weeks for GBM patients (37).

3.3. (^{131}I-chTNT-1/B MAb) Cotara

The necrotic core of malignant solid tumors contains histone H1 complexed to DNA, an antigen normally expressed intracellularly. Cotara is an immunotoxin consisting of ^{131}I-radiolabeled murine antibody specific for the histone H1-DNA complex (^{131}I-chTNT-1/B MAb). The antibody portion of the immunotoxin localizes the radioactive ^{131}I to tumor necrosis where it delivers a cytotoxic level of radiation to surrounding tumor cells. Of the 51 patients who enrolled in the phase I and II studies, six had recurrent anaplastic astrocytomas (12%), 37 had recurrent glioblastomas (73%), and eight had newly diagnosed glioblastomas (15%). Cotara concentrated at 1.0 mg/mL was infused through one or two catheters at 0.18 mL/h over 1 or 2 days. Clinical evaluation of patients was done using proton density, T1-weighted, and T2-weighted MRIs along with single photon emission computed tomography (SPECT) imaging. Adverse event data was collected on all study patients. Based on the NCI common toxicity scale, 18 patients (36%) experienced at least a grade 3 treatment-emergent neurologic or cognitive toxicity while only four patients (8%) experienced at least a grade 3 treatment-emergent systemic toxicity. Therapeutic efficacy was reported on a subset of 12 recurrent GBM patients who received doses within a therapeutic window defined by the investigators. All patients who received doses under the therapeutic window had progressed within 2 months of treatment and all patients over the therapeutic window experienced dose-limiting toxicity. Of the 12 patients evaluated for treatment efficacy, one had a partial response and six and stable disease 2 months after Cotara infusion. The median survival was 37.9 weeks (38).

3.4. Tf-CRM107 (TransMID)

A phase I clinical trial of infusion via CED of a recombinant targeted toxin, Tf-CRM107, was conducted in 1997 on 18 patients who demonstrated biopsy-proven recurrence of grade III and IV malignant brain tumors. Tf-CRM107 is a protein conjugate consisting of human transferrin (Tf) covalently linked to genetically modified diphtheria toxin (CRM107). Tf-CRM107 was infused through one to three catheters at a maximum rate of 4–10 μL/min in concentrations ranging from 0.1 to 1.0 μg/mL over 2–16 days. Patients were following clinically with T1-weighted MRIs. Nine of the 15 evaluable study patients (60%) showed a greater than 50% reduction in tumor volume as seen on MRI with maximal response occurring at variable time intervals following the first treatment, ranging from 1 to 14 months. Two patients had complete responses, comprising 22% of the responders and 13% of the overall study patients. The post-treatment, median survival was 74 weeks in the responders and 36 weeks in the non-responders. Overall, patients in the higher dose treatment groups were more likely to experience local toxicity due to peritumoral edema but were also more likely to respond to treatment (39).

A subsequent phase II study was completed in 2003 that evaluated the response to CED of Tf-CRM107 in 44 patients with recurrent or treatment-refractory grade III and IV gliomas across nine study sites. Tf-CRM107 was delivered through one or two catheters at a rate of 0.20–0.40 mL/h at a concentration of 0.67 μg/mL for 4–5 days until a volume of 40 mL was infused. As in the Phase I trial, patients were followed clinically with T1-weighted MRIs. Of the 34 evaluable patients, 12 (34%) had at least a partial response to therapy, defined as a greater than 50% reduction in tumor volume as assessed on MRI. Five study patients demonstrated a complete response, comprising 42% of the responders and 15% of the study group. The median survival of the group was 37 weeks with 13 patients surviving past the 1-year mark. All 44 enrolled patients received at least one CED infusion of Tf-CRM107 toxin and were therefore all evaluated for safety. Eight of the patients (18%) experienced cerebral edema responsive to steroids or mannitol. In the one case that resulted in a patient's death, the patient's family elected to withdraw care. The investigator in that case believed the cerebral edema could have been controlled with aggressive medical management (40). No systemic toxicity was seen in either the phase I or the phase II study.

3.5. NBI-3001 (IL-4-Pseudomonas Exotoxin)

In 2000, Rand et al. showed selective tumor toxicity of NBI-3001, a recombinant cytotoxin consisting of a fusion between IL-4 and a truncated form of Pseudomonas exotoxin (PE38). NBI-3001 was given via CED to nine patients with histologically confirmed recurrent grade IV gliomas (GBM) who previously failed external beam radiotherapy. NBI-3001 was infused through one to three catheters at a rate of 0.3–1.8 mL/h at concentrations ranging from 0.2 to 6.0 μg/mL over a period of 4–8 days. Seven of the nine patients underwent subsequent craniotomy for histologic tissue evaluation of post-infusion tumor. Tissue in six patients showed tumor necrosis and edema with preservation of normal brain, as shown in **Fig. 15.5** (41).

A subsequent study published in 2003 by Weber et al. examined the dose-related toxicity of NBI-3001. NBI-3001 was infused through one to three catheters at varying infusion rates in order to deliver 40–100 mL of drug at concentrations ranging from 6 to 15 μg/mL. Dose-escalation studies showed grade 3 or 4 drug-related toxicity in 39% of patients in all dosage groups based on NCI common toxicity criteria. The lowest dose group, 6 μg/mL × 40 mL, showed grade 3 or 4 toxicity in 22% of patients and was also determined to be the maximum tolerated dose. The most common adverse effects were seizures, headache, weakness, and cerebral edema. There were no noted systemic toxicities. The overall median survival was 8.2 months for all patients and 5.8 months for GBM patients with a 6-month survival rate of 52% and 48%, respectively (42). This is comparable to resection only for GBM progression which has an overall median survival of 5.3 months with a 6-month survival of 44% (43). One case report described a 27-year-old GBM patient who survived 3 years after recurrence after intratumoral infusion of NBI-3001 via CED (44).

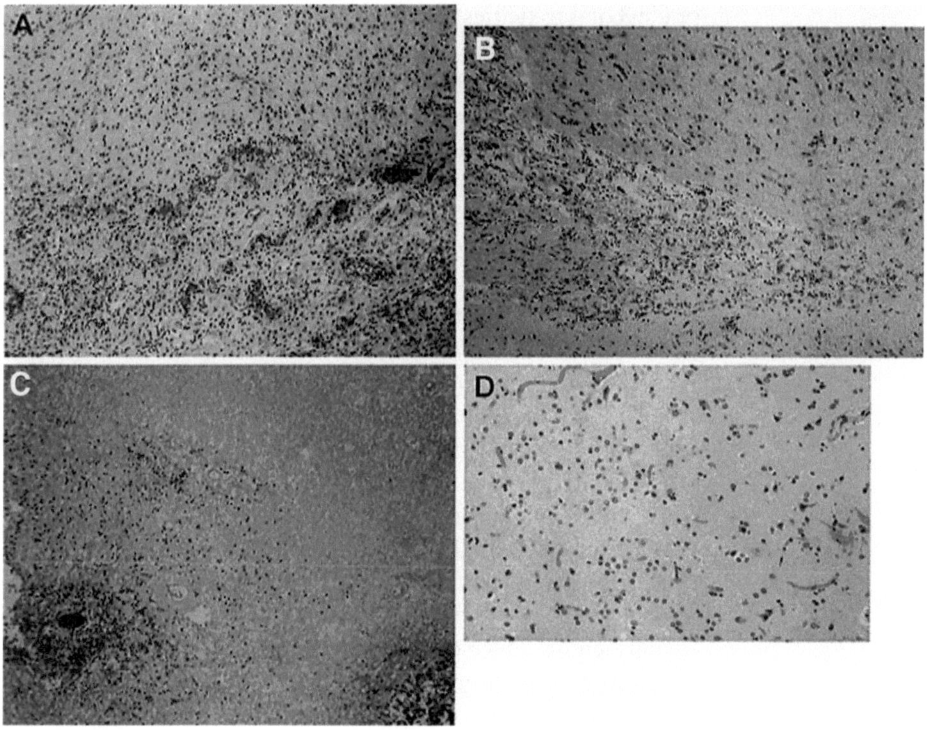

Fig. 15.5. Histologic evidence of tumor necrosis without damage to adjacent normal brain parenchyma after CED of the recombinant cytotoxin NBI-3001 for GBM. (**A**) H&E stain of tissue sections from patient 2, who received 0.2 mg/mL IL-4-toxin dose, which shows tumor necrosis but normal brain tissue shows no histological toxicity. (**B**) Tissue section from patient 3, who received 2 mg/mL of IL-4-toxin, shows necrotic tumor surrounded by normal brain tissue. Photomicrographs from patient 6, who received 6 mg/mL of the study drug, show tumor necrosis (**C**) and normal cortex with no histological damage (**D**). From Rand et al. (42), with permission.

3.6. Cintredekin Besudotox

The receptor for interleukin-13 (IL-13R) has been shown to be selectively overexpressed in malignant gliomas, making it a possible target for targeted therapy (45, 46). Cintredekin besudotox (CB) is a novel chimeric cytotoxin consisting of IL-13 conjugated to a truncated *Pseudomonas* exotoxin PE38QQR that was designed for direct intratumoral infusion via CED. A series of three phase I trials were conducted involving a total of 51 patients with progressive or recurrent grade III and IV primary gliomas. CB was infused through one or more catheters at a rate of 0.750 mL/h using concentrations of 0.25–1.0 μg/mL over a period of 4–6 days. Patients were scanned using T1-weighted MRI and SPECT imaging following CED infusion of ^{123}I-labeled human serum albumin (^{123}I-HSA). Overall median survival for all 51 study patients post-treatment was 45.9 weeks. Nine patients were progression-free for 1 year and seven patients progression-free for 2 years. The only reported NCI common toxicity scale grade 3 or 4 toxicities were headache, convulsion, and hemiparesis. Only hemiparesis was reported in more than one patient, affecting six patients (12%) (47).

An additional phase I trial looked at the efficacy of CB as part of the initial treatment regimen, along with standard external beam radiation therapy (EBRT) and temozolomide. The study group consisted of 22 patients with newly diagnosed high grade gliomas, 21 had glioblastomas and one had an anaplastic oligoastrocytoma. The patients were divided into four study groups – a lower CB concentration along with EBRT, with and without temozolomide, and a higher CB concentration along with EBRT, with and without temozolomide. CB was delivered through one or more catheters at a rate of 0.750 mL/h at concentrations ranging from 0.25 to 0.5 μg/mL for 96 h. Patients were scanned using T1-weighted and T2-weighted MRI along with SPECT imaging following CED ^{123}I-HSA infusion. At the time of publication, 11 of the initial 22 patients (50%) were still alive after a median follow-up of 44 weeks. Six patients (27%) remain progression-free after a median follow-up of 40 weeks. Twenty-eight adverse events were CB-related, eight (28%) of which were at least grade 3, and 34 adverse events were catheter-related, eight (24%) of which were at least grade 3 (48).

A phase III randomized clinical trial comparing treatment with intracerebrally infused CB via CED against Gliadel in patients with recurrent GBM following standard therapy, including surgical resection, radiation, and temozolomide, has been completed. The median survival after recurrence was 36.4 weeks for CB and 35.3 weeks for Gliadel. While the median survival for CB was comparable to Gliadel, there was no statistically significant advantage of CB over Gliadel. The lack of statistically significant survival advantage in the CB arm was attributed in part to inadequate drug distribution and improper catheter placement (49).

3.7. TP-38

Sampson et al. used CED for intratumoral infusion of TP-38, a chimeric protein resulting from the fusion of TGF-α and genetically modified Pseudomonas exotoxin (PE-38). The TGF-α domain of this tumor-specific toxin binds EGFR, which has been shown to be frequently upregulated in malignant gliomas (50). Of the 20 patients enrolled in the trial, one had a spindle cell metastasis and the other 19 had grade III or IV primary brain tumors. Additionally, the patients were stratified for dose escalation so that a maximally tolerated dose (MTD) and dose-limiting toxicities (DLT) could be determined. TP-38 was delivered through two catheters using concentrations ranging from 25 to 100 ng/mL at a rate of 0.8 mL/h over 50 h for a final infusion volume of 40 mL. Patients were scanned using T1-weighted and T2-weighted MRIs along with SPECT imaging following CED infusion of ^{123}I conjugated to human serum albumin ^{123}I-HSA. Two of the 15 patients with residual disease prior to the initiation of therapy (13%) showed radiographic response. Overall median survival following TP-38 infusion was 28 weeks for all patients, 20.1 weeks for patients with residual disease, and 33.0 weeks for patients with no residual disease. One patient had a complete response to treatment and was progression-free for 198 weeks (51). At the time of publication, that patient was still alive and had survived 260 weeks from the start of TP-38 therapy. The patient's pre- and post-treatment MRI scans are shown in **Fig. 15.6**. All treatment-related toxicities were neurologic and no signs of systemic toxicity were seen. Five study patients, each of whom had a pre-existing seizure disorder, experienced a seizure during drug infusion. DLT, defined as a grade 3 or 4 toxicity by the NCI common toxicity scale, were seen in two patients (52).

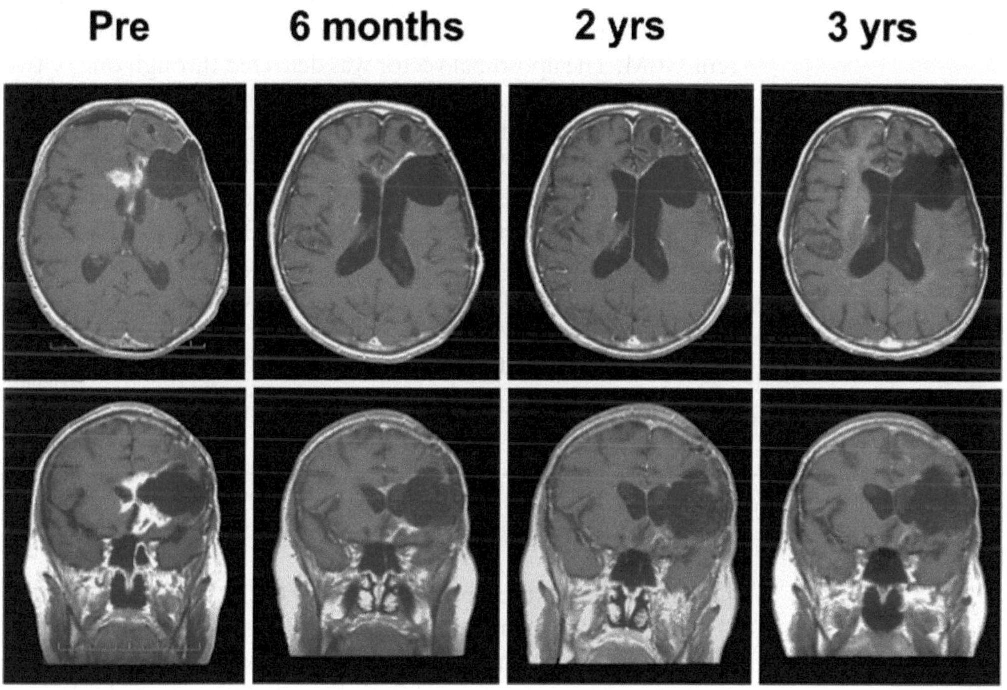

Fig. 15.6. Contrast-enhanced axial (*top panels*) and coronal (*bottom panels*) MR images showing extent of tumor prior to study treatment and at 6 months, 2 years, and 3 years after TP-38 infusion via CED. From Sampson et al. (51), with permission.

3.8. Liposome-Encapsulated Drug Delivery

Liposomal delivery of chemotherapeutics has shown great potential as a treatment modality for various cancers (53–56), and has been shown to increase a drug's sustained release, half-life, and therapeutic index (57). Similarly, liposomally mediated gene transfer has been shown to be easier to prepare and apply, less immunogenic, safer, and more stable at room temperature than viral vectors (58, 59). Liposomes have also been loaded with gadolinium to monitor CED drug distribution, as discussed below in the imaging section. Although liposomes containing antibody fragments, or immunoliposomes, have shown promise in the molecular targeting of cancer (60, 61), they have yet to be rigorously studied in CNS tumors.

Intratumoral injection of murine retroviral-vector-producing (M11) cells containing the herpes simplex type 1 thymidine kinase (HSV-*tk*) gene into patients who then undergo subsequent systemic administration of ganciclovir was first described by Ram et al. (62). Despite promising results from phase I/II clinical trials (63, 64), a follow-up phase III study showed no efficacy as an adjuvant to standard therapy (surgery and radiation) in 248 patients with newly diagnosed, previously untreated GBM, presumably due to limited delivery of the HSV-*tk* gene to tumor cells (65). More recently, Voges et al. studied the utility of CED of a liposomal formulation of the HSV-*tk* gene vector with subsequent intravenous (IV) administration of ganciclovir in eight patients with recurrent GBM. The liposomal vector was delivered through one or two catheters at rates of 0.025–0.6 mL/h until the total volume of 3.5–7.0 mL was delivered. Four days following completion of intracerebral infusion, 5 mg/kg ganciclovir was given IV for 1 h twice daily for 14 days. The patients were followed clinically with T1-weighted MRI and positron emission tomography (PET) scans. MRI showed focal tumor reduction in six patients (75%) and a greater than 50% decrease in overall tumor volume in two patients (25%). The overall median survival time for the eight patients was 28 weeks with 8 weeks as the median time to progression (66).

4. Imaging

Despite the many theoretical benefits of CED, it is critical to verify that the implementation of CED achieves its intended goal of actually infusing drugs at a therapeutic concentration over the appropriate target region. A number of different methods have been devised to monitor the region of infusion and assure that successful drug delivery is occurring. Non-invasive imaging is a critical component of these methods, and while some attempt to use imaging alone, other proposed methods have co-infused an imaging agent along with drug during CED. When an imaging agent is co-infused, this agent can then be monitored as a surrogate to estimate the region where drug is being delivered.

4.1. Diffusion MRI

Diffusion-weighted MRI (DWMRI) has been used for early detection of cancer response to chemotherapeutics (67–71) due to its ability to non-invasively differentiate between intracellular and extracellular water diffusion. In vitro studies have shown that intracellular water diffuses an order of magnitude slower than extracellular water (72, 73). This difference allows DWMRI to detect early effects of drugs on tumor structure and function at the cellular and tissue levels, such as changes in membrane permeability and cell lysis. Mardor et al. utilized DWMRI to monitor the response of CED paclitaxel in three patients and showed that drug infusion consistently resulted in a bright area, indicating increased water diffusion, 24–48 h after the start of infusion (74). A variable length of time after the appearance of the bright area, a dark region appeared within the bright area, presumptively denoting response to therapy which was detected at least 1 day before response detection by conventional T1-weighted and T2-weighted MRIs.

4.2. T2-Weighted MRI

Sampson et al. has demonstrated the ability to use T2-weighted MRIs co-registered with SPECT images to assess the volume of distribution of an infused cytotoxin, cintredekin besudotox for independent evaluation by three reviewers. The reviewers were able to assess, in complete agreement, which catheters provided intraparenchymal drug distribution and which catheters leaked infusions into subarachnoid or ventricular spaces. Every case in which the catheter was successfully placed resulted in a change in T2 signal. In the cases where there was no previous T2 signal abnormality, emergence of a new T2 signal showed the presence and geometry of intraparenchymal drug distribution. In the cases where previous T2 signal abnormality was present, change in signal shape and hyperintensity was observed. Sampson et al. concluded that T2-weighted MRI 48 h post-infusion was useful in determining the success of catheter placement and hence predicting the location and distribution of drug infusion (75).

4.3. Dynamic Contrast-Enhanced Imaging

Dynamic contrast-enhanced MRI has been described as an imaging modality to detect changes in regional cerebral perfusion since the early 1990s (76, 77). More recently, it has been used to track the properties of tracer molecules infused directly into the brain. Dynamic contrast-enhanced imaging can follow the course of these tracers from blood into neuronal tissue and back into blood. Based on molecular transport models, the permeability of normal brain and glioma tissue to a specific tracer molecule can be approximated. Then by performing multiple iterations of tracers with different molecular weights, models can be developed that predict permeability of high molecular weight molecules using lower molecular weight tracers (17). This may allow for the prediction of the geometry and distribution of a larger molecule (i.e., a cytotoxin) using the knowledge of the infusion properties of a small molecule that can be seen on imaging (i.e., gadolinium).

Direct labeling of drugs with gadolinium, while theoretically sound, is technically limited by the ability of imaging modalities to detect enhancement at non-toxic doses of drug. Usually the drug is delivered at too low a concentration for a direct labeling approach to be detected by current imaging techniques. Therefore, surrogates have been co-infused in an attempt to monitor those and assume that their distribution correlates to the actual distribution of drug delivery. If the co-infused surrogate tracer has similar characteristics to the delivered drug, then this is a viable technique. However, if the surrogate differs in size, charge, binding characteristics, rate of degradation, or numerous other characteristics, monitoring the area of tracer distribution may not be representative of the area of drug delivery.

4.4. Free Gd-DTPA

It is theoretically possible to use the small molecule Gd-DTPA to predict the convective boundaries of larger molecules, such as chemotherapeutics, cytotoxins, and immunotoxins. First, the differences between low molecular weight agents and high molecular weight agents needs to be considered. Low molecular weight molecules diffuse faster through porous neural tissue and are also removed faster in cerebral capillaries than high molecular weight molecules. These differences cause the concentration profiles of small and large molecules to become drastically different, especially at long intervals. However, by applying a series of measurements and subsequent calculations, we can compensate for these differences. Ideally, these calculations will result in overlapping concentration profiles that allow us to use a small molecule as a surrogate tracer for a large molecule. Since the small molecule Gd-DTPA is a commonly used imaging tracer, its volume of distribution can be easily visualized using conventional MRI.

On a short time scale of hours, the diffusion of low and high molecular weight agents do not differ significantly and are therefore approximately equal. In that time frame, the only adjustment needed is for the difference in removal rates, which can be measured in vivo using special computational algorithms on standard MR images acquired while infusing each molecule at a constant rate. If the ratio of the removal rates is a constant factor k such that the removal of the small molecule occurs k times faster than the removal of the large molecule, then compensation for this would require infusion of the small molecule at k times the concentration of the large molecule. In this theoretical setting, assuming the diffusivities of the two molecules are equivalent, the concentration profiles of the small and large molecules are now the same. Using this method of correction, Gd-DTPA can be co-infused with a larger molecule to actively monitor its volume of distribution. However, after several hours of infusion, the difference in diffusivities becomes significant and needs to also be considered. If compensation for only the difference in removal rate occurs, the higher diffusivity of Gd-DTPA compared to the high molecular weight agent will cause Gd-DTPA to have a wider concentration profile than the high molecular weight agent. Therefore, this difference also needs to be included in the correction. While the calculations become too complex for full explanation in this review, using these techniques it is nonetheless possible to determine the difference in diffusivities, such that the proper adjustments can be made and free Gd-DTPA can be used to model the drug distribution of high molecular weight compounds over long time periods.

An additional concern when using free Gd-DTPA is that it needs to be verified that the Gd-DTPA can be detected at the convection boundary. While the actual concentration is dependent on many factors, such as the manufacturer and MRI machine settings, it is generally in the range of nanomoles per voxel. Therefore if the tracer molecule is very sensitive, such as Gd-DTPA, it is possible to follow the outer boundary of a heavy molecule (i.e., cytotoxin) with a smaller tracer molecule (i.e., Gd-DTPA) for long times. The high sensitivity compensates for the fast removal rate. In fact studies have already shown, in a small number of cases and over short observation periods, that intracerebral co-infusion of Gd-DTPA is safe and visible on MRI (10, 78–80).

4.5. Albumin Conjugates

Nguyen et al. showed that iopanoic acid (IPA) bound to albumin can be used as a surrogate tracer for CT imaging, and also that albumin linked to Gd-DTPA can be used as a surrogate tracer for MRI (81). They infused IPA-albumin and Gd-DPTA-albumin and co-infused ^{14}C-labeled albumin into the putamen of rhesus monkeys and followed the distribution of infusate by subsequent CT and MRI imaging, respectively. Quantitative autoradiography (QAR) was used to measure the volume of ^{14}C-albumin distribution as a comparison to volumes calculated by CT and MRI. The investigators also proposed the use of Hounsfield units and degree of enhancement on MRI to extrapolate the concentration of infusate, using the linear relationship of iodine concentration with Hounsfield units and the pseudo-linear relationship between gadolinium concentration and enhancement on MRI. The relationship between gadolinium concentration and MRI enhancement is termed pseudo-linear because the true relationship between gadolinium concentration and MRI enhancement is bimodal (82), but in this study, the concentration of Gd-DTPA was limited to a range in which the correlation was approximately linear. While Gd-albumin has been shown to be an effective surrogate tracer for many CED drugs (81, 83), it is limited by its high cost and also the affinity of albumin for the catheter tubing. Because of albumin binding by the catheter, albumin-containing compounds must be infused at many-fold times the desired final intraparenchymal concentration. Sampson et al. showed the clinical utility of infusing ^{123}I-HSA via CED for subsequent SPECT imaging of the drug distribution (84). These studies were performed on human patients in conjunction with CED TP-38 for the treatment of malignant gliomas.

4.6. Gd-Liposomes

Saito et al. performed a series of studies on rats and primates in 2004 and 2005 which looked at MR imaging of Gd-loaded liposomes as a method for monitoring of CED liposomal drug delivery and distribution (85, 86). The animals were given CED of liposomal Gd (Lip/Gd) and liposomal DiI-DS, a fluorescent marker. After MRI images of the animals' brains were obtained, they were euthanized and necropsies were performed. The brain tissue was then histologically examined for the presence and distribution of the fluorescent probe DiI-DS. Results showed that the distribution of liposomes visualized on MRI was identical to the distribution seen on histology. Also, Lip/Gd and Lip/DiI-DS were infused into two rat brain tumor models, the C6 rat glioma and 9L-2 rat gliosarcoma models. Subsequent analysis showed the volume of distribution in the rat brain tumors was the same as the volume of distribution in the normal rat brains. Neither the rats nor primates showed any clinical toxicity as a result of Lip/Gd infusion.

5. Software Predictions

Since CED can result in suboptimal drug delivery if not performed correctly, visualization of the infused region is important, as is a method of predicting this distribution. Computer software that is currently available has been able to fuse images across different imaging modalities and therefore allow for better visualization of the regions of drug delivery. Additionally, to assist with treatment plans, computer software has been utilized that can model CED. These software programs allow for predictions of infused target areas and can also help detect anatomical features that could lead to poor drug delivery.

5.1. iPlan® Flow

In the clinical trial described above with seven patients who received CED of CB and ^{123}I-HSA, SPECT imaging was used to monitor the volume of distribution of CB. Studies in rats and primates have shown that labeled albumin is an effective surrogate tracer of CED cytotoxin distribution (87). ^{123}I-HSA was directly co-infused along with CB into the tumor beds of seven patients with recurrent malignant gliomas, and SPECT scans were taken at 24 and 48 h post-infusion. Analysis of the images using iPlan® Flow software designed by BrainLAB AG (Munich, Germany) showed SPECT signals were anisotropic with larger diameters in a parallel axis relative to the catheter tract, which is consistent with theoretical calculations and modeling (22). An example screen image from the iPlan software is shown in **Fig. 15.7**. Unfortunately, this visualization method is limited by the half-life of ^{123}I (13.2 h) and by the inherent resolution of SPECT imaging (84).

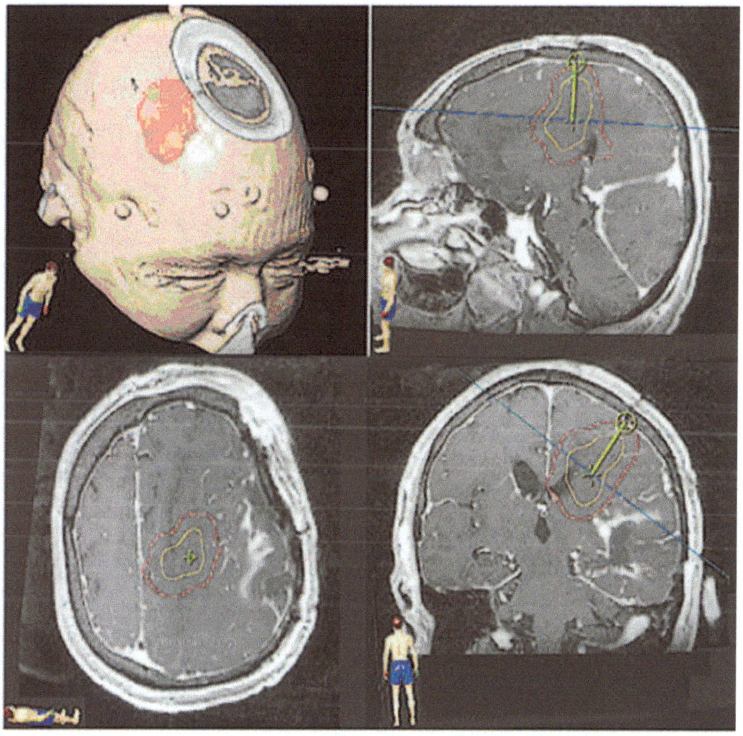

Fig. 15.7. Contrast-enhanced, T1-weighted MRI scans showing the catheter (*green line*) in various planes with overlay of the co-registered SPECT image showing radiotracer activity at the 50% isodose level (*yellow*) and at the 90% isodose level (*red*). Views shown are three-dimensional reconstruction (*top left*), obliques in plane with the catheter (*top right* sagittal and *bottom right* axial), and perpendicular to the plane of the catheter (*bottom left*). Note the large and characteristic shape of the distribution, which extends farther proximally from the catheter tip than it does distally. From Sampson et al. (84), with permission.

5.2. Sulcus Detection Algorithm

A further refinement of the BrainLAB AG (Munich, Germany) imaging software was the incorporation of a sulcus detection algorithm, for the purpose of predicting ineffective CED infusion secondary to leakage into the subarachnoid or intraventricular spaces. The locations of sulci are mapped automatically via an image processing algorithm applied to a three-dimensional T1-weighted MR image set. Sulci appear as thin surfaces of low image intensity in the T1-weighted volumes. Within the plane of the sulcus, image intensities are relatively constant, while perpendicular to the sulcus, a local minimum (trough) is seen. This observation forms the basis of a three-dimensional trough filter, which is sensitive to regions with high second derivative in one direction, and relatively constant second derivative in the perpendicular directions. Following the trough detection filter, the result is smoothed, thresholded, and thinned, using standard image processing techniques. The result is a binary 3D volume that serves as a sulcus map (**Fig. 15.8**), marking the locations deemed to contain a sulcus.

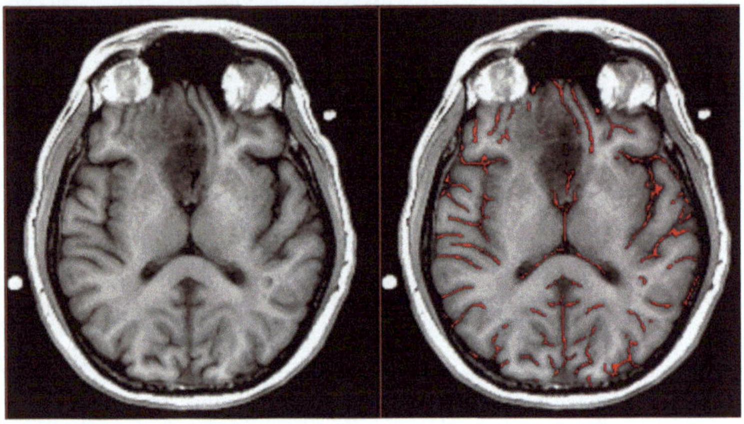

Fig. 15.8. Example of sulcus detection algorithm on a sample patient MRI scan. The *left image* shows a standard T1-weighted MRI. The *right image* shows the same MRI with the automatically computed sulcus map overlaid in *red*.

Sampson et al. tested the sulcus detection algorithm on a series of seven patients who were given CED cintredekin besudotox with a co-infusion of ^{123}I-labeled albumin and subsequently underwent MRI and SPECT scanning to determine drug distribution. The sulcus detection algorithm was then retrospectively applied to 14 evaluable catheters and showed 100% specificity and 71.4% sensitivity. Of the seven cases in which a sulcus was present, the algorithm detected five. Of the seven cases in which a sulcus was not present, the algorithm correctly identified the absence of sulci in all seven cases (88). In the same study, Sampson et al. also retrospectively developed a set of evaluation criteria to assess the adequacy of catheter positioning. The three criteria are as follows: (1) catheter tip depth of greater than 2.5 cm from the cortical surface, or from any deep sulci, or from the resection cavity wall if the catheter was inserted through the resection cavity, (2) catheter tip greater than 0.5 cm from any ependymal or pial surfaces, and (3) catheter tip greater than

0.5 cm from the resection cavity walls assuming the catheter was not inserted through the resection cavity. If all three criteria were fulfilled, the catheter was considered to be optimally placed and would be very unlikely to leak infusate into the subarachnoid space. If the first criterion was fulfilled but only one of the remaining two criteria was also fulfilled, the catheter was considered to be acceptably placed but not optimal. Finally, if the first criterion was not fulfilled, the catheter was considered inadequately placed regardless of the remaining criteria, and the probability of infusate leakage was predicted to be high.

6. Conclusion

CED has progressed significantly since its first description in 1994 by Bobo et al. (4). It has become an exciting area of active research for glioma therapy, and much more is now understood about its potential. Mathematical descriptions of the flow that is generated from CED have been determined, and these have helped to improve CED catheter design and have allowed for the development of software algorithms that can simulate drug distribution. Multiple drugs are currently being tested for administration via CED, and the encouraging results obtained so far certainly warrant further study. Furthermore, advances have been made to monitor drug delivery via novel imaging techniques, imaging molecule co-infusion, and computer simulations. Computers are now able to model CED in patients and assist with the planning of catheter placement, and a set of guidelines have even been published to aid in determining the optimal treatment plan and avoid excessive backflow or subarachnoid leakage (88). CED has allowed patients to be treated with drugs that previously were limited by intolerable systemic adverse effects or poor penetration of the blood–brain barrier and therefore, novel classes of anti-tumor therapeutics can now be pursued. While much work remains before CED can become a standard, widespread therapy, the progress to date has been promising and CED may ultimately prove to be a critical component of glioma therapy in the future.

References

1. Primary brain tumors in the United States statistical report. Central Brain Tumor Registry of the United States; 2007–2008.
2. Stupp, R., Mason, W. P., van den Bent, M. J. et al (2005) Radiotherapy plus concomitant and adjuvant temozolomide for glioblastoma. *N Engl J Med* (Mar 10), 352(10), 987–96.
3. Hegi, M. E., Diserens, A. C., Gorlia, T. et al (2005) MGMT gene silencing and benefit from temozolomide in glioblastoma. *N Engl J Med* (Mar 10), 352(10), 997–1003.

4. Bobo, R. H., Laske, D. W., Akbasak, A., Morrison, P. F., Dedrick, R. L., and Oldfield, E. H. (1994) Convection-enhanced delivery of macromolecules in the brain. *Proc Natl Acad Sci U S A* (Mar 15), 91(6), 2076–80.
5. Vogelbaum, M. A. (2005) Convection enhanced delivery for the treatment of malignant gliomas: symposium review. *J Neurooncol* (May) 73(1), 57–69.
6. Bankiewicz, K. S., Eberling, J. L., Kohutnicka, M. et al (2000) Convection-enhanced delivery of AAV vector in parkinsonian monkeys; in vivo detection of gene expression and restoration of dopaminergic function using pro-drug approach. *Exp Neurol* (Jul) 164(1), 2–14.
7. Oiwa, Y., Sanchez-Pernaute, R., Harvey-White, J., and Bankiewicz, K. S. (2003) Progressive and extensive dopaminergic degeneration induced by convection-enhanced delivery of 6-hydroxydopamine into the rat striatum: a novel rodent model of Parkinson disease. *J Neurosurg* (Jan) 98(1), 136–44.
8. Zirzow, G. C., Sanchez, O. A., Murray, G. J., Brady, R. O., and Oldfield, E. H. (1999) Delivery, distribution, and neuronal uptake of exogenous mannose-terminal glucocerebrosidase in the intact rat brain. *Neurochem Res* (Feb) 24(2), 301–05.
9. Lonser, R. R., Walbridge, S., Murray, G. J. et al (2005) Convection perfusion of glucocerebrosidase for neuronopathic Gaucher's disease. *Ann Neurol* (Apr) 57(4), 542–48.
10. Lonser, R. R., Schiffman, R., Robison, R. A. et al (2007) Image-guided, direct convective delivery of glucocerebrosidase for neuronopathic Gaucher disease. *Neurology* (Jan 23) 68(4), 254–61.
11. O'Connor, W. M., Davidson, B. L., Kaplitt, M. G. et al. (1997) Adenovirus vector-mediated gene transfer into human epileptogenic brain slices: prospects for gene therapy in epilepsy. *Exp Neurol* (Nov), 148(1), 167–78.
12. Fisher, R. S. and Ho, J. (2002) Potential new methods for antiepileptic drug delivery. *CNS Drugs* 16(9), 579–93.
13. Tenenbaum, L., Chtarto, A., Lehtonen, E., Velu, T., Brotchi, J., and Levivier, M. (2004) Recombinant AAV-mediated gene delivery to the central nervous system. *J Gene Med* (Feb) 6(Suppl 1), S212–S22.
14. Fisher, R. S. and Chen, D. K. (2006) New routes for delivery of anti-epileptic medications. *Acta Neurol Taiwan* (Dec) 15(4), 225–31.
15. Vogelbaum, M. A. (2007) Convection enhanced delivery for treating brain tumors and selected neurological disorders: symposium review. *J Neurooncol* (May) 83(1), 97–109.
16. Hornung, U. (ed) (1997) *Homogenization and Porous Media*. Springer, New York.
17. Raghavan, R., Brady, M. L., Rodriguez-Ponce, M. I., Hartlep, A., Pedain, C., and Sampson, J. H. (2006) Convection-enhanced delivery of therapeutics for brain disease, and its optimization. *Neurosurg Focus* 20(4), E12.
18. Saito, R., Krauze, M. T., Noble, C. O. et al (2006) Tissue affinity of the infusate affects the distribution volume during convection-enhanced delivery into rodent brains: implications for local drug delivery. *J Neurosci Meth* (Jun 30) 154(1–2), 225–32.
19. Guarnieri, M., Carson, B. S., Khan, A., Penno, M., and Jallo, G. I. (2005) Flexible versus rigid catheters for chronic administration of exogenous agents into central nervous system tissues. *J Neurosci Meth* (Jun 15) 144(2), 147–52.
20. Morrison, P. F., Chen, M. Y., Chadwick, R. S., Lonser, R. R., and Oldfield, E. H. (1999) Focal delivery during direct infusion to brain: role of flow rate, catheter diameter, and tissue mechanics. *Am J Physiol* (Oct) 277(4 Pt 2), R1218–R29.
21. Chen, M. Y., Lonser, R. R., Morrison, P. F., Governale, L. S., and Oldfield, E. H. (1999) Variables affecting convection-enhanced delivery to the striatum: a systematic examination of rate of infusion, cannula size, infusate concentration, and tissue-cannula sealing time. *J Neurosurg* (Feb) 90(2), 315–20.
22. Morrison, P. F., Laske, D. W., Bobo, H., Oldfield, E. H., and Dedrick, R. L. (1994) High-flow microinfusion: tissue penetration and pharmacodynamics. *Am J Physiol* (Jan) 266(1 Pt 2), R292–R305.
23. Krauze, M. T., Saito, R., Noble, C. et al (2005) Reflux-free cannula for convection-enhanced high-speed delivery of therapeutic agents. *J Neurosurg* (Nov) 103(5), 923–29.
24. Olson, J. J., Zhang, Z., Dillehay, D., and Stubbs, J. (2008) Assessment of a balloon-tipped catheter modified for intracerebral convection-enhanced delivery. *J Neurooncol* (Sep) 89(2), 159–68.
25. Barth, R. F., Soloway, A. H., and Fairchild, R. G. (1990) Boron neutron capture therapy of cancer. *Cancer Res* (Feb 15) 50(4), 1061–70.

26. Yang, W., Barth, R. F., Adams, D. M. et al (2002) Convection-enhanced delivery of boronated epidermal growth factor for molecular targeting of EGF receptor-positive gliomas. *Cancer Res* (Nov 15) 62(22), 6552–58.
27. Cunningham, J., Oiwa, Y., Nagy, D., Podsakoff, G., Colosi, P., and Bankiewicz, K. S. (2000) Distribution of AAV-TK following intracranial convection-enhanced delivery into rats. *Cell Transplant* (Sep–Oct) 9(5), 585–94.
28. Saito, R., Krauze, M. T., Noble, C. O. et al (2006) Convection-enhanced delivery of Ls-TPT enables an effective, continuous, low-dose chemotherapy against malignant glioma xenograft model. *Neuro Oncol* (Jul) 8(3), 205–14.
29. Noble, C. O., Krauze, M. T., Drummond, D. C. et al (2006) Novel nanoliposomal CPT-11 infused by convection-enhanced delivery in intracranial tumors: pharmacology and efficacy. *Cancer Res* (Mar 1) 66(5), 2801–06.
30. Saito, R., Bringas, J. R., Panner, A. et al (2004) Convection-enhanced delivery of tumor necrosis factor-related apoptosis-inducing ligand with systemic administration of temozolomide prolongs survival in an intracranial glioblastoma xenograft model. *Cancer Res* (Oct 1) 64(19), 6858–62.
31. Degen, J. W., Walbridge, S., Vortmeyer, A. O., Oldfield, E. H., and Lonser, R. R. (2003) Safety and efficacy of convection-enhanced delivery of gemcitabine or carboplatin in a malignant glioma model in rats. *J Neurosurg* (Nov) 99(5), 893–98.
32. Lidar, Z., Mardor, Y., Jonas, T. et al (2004) Convection-enhanced delivery of paclitaxel for the treatment of recurrent malignant glioma: a phase I/II clinical study. *J Neurosurg* (Mar) 100(3), 472–79.
33. Jennings, M. T. and Pietenpol, J. A. (1998) The role of transforming growth factor beta in glioma progression. *J Neurooncol* (Jan) 36(2), 123–40.
34. Frankel, B., Longo, S. L., and Ryken, T. C. (1999) Human astrocytomas co-expressing Fas and Fas ligand also produce TGFbeta2 and Bcl-2. *J Neurooncol* 44(3), 205–12.
35. Kjellman, C., Olofsson, S. P., Hansson, O. et al (2000) Expression of TGF-beta isoforms, TGF-beta receptors, and SMAD molecules at different stages of human glioma. *Int J Cancer* (May 20) 89(3), 251–58.
36. Schlingensiepen, R., Goldbrunner, M., Szyrach, M. N. et al (2005) Intracerebral and intrathecal infusion of the TGF-beta 2-specific antisense phosphorothioate oligonucleotide AP 12009 in rabbits and primates: toxicology and safety. *Oligonucleotides* (Summer) 15(2), 94–104.
37. Hau, P., Jachimczak, P., Schlingensiepen, R. et al (2007) Inhibition of TGF-beta2 with AP 12009 in recurrent malignant gliomas: from preclinical to phase I/II studies. *Oligonucleotides* (Summer) 17(2), 201–12.
38. Patel, S. J., Shapiro, W. R., Laske, D. W. et al (2005) Safety and feasibility of convection-enhanced delivery of Cotara for the treatment of malignant glioma: initial experience in 51 patients. *Neurosurgery* (Jun) 56(6), 1243–52 (discussion 1252–1253).
39. Laske, D. W., Youle, R. J., and Oldfield, E. H. (1997) Tumor regression with regional distribution of the targeted toxin TF-CRM107 in patients with malignant brain tumors. *Nat Med* (Dec) 3(12), 1362–68.
40. Weaver, M. and Laske, D. W. (2003) Transferrin receptor ligand-targeted toxin conjugate (Tf-CRM107) for therapy of malignant gliomas. *J Neurooncol* (Oct) 65(1), 3–13.
41. Rand, R. W., Kreitman, R. J., Patronas, N., Varricchio, F., Pastan, I., and Puri, R. K. (2000) Intratumoral administration of recombinant circularly permuted interleukin-4-Pseudomonas exotoxin in patients with high-grade glioma. *Clin Cancer Res* (Jun) 6(6), 2157–65.
42. Weber, F., Asher, A., Bucholz, R. et al (2003) Safety, tolerability, and tumor response of IL4-Pseudomonas exotoxin (NBI-3001) in patients with recurrent malignant glioma. *J Neurooncol* (Aug–Sep) 64(1–2), 125–37.
43. Brem, H., Piantadosi, S., Burger, P. C. et al (1995) Placebo-controlled trial of safety and efficacy of intraoperative controlled delivery by biodegradable polymers of chemotherapy for recurrent gliomas. The Polymer-brain Tumor Treatment Group. *Lancet* (Apr 22) 345(8956), 1008–12.
44. Rainov, N. G. and Heidecke, V. (2004) Long term survival in a patient with recurrent malignant glioma treated with intratumoral infusion of an IL4-targeted toxin (NBI-3001). *J Neurooncol* (Jan) 66(1–2), 197–201.
45. Debinski, W., Obiri, N. I., Powers, S. K., Pastan, I., and Puri, R. K. (1995) Human glioma cells overexpress receptors for interleukin 13 and are extremely sensitive to a novel chimeric protein composed of interleukin 13 and pseudomonas exotoxin. *Clin Cancer Res* (Nov) 1(11), 1253–58.

46. Debinski, W., Gibo, D. M., Hulet, S. W., Connor, J. R., and Gillespie, G. Y. (1999) Receptor for interleukin 13 is a marker and therapeutic target for human high-grade gliomas. *Clin Cancer Res* (May) 5(5), 985–90.
47. Kunwar, S., Prados, M. D., Chang, S. M. et al (2007) Direct intracerebral delivery of cintredekin besudotox (IL13-PE38QQR) in recurrent malignant glioma: a report by the Cintredekin Besudotox Intraparenchymal Study Group. *J Clin Oncol* (Mar 1) 25(7), 837–44.
48. Vogelbaum, M. A., Sampson, J. H., Kunwar, S. et al (2007) Convection-enhanced delivery of cintredekin besudotox (interleukin-13-PE38QQR) followed by radiation therapy with and without temozolomide in newly diagnosed malignant gliomas: phase 1 study of final safety results. *Neurosurgery* (Nov) 61(5), 1031–37 (discussion 1037–1038).
49. Sawyer, A. J., Piepmeier, J. M., and Saltzman, W. M. (2006) New methods for direct delivery of chemotherapy for treating brain tumors. *Yale J Biol Med* (Dec) 79(3–4), 141–52.
50. Wong, A. J., Bigner, S. H., Bigner, D. D., Kinzler, K. W., Hamilton, S. R., and Vogelstein, B. (1987) Increased expression of the epidermal growth factor receptor gene in malignant gliomas is invariably associated with gene amplification. *Proc Natl Acad Sci U S A* (Oct) 84(19), 6899–903.
51. Sampson, J. H., Reardon, D. A., Friedman, A. H. et al (2005) Sustained radiographic and clinical response in patient with bifrontal recurrent glioblastoma multiforme with intracerebral infusion of the recombinant targeted toxin TP-38: case study. *Neuro Oncol* (Jan) 7(1), 90–96.
52. Sampson, J. H., Akabani, G., Archer, G. E. et al (2008) Intracerebral infusion of an EGFR-targeted toxin in recurrent malignant brain tumors. *Neuro Oncol* (Jun) 10(3), 320–29.
53. Gregoriadis, G., Wills, E. J., Swain, C. P., and Tavill, A. S. (1974) Drug-carrier potential of liposomes in cancer chemotherapy. *Lancet* (Jun 29) 1(7870), 1313–16.
54. Harrington, K. J. (2001) Liposomal cancer chemotherapy: current clinical applications and future prospects. *Expert Opin Investig Drugs* (Jun) 10(6), 1045–61.
55. Maruyama, K. (2000) In vivo targeting by liposomes. *Biol Pharm Bull* (Jul) 23(7), 791–99.
56. Voinea, M. and Simionescu, M. (2002) Designing of "intelligent" liposomes for efficient delivery of drugs. *J Cell Mol Med* (Oct–Dec) 6(4), 465–74.
57. Moog, R., Burger, A. M., Brandl, M. et al (2002) Change in pharmacokinetic and pharmacodynamic behavior of gemcitabine in human tumor xenografts upon entrapment in vesicular phospholipid gels. *Cancer Chemother Pharmacol* (May) 49(5), 356–66.
58. Zhu, J., Zhang, L., Hanisch, U. K., Felgner, P. L., and Reszka, R. (1996) A continuous intracerebral gene delivery system for in vivo liposome-mediated gene therapy. *Gene Ther* (Jun) 3(6), 472–76.
59. von Eckardstein, K. L., Patt, S., Zhu, J., Zhang, L., Cervos-Navarro, J., and Reszka, R. (2001) Short-term neuropathological aspects of in vivo suicide gene transfer to the F98 rat glioblastoma using liposomal and viral vectors. *Histol Histopathol* (Jul) 16(3), 735–44.
60. Park, J. W., Hong, K., Kirpotin, D. B., Meyer, O., Papahadjopoulos, D., and Benz, C. C. (1997) Anti-HER2 immunoliposomes for targeted therapy of human tumors. *Cancer Lett* (Oct) 14 118(2), 153–60.
61. Park, J. W., Hong, K., Kirpotin, D. B., Papahadjopoulos, D., and Benz, C. C. (1997) Immunoliposomes for cancer treatment. *Adv Pharmacol* 40, 399–435.
62. Ram, Z., Culver, K. W., Oshiro, E. M. et al (1997) Therapy of malignant brain tumors by intratumoral implantation of retroviral vector-producing cells. *Nat Med* (Dec) 3(12), 1354–61.
63. Klatzmann, D., Valery, C. A., Bensimon, G. et al (1998) A phase I/II study of herpes simplex virus type 1 thymidine kinase "suicide" gene therapy for recurrent glioblastoma. Study Group on Gene Therapy for Glioblastoma. *Hum Gene Ther* (Nov 20) 9(17), 2595–604.
64. Shand, N., Weber, F., Mariani, L. et al (1999) A phase 1–2 clinical trial of gene therapy for recurrent glioblastoma multiforme by tumor transduction with the herpes simplex thymidine kinase gene followed by ganciclovir. GLI328 European-Canadian Study Group. *Hum Gene Ther* (Sep 20) 10(14), 2325–35.
65. Rainov, N. G. (2000) A phase III clinical evaluation of herpes simplex virus type 1 thymidine kinase and ganciclovir gene therapy as an adjuvant to surgical resection and radiation in adults with previously untreated glioblastoma multiforme. *Hum Gene Ther* (Nov 20) 11(17), 2389–401.
66. Voges, J., Reszka, R., Gossmann, A. et al (2003) Imaging-guided convection-enhanced delivery and gene therapy of glioblastoma. *Ann Neurol* (Oct) 54(4), 479–87.

67. Chenevert, T. L., McKeever, P. E., and Ross, B. D. (1997) Monitoring early response of experimental brain tumors to therapy using diffusion magnetic resonance imaging. *Clin Cancer Res* (Sep) 3(9), 1457–66.
68. Chinnaiyan, A. M., Prasad, U., Shankar, S. et al (2000) Combined effect of tumor necrosis factor-related apoptosis-inducing ligand and ionizing radiation in breast cancer therapy. *Proc Natl Acad Sci U S A* (Feb 15) 97(4), 1754–59.
69. Gupta, R. K., Sinha, U., Cloughesy, T. F., and Alger, J. R. (1999) Inverse correlation between choline magnetic resonance spectroscopy signal intensity and the apparent diffusion coefficient in human glioma. *Magn Reson Med* (Jan) 41(1), 2–7.
70. Chenevert, T. L., Stegman, L. D., Taylor, J. M. et al (2000) Diffusion magnetic resonance imaging: an early surrogate marker of therapeutic efficacy in brain tumors. *J Natl Cancer Inst* (Dec 20) 92(24), 2029–2036.
71. Geschwind, J. F., Artemov, D., Abraham, S. et al (2000) Chemoembolization of liver tumor in a rabbit model: assessment of tumor cell death with diffusion-weighted MR imaging and histologic analysis. *J Vasc Interv Radiol* (Nov–Dec) 11(10), 1245–1255.
72. Pilatus, U., Shim, H., Artemov, D., Davis, D., van Zijl, P. C., and Glickson, J. D. (1997) Intracellular volume and apparent diffusion constants of perfused cancer cell cultures, as measured by NMR. *Magn Reson Med* (Jun) 37(6), 825–832.
73. Van Zijl, P. C., Moonen, C. T., Faustino, P., Pekar, J., Kaplan, O., and Cohen, J. S. (1991) Complete separation of intracellular and extracellular information in NMR spectra of perfused cells by diffusion-weighted spectroscopy. *Proc Natl Acad Sci U S A* (Apr 15) 88(8), 3228–3232.
74. Mardor, Y., Roth, Y., Lidar, Z. et al (2001) Monitoring response to convection-enhanced taxol delivery in brain tumor patients using diffusion-weighted magnetic resonance imaging. *Cancer Res* (Jul 1) 61(13), 4971–4973.
75. Sampson, J. H., Raghavan, R., Provenzale, J. M. et al (2007) Induction of hyperintense signal on T2-weighted MR images correlates with infusion distribution from intracerebral convection-enhanced delivery of a tumor-targeted cytotoxin. *AJR Am J Roentgenol* (Mar) 188(3), 703–709.
76. Edelman, R. R., Mattle, H. P., Atkinson, D. J. et al (1990) Cerebral blood flow: assessment with dynamic contrast-enhanced T2*-weighted MR imaging at 1.5 T. *Radiology* (Jul) 176(1), 211–220.
77. Rempp, K. A., Brix, G., Wenz, F., Becker, C. R., Guckel, F., and Lorenz, W. J. (1994) Quantification of regional cerebral blood flow and volume with dynamic susceptibility contrast-enhanced MR imaging. *Radiology* (Dec) 193(3), 637–641.
78. Lonser, R. R., Warren, K. E., Butman, J. A. et al (2007) Real-time image-guided direct convective perfusion of intrinsic brainstem lesions. Technical note. *J Neurosurg* (Jul) 107(1), 190–197.
79. Mardor, Y., Rahav, O., Zauberman, Y. et al (2005) Convection-enhanced drug delivery: increased efficacy and magnetic resonance image monitoring. *Cancer Res* (Aug 1) 65(15), 6858–6863.
80. Murad, G. J., Walbridge, S., Morrison, P. F. et al (2007) Image-guided convection-enhanced delivery of gemcitabine to the brainstem. *J Neurosurg* (Feb) 106(2), 351–356.
81. Nguyen, T. T., Pannu, Y. S., Sung, C. et al (2003) Convective distribution of macromolecules in the primate brain demonstrated using computerized tomography and magnetic resonance imaging. *J Neurosurg* (Mar) 98(3), 584–590.
82. Wang, S. C., White, D. L., Pope, J. M., and Brasch, R. C. (1990) Magnetic resonance imaging contrast enhancement versus tissue gadolinium concentration. *Invest Radiol* (Sep) 25(Suppl 1), S44–45.
83. Lonser, R. R., Walbridge, S., Garmestani, K. et al (2002) Successful and safe perfusion of the primate brainstem: in vivo magnetic resonance imaging of macromolecular distribution during infusion. *J Neurosurg* (Oct) 97(4), 905–913.
84. Sampson, J. H., Brady, M. L., Petry, N. A. et al (2007) Intracerebral infusate distribution by convection-enhanced delivery in humans with malignant gliomas: descriptive effects of target anatomy and catheter positioning. *Neurosurgery* (Feb) 60(2 Suppl 1) ONS89-98; discussion ONS98–89.
85. Saito, R., Bringas, J. R., McKnight, T. R. et al (2004) Distribution of liposomes into brain and rat brain tumor models by convection-enhanced delivery monitored with magnetic resonance imaging. *Cancer Res* (Apr 1) 64(7), 2572–2579.
86. Saito, R., Krauze, M. T., Bringas, J. R. et al (2005) Gadolinium-loaded liposomes allow for real-time magnetic resonance imaging of convection-enhanced delivery in the primate brain. *Exp Neurol* (Dec) 196(2), 381–389.

87. Murad, G. J., Walbridge, S., Morrison, P. F. et al (2006) Real-time, image-guided, convection-enhanced delivery of interleukin 13 bound to pseudomonas exotoxin. *Clin Cancer Res* (May 15) 12(10), 3145–3151.

88. Sampson, J. H., Raghavan, R., Brady, M. L. et al (2007) Clinical utility of a patient-specific algorithm for simulating intracerebral drug infusions. *Neuro Oncol* (Jul) 9(3), 343–353.

Subject Index

K.K. Jain (ed.), *Drug Delivery to the Central Nervous System,* Neuromethods 45,
DOI 10.1007/978-1-60761-529-3, © Humana Press, a part of Springer Science+Business Media, LLC 2010

N

O

P

R

S

T

U

V

MIX
Papier aus verantwortungsvollen Quellen
Paper from responsible sources
FSC® C105338

If you have any concerns about our products,
you can contact us on
ProductSafety@springernature.com

In case Publisher is established outside the EU,
the EU authorized representative is:
Springer Nature Customer Service Center GmbH
Europaplatz 3, 69115 Heidelberg, Germany

Printed by Libri Plureos GmbH
in Hamburg, Germany